The Gliocentric Brain

Wolfgang Walz

The Gliocentric Brain

Phenotype Plasticity of the Damaged Brain

Springer

Wolfgang Walz
Psychiatry
University of Saskatchewan
Saskatoon, SK, Canada

ISBN 978-3-031-48104-8 ISBN 978-3-031-48105-5 (eBook)
https://doi.org/10.1007/978-3-031-48105-5

Cover image from M. Pekny & M. Pekna 2014. Astrocyte reactivity and reactive gliosis: costs and benefits. Physiol Rev 94: 1077-98. Copyright permission obtained from American Physiological Society.

This Springer imprint is published by the registered company Springer Nature Switzerland AG
The registered company address is: Gewerbestrasse 11, 6330 Cham, Switzerland

Paper in this product is recyclable.

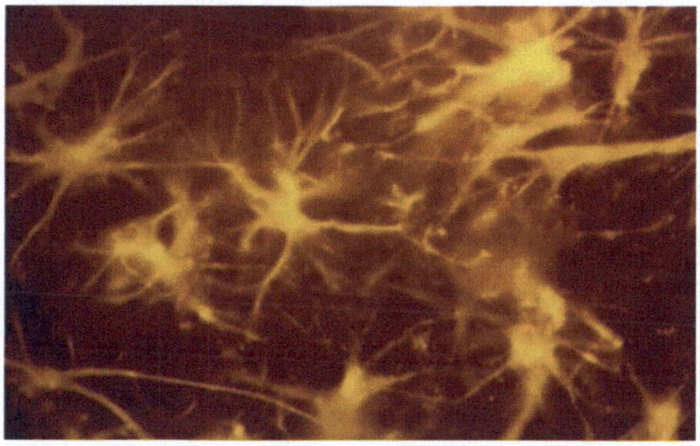

Preface

My first glial cell experiment took place on November 2, 1976 at the start of my undergraduate thesis (*Diplomarbeit*) in the laboratory of Wolf-Rüdiger Schlue at the Universität Konstanz. The thesis characterized the electrical properties of neuropile glial cells in the medicinal leech [1]. At that time glial cells were seen at best as passive participants in neuronal homeostasis [2], if not simply as a filling (glia meaning "glue" in ancient Greek) between neuronal elements, despite the fact that astrocytes were known to constitute a giant syncytium within the brain [3]. At first this research was received with some skepticism, and it needed some persistence to gain acceptance. Captivated by this grouping of cells, I dedicated my professional life to their investigation. With the support of un-interrupted external funding for 31 years at the University of Saskatchewan, together with my laboratory team, we used various preparations to investigate the function of these cells in health and disease, including cell cultures, brain slices and in vivo rodent brains. Today the perception of glial cells has changed radically. Glial cells are not only seen to be central homeostatic players in the nervous system, making the function of the brain as an organ possible, but are also acknowledged to participate in all major signaling processes as equal partners with neurons [4–6]. This acknowledgment was a long time coming.

Glial cells do not act in isolation. Neurons are recognized as the major backbone in circuit signaling and integrative processing and must be considered central to any honest and unbiased approach to brain function. However, in the adult brain, neurons are mainly in a predetermined fixed position within a clearly prescribed circuit. The removal or loss of any neuron comes at a price. Any room to maneuver during pathological challenges to the brain is very restricted and neurons must be assisted in their survival at all costs. This is where the critical importance of glial cells come into play. Their plasticity allows them to change phenotypes and position, to start proliferating and even to be removed according to need. The primary purpose of this plasticity is to prioritize neuronal signaling. Even after damage to neuronal circuits, the priority of glial cells is to repair and reconstitute these circuits to compensate for lost wiring connections. It can be argued therefore that it is a moot question if glial cells or neurons are more important for successful brain function. Healthy brain circuits are the primary goal, with all cell types cooperating and acting in an integrative system to ensure brain plasticity during all challenges, whether physiological or pathological. This monograph focuses on integrating the plasticity of non-neuronal cells, especially glial cells, in normal function and during major diseases of the brain as an integrated organ.

Victoria, BC, Canada Wolfgang Walz

References

1. Walz W. Do neuronal signals regulate potassium flow in glial cells? Evidence from an invertebrate central nervous system. J Neurosci Res. 1982;7(1):71–9.
2. Kuffler SW, Nicholls JG. The physiology of neuroglial cells. Ergeb Physiol Biol Chem Exp Pharmakol. 1966;57:1–90.
3. Watson WE. Physiology of neuroglia. Physiol Rev. 1974;54(2):245–71.
4. Liu Y, Shen X, Zhang Y, Zheng X, Cepeda C, Wang Y, et al. Interactions of glial cells with neuronal synapses, from astrocytes to microglia and oligodendrocyte lineage cells. Glia. 2023;71(6):1383–401.
5. Hasel P, Aisenberg WH, Bennett FC, Liddelow SA. Molecular and metabolic heterogeneity of astrocytes and microglia. Cell Metab. 2023;35(4):555–70.
6. Rasmussen RN, Asiminas A, Carlsen EMM, Kjaerby C, Smith NA. Astrocytes: integrators of arousal state and sensory context. Trends Neurosci. 2023;46(6):418–25.

Contents

1 Introduction ... 1

2 The Brain as an Organ ... 7

3 Glial Cells During the Life Cycle ... 29

4 Reactive Microglia and Astrocyte Phenotype Transitions: A Framework ... 59

5 Transition of Microglia to Reactive States ... 67

6 Reactive Astrocytes ... 77

7 Neuroinflammation ... 83

8 The Brain and the Immune System ... 91

9 Viral and Bacterial Infections ... 101

10 Autoimmune Diseases ... 113

11 Adult Glial Cell Proliferation and Neurogenesis ... 125

12 Glioma ... 135

13 Neurodegenerative Disorders ... 151

14 Vascular Diseases ... 167

15 Seizures ... 183

16 Traumatic Brain Injury ... 195

17 Major Psychiatric Disorders ... 207

18 Glia in Recovery Processes and Repair ... 231

19 Glial Phenotype Plasticity ... 241

Index ... 247

Chapter 1
Introduction

Abstract Neuronal somata are in a more or less fixed position within their circuit. During their lifetime, neurons acquire their specific function within a circuit due to their previous activity. Although there is some plasticity, the neuronal loss can therefore not easily be compensated. Glial cells have far greater plasticity to change phenotypes and will adjust to injury by changing function and morphology. They migrate and proliferate according to circumstances and depending on an accurate reading of the damage signals. Obviously, glial cells are more able to abandon their signal and homeostatic functions than neurons. In injury, their main goal is the integrity of the brain parenchyma and to protect neuronal circuits as much as possible. They are also able to initiate and participate in the recovery and repair processes in order to adapt surviving circuits to new tasks.

Keywords Astrocytes · Blood · brain barrier · Brain injury · Brain swelling · Gastrointestinal tract · Microglia · Neuron doctrine · skin

In the last 20 years, neuroscience has undergone an expansion from the neuron doctrine toward a view that is more inclusive of other cellular players. Furthermore, these non-neuronal cell types are now known to exhibit a larger plasticity of their phenotypes than neuronal elements, especially during pathological events. Neurons are locked into a circuit, and any removal would cause a major functional upheaval. To be sure, neurons that survive injury exhibit some plasticity in creating compensatory circuits, but their ability to switch phenotypes is dwarfed by the glial cell plasticity. The notion of a gliocentric brain was first pointed out by Nedergaard in 2009 to underline the fact that certain neurological diseases are due to a failure of the glial cells rather than the neuronal elements and that treatment options should focus on this group of cells [1]. This gliocentric view of the brain has been criticized, of course [2]. It is not the purpose of this monograph to replace one doctrine with another one; therefore, I will try and present as much as possible a balanced and critical view. Still, this monograph brings together large amounts of material in support of the function of the brain as an organ. Naturally, this approach will delegate the neurons as crucial yet not only players. The most important message of this monograph will be that there is no central or commanding element in the brain.

W. Walz, *The Gliocentric Brain*, https://doi.org/10.1007/978-3-031-48105-5_1

Rather, there are many participants whose integration with each other assures the smooth operation of this organ. The gliocentric approach is especially justified by the tremendous plasticity of the non-neuronal cells. This plasticity is most pronounced during and after damage to the central nervous system (CNS), when these glial cells change their phenotype depending on the context. In the pathological context that requires it, the glial cell function can be remodeled so much that it appears that the cells transitioned into a new cell type.

In this monograph, I will integrate the role of the various glial cells in the healthy and adult brain. I will expand into blood flow regulation, blood–brain barrier, and the brain's own immune system as well as the interactions with peripheral immune system components. Originally, glial cells were seen as important homeostatic players in the central nervous system (CNS). This view expanded in the last 15 years to include all glial cells in information processing and signaling. The view is now that astrocytes (and maybe microglia) play a large role in synchronization, pattern generation, and long-range signaling of brain activity [3, 4]. However, glial plasticity is most prominent after damage to the CNS and the interactions with the immune system. A large part of the monograph will therefore focus on neurological and psychiatric pathologies. To understand the reaction of the brain components in injury better, it is best to compare it with those of some other organs that are frequently challenged by damaging events, such as skin and the gastrointestinal (GI) system.

All organs in the adult mammalian body are subject to damage. This damage is due to various causes, such as physical external injury, bacterial and viral infections, as well as internal events, such as reduced blood flow, hemorrhage, or cancerous growth. Different organs face different challenges and are therefore equipped differently to cope with the challenges. The skin has a huge area and covers most of the body's external surface. In humans, there is no protective layer like fur on the outward side. This organ is therefore subject to a high likelihood of physical damage and penetration with accompanying infections. It is imperative that this organ is well equipped to deal with these challenges and to reconstitute the tissue after such an insult [5]. This reconstitution must occur as fast as possible to uphold the integrity of the body's surface barrier [6]. The epithelial cell lining of the gastrointestinal (GI) tract is facing a similar problem. The lumen of the GI tract is technically outside the body; thus, this epithelial cell layer can be regarded as protecting the body against an external environment. It too has a huge area despite the basic cylindrical structure of the GI canal. The 6-meter-long human small intestine has extensions known as folds of Kerckring, villi, and microvilli, which increase the surface by a factor of 600. This extensive epithelial area faces an environment that is usually even more hostile than the environment faced by the epithelial layer of the skin. Destructive digestive enzymes, extremely acidic pH values, and high density of various bacteria occur along the GI compartments. In addition, there are strong mechanical forces and osmotic gradients acting on the GI epithelia. It is a normal occurrence for cells of this epithelial layer to be mechanically sheared off and replaced by new cells without giving bacteria much of a chance to penetrate the internal environment of the body. Consequently, the lining of the GI tract is closely intertwined with the immune system. The immune system has specializations within the lining to fight viral and bacterial intrusions. These local immune responses are far more developed than those of the skin epithelial layer [7, 8].

The brain on the other hand is facing similar challenges but in a different context. While skin and GI tissue can usually be reconstructed to their original status after damage by recreating the same architecture, every time there has been damage and using the proliferative capacity of stem cells, the brain is in a different situation. It is concerned with information processing and as such must exhibit learning and memory. During these processes, specific synaptic connections between neurons are strengthened, whereas others are weakened and lost. This information is acquired and not stored as genetic code, and it is part of the CNS modus of operation. This means that after the destruction of some of the circuits, they cannot simply be reconstructed from genetic memory or positional information as their specific connections were based partly on previous use. There is some flexibility of cause, called plasticity, but the underlying problem is not easily addressed when reconstructing lost brain tissue. The strategy developed during evolution was to protect the brain heavily from external threats such as mechanical damage and infection. This was done by encasing the brain in a rigid solid bone structure, the skull. It is also the only organ to have a blood–brain barrier that adds an additional layer of protection. Yet there are problems, as the skull is rigid and does not allow much tolerance for cell swelling or edema. Any larger energy reserves, which could be stored in the brain parenchyma, like glycogen are therefore ruled out. The use of glycogen as an energy substrate will increase the osmolarity of the cytoplasm and needs tolerance to cell swelling, which is not possible because of the rigid skull encasing and the sensitivity of electrical signals to the concentration of ions and signal substances in the interstitial space. Any water movements across cell membranes will alter the volume of the interstitial space between brain cells and cause a change in the concentration of these substances. This will alter the electrical field across neuronal membranes and in turn the electrical signal processing. The brain also uses a disproportional amount of energy in relation to its weight at any time (20% of the body's energy consumption although it only represents 2% of its weight). Because of all these factors, the brain is dependent on a high blood flow rate to provide oxygen and glucose and to remove carbon dioxide. Thus, due to its lack of energy storage, the brain is dependent on a high rate of blood flow at any time. Reduction of blood flow to an area of the brain will lead to neuronal electrical dysfunction and ion gradient breakdown within seconds. Another problem is that the low tolerance for fluctuations of ions, pH, and signal molecules will result in severe consequences of a breach in the blood–brain barrier. It can lead to dysfunctional circuits including epileptic seizure discharges. Therefore, the brain, although reasonably protected against outside mechanical impact and injury, is highly vulnerable to internal disruptions due to its dependence on a high rate of blood flow [9]. This is despite a highly sophisticated blood–brain barrier and the fact that cells of the immune system reside directly in the brain parenchyma and constantly survey the environment for damage. In addition to all these challenges, the brain is also subject to another threat, not experienced in this intensity by other organs: with advancing age, the neuronal circuits are highly vulnerable to neurodegenerative processes [10]. Options for the reconstruction of the original tissue with its specific neuronal synaptic connections are limited as the integrity of the blood–brain barrier function has a high priority during repair processes and the simple recreation of new neurons with synaptic contacts lacks the memory traces laid down by earlier experiences. There is another problem with reconstituting brain tissue after injury at least in the CNS of higher mammals: stem cells with the capability of creating new neurons, while existing, have a limited capacity to lead to new neurons, which could be integrated into functional circuits.

However, neurons in the central nervous system are not working in isolation. Throughout their life cycle from development to recovery from injury, neurons form a functional unit and close partnership with various other cells from neuroepithelial lineage and from the immune system. These cells are various forms of glial cells, which are involved in the macro- and micro-support systems of neurons and the establishment of neuronal circuits with interacting modulating glial partners. They are also reacting to various degrees of injuries and damage by adjusting their phenotype to ensure the survival of essential signal processing. This adjustment of phenotypes of the glial support around neurons after functional disturbance is a phenomenon called reactive gliosis. Moreover, it is also observed during normal aging, albeit with less intensity. Crucial is the microglial–astrocytic communication axis. It is complicated by the occurrence of various degrees of neuroinflammation, and it may involve oligodendrocytes and oligodendrocyte progenitor cells as well as invading cells of the immune system. Its impact on neuronal health is not only protective but can also involve a toxic response detrimental to neuronal survival and regeneration. As will be seen, the reactive gliotic response is a continuation of the interactions between neurons and their cellular and non-cellular environment during neuronal genesis and development as well as of the homeostatic and signaling interactions in the adult organism.

References

1. Takano T, Oberheim N, Cotrina ML, Nedergaard M. Astrocytes and ischemic injury. Stroke. 2009;40(3 Suppl):S8–12.
2. Halliday GM. Re-evaluating the glio-centric view of multiple system atrophy by highlighting the neuronal involvement. Brain J Neurol. 2015;138(Pt 8):2116–9.
3. Robertson JM. The gliocentric brain. Int J Mol Sci. 2018;19(10):3033.
4. Pacholko AG, Wotton CA, Bekar LK. Astrocytes-the ultimate effectors of long-range neuromodulatory networks? Front Cell Neurosci. 2020;14:581075.
5. Palmieri B, Vadala M, Laurino C. Review of the molecular mechanisms in wound healing: new therapeutic targets? J Wound Care. 2017;26(12):765–75.
6. Sorg H, Tilkorn DJ, Hager S, Hauser J, Mirastschijski U. Skin wound healing: an update on the current knowledge and concepts. Eur Surg Res Europaische Chirurgische Forschung Recherches Chirurgicales Europeennes. 2017;58(1–2):81–94.
7. Gehart H, Clevers H. Tales from the crypt: new insights into intestinal stem cells. Nat Rev Gastroenterol Hepatol. 2019;16(1):19–34.

8. Allaire JM, Crowley SM, Law HT, Chang SY, Ko HJ, Vallance BA. The intestinal epithelium: central coordinator of mucosal immunity. Trends Immunol. 2018;39(9):677–96.
9. Attwell D, Laughlin SB. An energy budget for signaling in the grey matter of the brain. J Cereb Blood Flow Metab. 2001;21(10):1133–45.
10. de la Torre JC. Are major dementias triggered by poor blood flow to the brain? Theoretical considerations. J Alzheimer's Dis. 2017;57(2):353–71.

Chapter 2
The Brain as an Organ

Abstract Neurons are highly specialized for circuit activity. Their unique structure and function make them very vulnerable to small fluctuations in the composition of the extracellular space. A very important process is synaptic integration, which is the most energy-consuming process of the neuronal circuit. As the synapses are very often far from the cell body, a dynamic energy supply is a problem. For these and other challenges, the brain has an elaborate system of support or glial cells which in addition can act as partners in signaling pathways. The major group of glia are astrocytes, which have their own elaborate structure with domains and constitute a syncytium. They also cover the blood–brain barrier with endfeet. Their architecture destines them to be the prime partner of neurons in homeostasis and signaling. Another glial cell group is oligodendrocytes and their precursor cells. Oligodendrocytes specialize for myelination, whereas oligodendrocyte precursor cells are more complicated as they are on the receiving end of neuronal presynaptic endings. Finally, microglia are macrophages trapped in the brain after the establishment of the blood–brain barrier. They adapted to this peculiar environment by not only playing the part of a resident macrophage waiting for pathological challenges but also being involved in neuronal signaling. Three fluid systems are irrigating the brain. The blood supply thins out into fine processes, whose endothelial cells are the location of specific exchange systems and of the blood–brain barrier. The cerebrospinal fluid is created by the choroid plexus and migrates by bulk flow. It is in free exchange with the parenchyma and empties into the lymph and venous system, depending on pressure gradients. The glymphatic system is a convection system, from the perivascular arterial space through the glial syncytium and extracellular space toward several exit passages, venous system, and lymphatic system and on the way possibly mixing with cerebrospinal fluid. It is dependent on aquaporin channels and the dimensions of the extracellular space. That space is almost doubled during sleep, which eases the removal of waste products like β-amyloid peptide.

Keywords Astrocyte endfeet · Blood–brain barrier · Brain blood circulation · Brain capillaries · Brain extracellular space · Brain fluid dynamics · Cerebrospinal fluid · Choroid plexus · Glymphatic system · Neuronal energy demands

W. Walz, *The Gliocentric Brain*, https://doi.org/10.1007/978-3-031-48105-5_2

2.1 Introduction

Neurons and their circuits form the basis of signal processing in the central nervous system. The main communication channels between neurons and neurons and their effector organs are defined contacts, the chemical and electrical synapses. This mode of signaling is called wire transmission. There are additional modes for communication (for example, volume transmission [1]), but the synapses are the major mechanism by which neurons of a circuit contact each other. Neurons are highly differentiated cells that depend on the maintenance of optimal conditions of their environment for function and survival. In fact, neuronal circuits in the central nervous system are not capable of functioning properly without an elaborate support system of non-neuronal cells. The densely packed architecture of the brain tissue and the huge energy requirement of the neuronal machinery determine this imperative. These non-neuronal cells and their processes are intimately related to the neuronal components, often wrapping tightly around them. This proximity is important for interactions with, and support of, neurons on one hand, while at the same time restricting the dimensions of the extracellular volume around the neurons and their processes on the other hand. This leads to both morphological and functional complexity. Neurons can establish functional circuits in various cell cultures and ex vivo systems without support cells. However, in this case, the experimenter takes on the role of surrogate for the support cells by adjusting and maintaining the neuronal environment.

Short-term neuronal signal processing and integration relies on precise changes in ionic conductances of the cell membrane to generate action and synaptic potentials. The time courses of these conductance changes are very sensitive to the external concentration of a variety of ions and chemical compounds, many of them released by the neurons themselves during periods of high or even modest activity. The constrained dimensions of the extracellular space around the neurons facilitate a rapid build-up of such neuronally derived neuroactive compounds.

Neurons are very sensitive to external changes and less than optimal supply or clearance processes of most components in their immediate environment. Their major function is information processing, which relies on ionic current densities across their membranes. Even small changes in the microenvironment which do not threaten neuronal survival can still distort the neuronal signals and lead to detrimental impacts on the survival of the whole organism. This prompts the question: if supporting cells are so important strategically to protect neuronal integrity and functioning, do they also influence neuronal signal processing?

This chapter explores the cellular structures that surround the neuronal circuits and ensure the stability of neuronal function in the mature and healthy mammalian brain. It begins with the identification of neuronal demands for support, then it introduces the neuronal microenvironment and its main functional components. It concludes by covering the macrostructure of the brain that facilitates fluid exchange, and the rapid transport of fluid and solutes to and from the neuronal environment. The chapter emphasizes therefore the function of the brain as an assembly of specialized cells that are organized to promote the optimal signal output for various challenges to the individual organism. The chapter highlights the somewhat underappreciated concept of the brain as a smooth functioning entity with optimal supply and metabolic systems to ensure highly flexible and versatile signal processing systems.

2.2 Neurons and Their Demands of Their Microenvironment

Neurons are excitable cells. This means they are differentiated to conduct action potentials and to create synaptic currents and release the content of synaptic vesicles. From a functional point of view, neurons in the central nervous system (CNS) have several issues.

1. They have high energy demands. This is comparable with that of heart muscle and skeletal muscle cells during contraction as well as that of absorptive cells in the intestine during digestion and absorption. This inevitably results in the necessity for efficient removal of the waste products of aerobic metabolism: carbon dioxide and water.
2. Synaptic activity means the major sites of energy consumption in neurons are the pre- and postsynaptic endings. These synapses are quite often far removed from the cell body and therefore the nucleus. They are connected to the metabolic machinery of the cell body by a dendritic or axonal process with a small diameter and often several centimeters long. A compensating mechanism is needed which will avoid undue delays of energy supplies during times of intense synaptic activity.
3. Electrical signals are based on transmembrane ion fluxes. Some of these ions, such as potassium, calcium, and hydrogen, change their extracellular concentrations in relation to the intensity of the neuronal action potential frequency and synaptic potentials [2]. These ion concentrations influence the opening and closing kinetics of the many gates of the voltage-gated channels (see below for further details). This means any prolonged activity requiring net transmembrane flux of those ions proximate to the neuronal cell membrane has the potential to change the excitability properties of the neurons. Similar issues arise when the tight synaptic cleft is flooded with neurotransmitters, which have to be removed or deactivated quickly in order to retain their signal function. In summary, regardless of the cause, changes in extracellular ion concentration result in feedback which modulates synaptic activity. Therefore, astrocytes, capable of changing the concentration of extracellular ions, are well positioned to stabilize neuronal activity.
4. The issues described above make neurons and their components acutely dependent on their extracellular environment. They need specialized cells that assist in short- and long-term maintenance. The brain is a complicated organ with an extensive blood supply serving densely packed neurons and supporting or satellite cells. It is encased in a rigid skull for physical protection. The extracellular space of the brain acts functionally like a discrete fluid compartment with rela-

tively narrow dimensions. This exacerbates the issues outlined in points 1–3 above. Molecules released by active neurons tend to accumulate on the outside surface of the neuronal cell membrane because of the short diffusion paths and narrow dimensions of the extracellular space.

Key to neuronal excitability are the neuronal membrane conductances for ions. These depend on both the driving forces and various inside and outside signal substance concentrations. As neuronal activity is dynamic, with a high and variable energy consumption, it is a challenge to keep the environment of the extracellular space constant. This requirement for a close-to-constant internal milieu demands mechanisms that rapidly remove excess solutes and replenish molecules and ions that are taken up by neurons. Most of these ions and molecules that vary in their extracellular concentrations during neuronal activity modulate neuronal membrane mechanisms [2, 3]. This would lead to a detrimental feedback process which would distort the original signal. Therefore, compensatory mechanisms are needed that keep the extracellular changes evoked by neuronal activity within physiologically tolerable limits.

Neurons are some of the largest consumers of energy in the body. Unlike skeletal muscle and digestive epithelium, but more like cardiac cells, they are constantly active without resting phases. Indeed, the energy consumption of the brain hardly differs between behaviorally active and resting states [4]. The pattern of energy consumption is similar to that of continuously active cardiac muscle cells [5]. Even during sleep, there is heavy neuronal traffic. In the resting body, the human brain (with about 2% of body weight) is responsible for 20% of energy metabolism, 15% of cardiac output, 20% of oxygen, and 25% of glucose consumption [6]. Within the brain, the neurons consume 75–80% of the brain energy production with the remainder going to non-neuronal structures [7]. Within the neuron, about 80% of energy consumption is used for synaptic signaling. In myelinated neurons, the conduction of action potentials is very energy efficient compared with synaptic events. There are two main processes of energy consumption at the synapse: (1) Postsynaptic potentials require adenosine triphosphate (ATP) consumption to restore transmembrane ion gradients by transport mechanisms. (2) Vesicle recycling at the presynaptic ending appears to be the major user of ATP. The fact that the location of the major energy sink of neurons is the synapse means that for many neurons, the site of the major ATP consumption is far removed from the cytoplasm of the cell body [8].

The accumulation of neuroactive substances is aggravated by the shrinking of the volume of the extracellular space during different functional states (see below). Prime candidates for such neuroactive substances are potassium, calcium, and hydrogen ions as well as glutamate molecules. Potassium ions are released by action potential propagation. It has been estimated that one action potential can raise the extracellular potassium by 1 mM above the resting level of 2.7–3.2 mM in normal extracellular space (i.e., the space not being artificially enlarged by local damage caused by the tip of the extracellular microelectrodes) if compensatory clearance mechanisms by astrocytes are discounted [3]. Repeated neuronal activity including high-frequency discharges can increase the potassium concentration up to 12 mM

[9]. This is a problem as extracellular potassium ions modulate some channel activity and transmitter release [10]. An opposite dilemma arises when extracellular calcium falls from 1.2 to 0.8 mM during neuronal activity [11]. Prolonged neuronal activity leads to a marked external alkalinization of the extracellular space. This alkalinization has two sources: H^+ shifts into cells caused by glutamatergic transmission and bicarbonate shifts into the extracellular space caused by GABA (gamma-aminobutyric acid) mediated transmission. While there are resulting internal pH changes, external alkalinization alone influences ion channel kinetics due to interaction with charges on the surface proteins of neurons [12].

The synaptic cleft is even narrower than the extracellular space around the soma or unmyelinated axon. The signal function of neurotransmitter and neuromodulator release at the chemical synapse is dependent on these molecules being rapidly removed or inactivated after receptor interaction. Diffusion to neighboring synapses can be an additional problem [13]. Thus, there is a need to regulate the life span of neuroactive substances in the neuronal microenvironment. These examples illustrate the dependence of neuronal function on the homeostatic properties of their environment. If the external space around neurons was unlimited and effective diffusion or convection flow prevented the accumulation of substances around the neuronal membrane, these problems would not exist. However, the need to replenish the lost substances or their precursors would still exist. The reality is very different from the ideal case outlined above. A very narrow extracellular space with reduced diffusion results in major changes in ions and other neuroactive substances, requiring active regulation of extracellular fluid composition for proper function and the fine-tuning of signaling.

2.3 Extracellular Space

All tissues in the body are densely packed and have a narrow extracellular space, also called the interstitial space. However, the percentage of interstitial fluid volume in relation to the overall space of different organs differs. Skin and lung have a relative interstitial fluid volume which is about four times as large as for muscle, kidney, and brain [14]. This relatively small volume of brain extracellular space is not surprising, given its dense vascular supply, large number of support cells, and extensive neuronal morphology with myelinated and unmyelinated axons. In the body, interstitial fluid is involved in the generation of lymph and has a complicated architecture. It was recently proposed that this fluid space is dynamic and acts as a shock absorber [15]. The situation in the brain is somewhat different but no less complicated. First of all, the term "extracellular space" is a misnomer. It is, by convention, used in other organs to mean interstitial space, blood vessels, as well as lymph vessels. In the CNS (central nervous system), the term extracellular space is used differently, which refers to the interstitial space alone. It had been assumed that the percentage of the brain volume that is occupied by the extracellular space (ECS) is about 20–22%. This ratio has been recently revised by the work of Nedergaard's laboratory. Notably, 20% is the volume percentage of anesthetized animals; however, in awake animals, the ratio is considerably smaller, accounting for about 13% [16]. The blood volume is 2–4.7% depending on the brain region [17]. The ECS has a complicated structure and separates all cellular structures from each other except for gap junctions. The average width is not uniform and depends on fixation techniques, the method of anesthesia, and the brain region. At least in anaesthetized animals, it has been estimated to be 30–65 nm [18, 19]. It is narrower in the synaptic cleft and between the innermost myelin sheet and the axon membrane. Moreover, the ECS is not a uniform narrow space filled with water and small molecules. Three factors contribute to this complicated three-dimensional morphology.

1. The existence of dead-space microdomains: astrocytic processes, cell invaginations, dead-end pores, and the like [20].
2. In analytical models, the geometrical 3D arrangements of the cellular elements are more comparable with convex polyhedral than cubes [20].
3. The fluid that fills the ECS has a high viscosity which is not compatible with small solutes dissolved in water. Long-chain macromolecules dominate and create a so-called extracellular matrix (see below).

All three factors are obstacles to free diffusion away from cell surfaces. Diffusion along a straight line yields a tortuosity of 1.0, while the tortuosity of the ECS of the brain has been estimated to be approximately 1.6. The complicating factors cause an increase in the path length of any molecule that is able to diffuse [21]. The percentage of the brain volume that the ECS occupies is variable. As pointed out above, it is about 13% in the awake adult animal. Sleep or anesthesia increases the space by about 23%. It is about twice this percentage in the developing nervous system [21]. The ECS volume fraction varies between brain regions: for example, in the cat cortex, it is double the size in the molecular layer, stratum radiatum, and oriens, compared with the pyramidal layer [22]. It is reduced during neuronal activity in an intensity-dependent manner, with epileptic activity having an even larger effect on ECS shrinkage [22]. Anoxia/ischemia leads to a near collapse of the relative extracellular space to around 7% of the brain volume (in the anesthetized animal), a reduction that is reversible if the anoxia/ischemia phase does not last too long [23].

As indicated above, the ECS is not an "empty" water-filled space. Elongated molecules extend into this space and these molecules, associated with different cells, can make contact with each other. This extracellular matrix forms a microenvironment around cells with a surplus of negative charges, reduced diffusion rates, and a pH (potential of hydrogen, a measure of acidity, or alkalinity of a solution) value different from the ECS as a whole. Most of these molecules have transmembrane and intercellular signal function in addition to serving as structural support. The extracellular matrix is prevalent in connective, skin, and bone tissue. The basement membrane of epithelial cells is a specialized form of extracellular matrix [24]. The major components are synthesized within cells and subsequently exported by exocytosis. They consist of elongated chains of proteoglycans or polysaccharides, such as collagen, elastin, hyaluronic acid, and fibronectin. However, the composition of the matrix in the brain is different from that of the rest of the body. It has a hyaluronic acid backbone, and this backbone is synthesized in the cell membrane. Hyaluronic acid can be up to 25,000 repeats of disaccharides. Attached to this backbone are chondroitin sulfate proteoglycans (PSGs) and tenascins [25]. All cell types in the brain parenchyma are capable of synthesizing and assembling the molecular components of the extracellular matrix [26]. The extracellular matrix is functionally heavily involved in the development and maintenance of a healthy adult brain and in the aftermath of brain injuries. During development, the matrix is involved in cell migration, neurite outgrowth, and synaptogenesis. In the adult brain, its major functions are in structural stability, synaptic plasticity, and neuronal–astrocytic interactions. After injury, the matrix is heavily involved in gliosis and limiting the regeneration of axons [26]. In addition, there are at least two specialized functions of the brain extracellular matrix. First, around the blood–brain barrier, the matrix is somewhat differently organized from the parenchymal ECS matrix [27]. Second, on a subset of interneurons, perineuronal nerve nets are found and play an important role in synapse stabilization and plasticity [28].

2.4 Blood Supply

The microvascular supply system of the brain accounts for roughly 2–5% of its volume depending on the brain region [17, 29]. The blood flow in the parenchyma occurs in small arterioles (diameter 8–50 μm), capillaries (below 8 μm), and small venules [30]. The average distance between a neuronal cell body and the capillary is roughly 10–15 μm. The length of the total cerebral vasculature in the human brain is about 700 km [31]. The volume rate of the blood flow (velocity vs. volume flow) varies in different brain regions, being highest in gray matter and lowest in white matter. There is autoregulation of the blood vessel diameter to keep blood pressure changes within limits. Interestingly, partial carbon dioxide pressure is a more potent regulator of blood flow than oxygen tension: if carbon dioxide partial pressure is increased, cerebral blood flow is dramatically increased due to a corresponding reduction in vascular resistance. This means that the removal of this aerobic metabolism product is as important as the delivery of oxygen or glucose [32]. The barrier between blood and brain parenchyma is unique. It shields the brain from fluctuations of ions and metabolites in the blood. It also allows the brain to regulate its extracellular space (really the interstitial space) composition as it restricts the diffusion of substances that cannot cross lipid layers. The major component of the barrier is the endothelial cell layer. The cells of this layer are connected by tight junctions, ensuring that transport can normally only occur by diffusion through the endothelial cell membranes or by selective transport mechanisms across these membranes. On the parenchymal side of the endothelial barrier is a basement membrane (basal lamina). However, this membrane does not play a role as a physical barrier. The basal lamina is covered by astrocytic endfeet. Although these endfeet interact with endothelial cells, they, like the basement membrane, are not a functional barrier and the space between these endfeet is freely available for diffusion [33]. Pericytes and smooth muscle cells are in the space between the basal lamina and astrocytic endfeet. Whereas smooth muscle cells are in the medial layer of arteries and venules, multi-functional pericytes wrap around endothelial cells of capillaries. Pericytes appear to be an important signal transmission station between neurons, astrocytes, and endothelial cells as their absence leads to a dysfunctional neurovascular unit. They are also important for angiogenesis [34]. Another important feature of the blood–brain interphase is the choroid plexus. It is the site of the production of the cerebrospinal fluid, and it is highly vascularized as it has five times the normal blood flow rate. The surface area of the choroid plexus represents up to half the surface area of all brain endothelial cells together due to the presence of microvilli (2–5 m^2 compared with 10 m^2 in the human brain). It does not consist of endothelial cells but is a cell layer derived from ependymal cells [35]. However, in contrast to the ependymal cell layer separating cerebrospinal fluid and interstitial fluid, these ependymal cells of the choroid plexus are connected by tight junctions. The cells of the choroid plexus have a high mitochondria density allowing for the active accumulation of solutes. The accumulation of solutes, and the resultant increase in osmotic pressure, draws water into the ventricles to form the cerebrospinal fluid. This subject will be addressed in more detail later in this chapter.

2.5 Glial Cell Relationships and Lineage

Historically, the term "glial cell" was used as a general term covering all non-neuronal cells within the central nervous system. The cells within the parenchyma and its lining (except the endothelial cells around the capillaries) and the cells with close contact with neurons are all termed glial cells, thus, including astrocytes, oligodendrocytes, microglia, ependymal cells, tanycytes, and NG2 cells (nerve-glial antigen 2 cells, polydendrocytes, or oligodendrocyte precursors). However, even within a glial cell type like astrocytes, there is a huge heterogeneity of functional properties as will be shown later. Although occurring only in the nervous system, these glial cell types account for 20% of all human body cells [36]. What is their function? What is the relationship between these glial cell types? What is their familial lineage? The definition of a glial cell and subtypes and how they relate to each other functionally and by lineage is crucial.

New genetic tools have facilitated the investigation of the function and lineage of these glial cells [37]. Fluorescent protein labeling in transgenic mice, viral injections, or deoxyribonucleic acid (DNA) fluorescent reporter vectors such as *Brainbow* or *StarTrack* have enabled the tracking of the fate of these cells from the progenitor to the final differentiated product [38, 39]. Lineage relationships, migration routes, final destinations, and changes in functional properties can now be delineated much more convincingly than ever before.

Astrocytes form the largest group of glial cells. They are very heterogeneous. The main distinction within the category of parenchymal astrocytes is the separation into fibrous and protoplasmic astrocytes [40]. Fibrous astrocytes are situated in white matter and have a morphology that accommodates tract architecture: they are elongated along fiber tracts. Protoplasmic astrocytes are evenly spaced within gray matter and have a completely different morphology. They have a huge number of fine and complex multi-directional processes and occupy specific domains in the parenchyma. There is no "empty" space that is not occupied by such an astrocytic domain. The overlap between astrocytic domains is negligible. Astrocytes send processes to the blood capillaries which they cover completely with endfeet. Other astrocytic processes reach the perisynaptic space between neurons and are close to dendritic spines. In rodents, one astrocytic domain can cover 20,000–120,000 synapses. In humans, one domain can encompass 270,000 to 2 million synapses [41]. The processes of adjacent astrocytes are joined by gap junctions. These junctions facilitate strong electrical coupling, so much so, that astrocytic domains are joined together as a functional syncytium. The "gold standard" for astrocyte identification was, and is, that they express glial fibrillary acidic protein (GFAP), an intermediary filament. It has been generally assumed that brain cells that express this protein are astrocytic in nature. However, there are problems with this assumption. For example, astrocytes in the cerebral cortex express GFAP at or below the detection limit [42]. Neural stem cells in the subventricular zone (SVZ) also express GFAP, and there is an ongoing controversy over the astrocytic nature of these cells (see below). Another frequently used astrocyte marker is S100β (calcium-binding protein B). Astrocytes also express glutamate transporter proteins (GLAST or glutamate-aspartate transporter and GLT-1 or excitatory amino acid transporter 2 or EAAT2). These markers can be used in combination with each other to verify the astrocytic nature of a cell under investigation [43]. However, there are differences in the expression of these proteins in major CNS regions like spinal cord versus cerebral cortex and even within closely related brain regions. This is especially obvious if one looks at potassium channels. While astrocytes are not capable of creating action potentials, they do possess a variety of voltage-gated channels. The prevalence and specific subtypes of inverted rectifier potassium channels differ during development and vary from one brain region to another. Similar situations exist for other voltage-gated channels. Populations of astrocytes in the different layers and columns of the cerebral cortex have different functional identities [44]. This regional variation of functional properties has caused some investigators to refer to these functional groupings as "astromeres" [45]. An astromere is considered a group of regionally specialized astrocytes with functional properties tailored to the needs of, and interaction with, surrounding neurons.

In human and non-human primates, astrocytic heterogeneity is different from that in other mammals. Human fibrous and protoplasmic astrocytes have a different morphology and density. Human fibrous astrocytes are also larger, with a more complex morphology [46]. Human protoplasmic astrocytes have a larger cell body and more processes. They are also organized in domains, but the field of overlap between adjacent domains is larger. There are also two unique classes of astrocytes in the human cerebral cortex not seen in rodents: interlaminar and varicose projection astrocytes [41]. The interlaminar astrocytes are in layer 1 of the cerebral cortex and extend a few processes with large diameters to the pial glial limitans and to deeper layers. The varicose projection astrocytes are in layers 2–4 and have just a few processes that extend to more distant regions. The processes have regularly spaced varicosities. While the function of these unique human cells is not yet clear, they likely indicate a more complex astrocytic neuronal interaction. Because they are unique to humans, rodent models cannot be used to investigate their function.

In the adult mammalian brain, two groups of cells create new neurons: cells in the subgranular zone (SGZ) in the dentate gyrus area of the hippocampus and cells in the subventricular zone (SVZ) adjacent to the lateral ventricles. In the SGZ, there is a population of slowly dividing astrocytes with a resemblance to radial glia. Their progenitors differentiate into granular neurons, migrating in stages [47]. However, the SVZ is the far larger neuron-generating site, where there are two populations of neural stem cells in the wall of the lateral ventricles, next to the ependymal cells. There is a slower and a faster dividing population of these stem cells. The faster dividing subpopulations derive from the slower proliferating one. This is in all likelihood to keep the number of cell divisions of the slower dividing group low in order to prevent mutations during DNA (deoxyribonucleic acid) copying, thus preserving their integrity during the lifetime of the organism as much as possible. Most of the faster dividing group differentiates into neuroblasts and then differentiates again along the migration route to the olfactory bulb to replace olfactory neurons. However, not all faster proliferating neural precursor cells (NPCs) turn into neurons. Some differentiate into oligodendrocytes and astrocytes. It has been shown that there is a regional specialization: the progenitors of the lateral wall lead to more neurogenesis, whereas those in the septal wall lead to more gliogenesis [48]. This has been demonstrated for the rodent brain, and there are indications that the human SVZ performs in a similar fashion [47].

In addition to this complex relationship, there are specialized astrocytes other than those associated with compartmental borders where properties and morphology are different, but not fundamentally so. Most prominent are the Muller cells of the retina, the Bergmann glia of the cerebellum, and the pituicytes of the neurohypophysis. Velate astrocytes in the cerebellar cortex are related to Bergmann glia [49].

NG2 glia are named after a specific antigen (neuron-glial antigen 2) and are also called oligodendrocyte progenitor cells (OPCs) or polydendrocytes in order not to name them after a marker [50]. NG2 glia are homogenously distributed throughout the whole adult brain. However, their cellular properties suggest different cell subsets [51]. Their function in the white matter of the brain is clearly the generation of myelinating oligodendrocytes. They have an additional function other than serving as OPCs in white matter. It has been shown that NG2 glia form synapses with neurons in some brain regions [52]. The neuronal partner forms the presynaptic part while the glial partner forms the postsynaptic part, leading to a signal transfer from neuron to glia. Various transmitters are used, and miniature excitatory synaptic potential has been recorded in NG2 glia as a response to neuronal stimulation. The function of this signal system is unknown. Oligodendrocytes are the myelinating cells of the CNS. The number of myelin sheets, their thickness, and the circuit specificity are important for action potential processing along an axon.

Oligodendrocytes receive a multitude of neuronal signals. Recently, evidence has been collected for neuronal activity-regulated myelination [53].

Microglia are not related to astrocytes or oligodendrocytes. They are cells of the immune system, the resident macrophages of the CNS. They originate from erythromyeloid progenitor cells [54] and during human development migrate into the brain. They are eventually trapped there by the formation of the blood–brain barrier. Microglia participate in developmental processes like synaptic pruning. In the adult brain, their density is decreased, and they become ramified resting microglia. They demonstrate self-renewal and are activated in pathological states [55]. However, there are indications that they are also participants in normal, non-pathological processes of the nervous system and not just silent partners, only activated by injury or damage [56].

2.6 Brain Fluid Compartments and Their Borders

The arterioles are the major site of regulation of blood flow velocity due to the setting of the vascular tone by smooth muscle. Blood flows from arterioles through capillaries and then to venules. All brain blood vessels are surrounded by a single layer of endothelial cells with tight junctions between them. The endothelial cells of the capillaries are the site of gas exchange and uptake of glucose and other major nutrients. The tight junctions are the physical basis of the blood–brain barrier. Such a barrier is unique to the brain and its functional implication is, that unless substances are lipid soluble, the only way they can cross from blood to brain parenchyma is by specific transport processes through endothelial cells. Venules are the preferred entrance site of leukocytes during brain injury. Only a small number of leukocytes enter through arterioles, and virtually none through capillaries [57]. Venules join eventually to form the larger veins. These veins in turn collect into larger vessels on the surface of the brain in the subarachnoid space. They then cross the arachnoid mater into the dural venous sinuses. From there the venous blood drains into the major veins of the body. The venous drainage system of the brain is not parallel to the arterial supply system [58].

There is another compartment in the brain, which shares some features with the body's lymph system. This is the cerebrospinal system containing the cerebrospinal fluid (CSF). The cerebrovascular system is mainly located in the four cerebral ventricles. The cerebral spinal fluid is created in the choroid plexus. Each ventricle has such a choroid plexus. In the plexus vasculature, the endothelial cells lack tight junctions. Instead, a form of ependymal cells (see below) with tight junctions is found on the surface facing the ventricles. These cells create the CSF by secretion of ions into the ventricular space. The resulting increase in the osmotic pressure inside the ventricles is the cause of water flow into the CSF from blood. Thus, the active transport of ions and the subsequent passive flux of water by the choroid plexus is the cause of CSF bulk flow inside the ventricles. The choroid plexus is not as active as the endothelial cells of the capillaries in the extraction of nutrients into the brain. However, it is a main site of ascorbate and folate uptake and likely some other substances. The CSF exits the brain into the subarachnoid space and from there mainly into venous blood sinuses at the arachnoid granulations [59]. However, lately, evidence accumulated for another exit pathway: drainage into the meningeal lymphatic system (see below).

There are two kinds of glial cells that form the borders of the different brain compartments. Those lining the ventricular surfaces and those lining the surface of the brain just under the pia mater. Ependymal cells are glial cells that line the border between parenchyma and ventricles and therefore between extracellular fluid and CSF. They have features of epithelial cells. Microvilli and cilia are present on the ventricular side. Gap junctions between ependymal cells facilitate the synchronization of the beating cilia into rhythmic movement, assisting the streaming of CSF. Gap junctions also extend to the astrocytes within the parenchyma. It was therefore speculated that this might indicate the existence of a "panglial syncytium" [60]. There are no tight junctions, and the space between the ependymal facilitates the free diffusion of large molecules. There is a basement membrane on the parenchymal side of the ependymal cell layer. Despite the lack of tight junctions, this could be a partial barrier that is regulated, as there are adherens junctions in this cell layer [61]. Choroid plexus cells are a form of ependymal cells and are continuous with the ependymal cell layer. They line special surfaces on the four ventricles and are connected by tight junctions (unlike the ependymal cells). As discussed above, they are responsible for the creation of the CSF.

Tanycytes have features of radial glial cells and astrocytes. There appear to be four types with different functions, some of which are involved in interactions between CSF, portal blood, and hypothalamus [62]. They line the floor of some ventricles, and some are associated with circumventricular organs and extend their processes into the hypothalamus. They are actively involved in energy metabolism and release neuropeptides into the blood as well as sense the glucose concentration in the blood [63]. Other ependymal cells with specialized functions form similar barriers with specialized functions around circumventricular organs [64].

The glia limitans is a thin barrier formed by the endfeet of marginal glia just below the pia mater surface of the brain. Marginal glia are also called surface astrocytes or layer 1 astrocytes [65]. The pia mater itself is made up of fibroblasts. Between the astrocytic endfeet and the pia mater is a basement membrane. Marginal glia have cell bodies close to the pia mater so their endfeet do not extend far from the cell bodies at the pia mater. However, on their opposite side, these marginal glia extend long processes into the brain parenchyma. The glia limitans (endfeet of the marginal glia with basement membrane) covers the whole innermost surface of the brain with two exceptions: the olfactory bulb which is on the migration route of differentiating neuroblasts, and the cerebellum where the glia limitans is made up of foot-plates of Bergmann glia [66]. The function of these cells is not clear. They are likely involved in controlling access of immune cells, mechanical buffering, and osmoregulation.

2.7 Brain Fluid Dynamics

There are three different circulatory systems in the brain, each of them with a different functional emphasis (see Fig. 2.1). However, there is also some overlap in their functions.

2.7.1 *Brain Blood Supply*

Arterial, oxygen-rich blood enters the brain via a pair of internal carotid arteries and a pair of cerebral arteries. The pair of cerebral arteries join to form the basilar artery. Arterial blood from all three arteries is mixed upon entering the circle of Willis. This anastomosis of the supply arteries ensures that an occlusion of any of the four arteries entering the brain does not have disastrous consequences. Three pairs of cerebral arteries branch off from the circle of Willis (anterior, middle, and posterior cerebral arteries). They then branch further to distribute arterial blood to the brain. Several cerebellar arteries branch from the basilar artery out before the blood enters the circle of Willis. The arterial blood supply separates finally into small arterioles, which are contractile due to smooth muscles. The cerebral arteries run along the cortical surface in the subarchnoid space. They then penetrate the brain surface through the pia mater as pial vessels and transition into arterioles. They then turn into capillaries which are the site of gas exchange and uptake of nutrients. The capillaries are just large enough for red blood cells to squeeze through, which facilitates

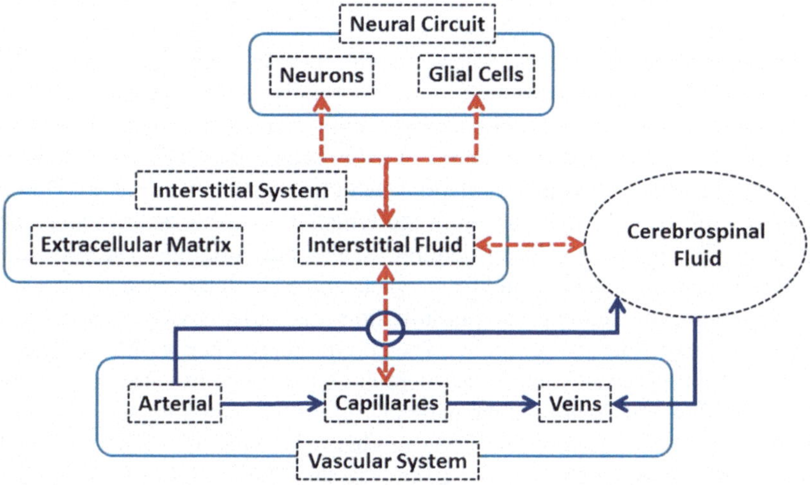

Fig. 2.1 Three compartments of the brain and their substance exchanges. The blue solid line represents the "one-way" exchange of substances, and the red dashed line represents interactive exchanges. (From Ref. [74] with permission. Copyright Elsevier Publishers)

gas exchange by reducing the length of the diffusion pathway. Finally, the small venules receive relatively oxygen-poor and carbon dioxide-rich blood from the capillaries. Arterioles, capillaries, and venules are covered by endothelial cells with tight junctions. These are the site of the blood–brain barrier. The endothelial cells can selectively transport substances in or out of the brain as the diffusion pathway is only available for lipophilic substances. Most of the uptake of substances by the endothelial cells is in the capillary section. The arterioles can change their diameter due to smooth muscle contraction or relaxation. They therefore have a major role in blood supply regulation. The small venules are the major site of entry for cells from the immune system during pathological conditions [30–32, 57].

2.7.2 Cerebrospinal Fluid

The cerebrospinal fluid or ventricular system is a dynamic fluid exchange device with similarities to the lymphatic system. The fluid is created by the choroid plexus, a group of specialized ependymal cells, which are on the ventral surface of each of the four ventricles (see Fig. 2.2). The endothelial cells of the capillaries do not contribute to the CSF [59]. The cells of the choroid plexus transport ions and other substances into the ventricular cavities, creating an osmotic gradient that causes water molecules to diffuse from the blood into the cerebrospinal cavity. This continuous diffusion of water at the choroid plexus creates a pressure gradient within the cerebrospinal fluid cavity which is responsible for the CSF bulk flow [35]. It is supported by coordinated movements of the cilia of the ependymal cells. In humans, the CSF volume is about 150 ml and the flow rate is about 0.4 ml/min. The direction of the flow is from the two lateral ventricles through the interventricular foramina and further into the third ventricle. From there, the CSF flows through the cerebral aqueduct to the fourth ventricle where it splits into two pathways: one down the central canal of the spinal cord (ending up eventually in the cervical lymph node) and the other flows via three small openings (foramina) into the subarachnoid space. If the CSF pressure exceeds the venous blood pressure, the CSF enters the venous sinuses through the arachnoid granulations, which act as one-way valves. However, a major part of the CSF will end up in the lymphatic vessels of the subarachnoid space if the CSF pressure is lower than the venous pressure [67]. The CSF flow in the ventricles is pulsatile back and forth synchronized with the cardiac cycle with a resulting slow net movement. The pulsatile flow leads to a mixing of the CSF [68]. Ependymal cells border the ventricles. They do not represent a significant barrier to diffusion into the parenchyma. Thus, there is an exchange of substances between extracellular fluid (ECF) around the parenchymal cells (including neurons) and the CSF. There are no major concentration gradients between ECF and CSF and no substances are sequestered in either compartment. The choroid plexus transports inorganic ions such as sodium, chloride, and bicarbonate to create the osmotic pressure difference. It also transports trace metals (manganese), ascorbate, folate, and

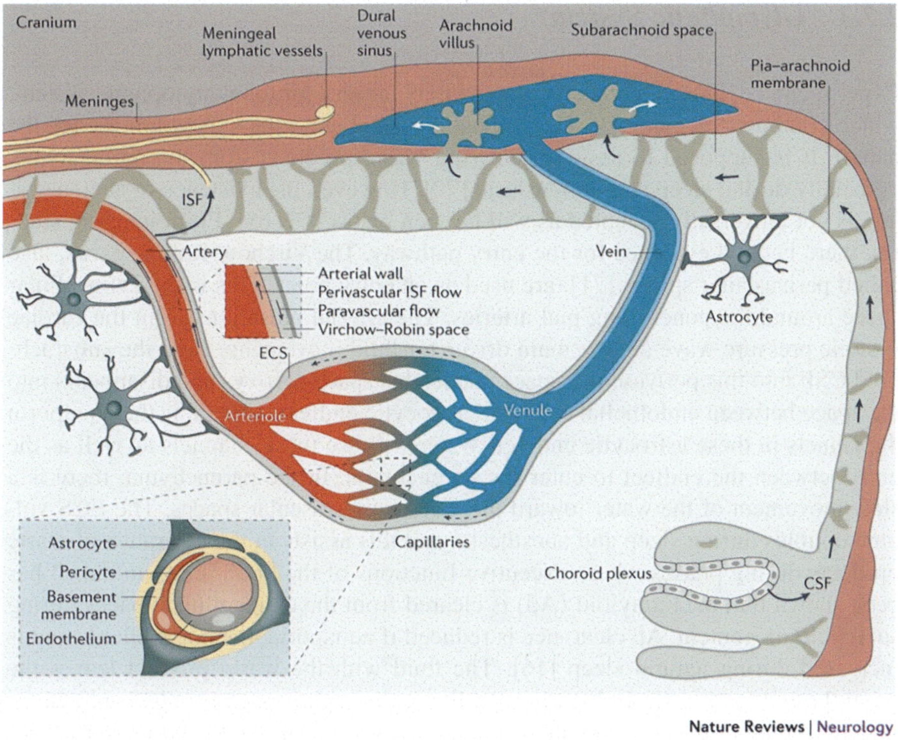

Fig. 2.2 Brain capillaries are a key site of the blood–brain barrier (BBB). The capillary cross section (large inset) shows a tightly sealed endothelium, which shares a common basement membrane with pericytes and astrocyte end-feet wrapping around the capillary wall. The arterial cross section (small inset) shows the perivascular flow of interstitial fluid (ISF) through the arterial wall in the opposite direction to blood flow; paravascular flow might also occur in the same direction as blood flow. CSF is produced by the choroid plexus and flows from brain ventricles into subarachnoid spaces, draining into the meningeal lymphatic system and/or venous blood through the arachnoid villi. ISF can exchange with CSF in the ventricles (not shown) and subarachnoid spaces. ECS, extracellular space. (From Ref. [75] with permission. Copyright Springer Nature)

nucleosides as well as some peptides (leptin). In addition, the choroid plexus synthesizes some substances and releases them into the CSF (brain-derived neurotrophic factor). Thus, the choroid plexus is an important uptake system in addition to the endothelial cells of the capillaries. The lack of a diffusion barrier at the CSF–ECF interface facilitates a mixture of these two distribution systems. The circadian rhythm hormone melatonin is synthesized in the pineal gland and secreted into the CSF at the third ventricle. It is then distributed through all the brain fluid spaces [69]. There is also some evidence that the CSF flow system removes certain metabolic products from the ECF into the venous blood and lymph by passive distribution (copper, riboflavin).

2.7.3 Glymphatic System

Experimental evidence has accumulated lately for an additional glymphatic system, which combines elements of the lymphatic system and glial syncytium (hence the name). It is supposed to assist in clearing waste products from the parenchyma, especially during sleep and non-arousal [70]. However, the existence of such a system is not universally accepted as will be seen below [59, 68]. From imaging studies, there is good evidence for the entry pathway. The Virchow-Robin spaces, also called perivascular spaces [71], are used as an entry point. This is the extracellular space around the penetrating pial arteries. The arterial pulsations from the cardiac systolic pressure wave are the main driver for fluid movements from the subarachnoid CSF into this perivascular space and as this space narrows and disappears into the space between endothelial cells and astrocytic endfeet. Crucial is the aquaporin 4 channels in these astrocytic endfeet. Water will use these channels as well as the gaps between the endfeet to enter the parenchyma. In the parenchyma, there is a slow movement of the water toward the venous perivascular spaces. The ECS volume doubles during sleep and anesthesia, and this assists in the clearance of waste products during phases where executive functions of the brain are minimal. It has been shown that beta-amyloid (Aβ) is cleared from the extracellular spaces during such fluid movement. Aβ clearance is reduced if aquaporin 4 is not available and is increased during natural sleep [16]. The fluid with the waste product leaves the brain through astrocytic endfeet around the perivascular space venules and venous system. As the flow is outside the venous blood vessels, it would end up in the subarachnoid space, partly being recycled through glymphatic uptake, partly flowing into the venous system through arachnoid villi and partly entering the meningeal lymphatic system [72]. While the entrance of fluid, its clearance function, and the importance of sleep are well documented for the glymphatic system, imaging the exit pathway is a problem. Experimental fluorescence tracers entering the brain with this pathway can be imaged. However, they become too diluted to establish a clear proof of exit. There are also questions as to whether arterial pulsations are large enough to drive the exit system, and whether other mechanisms could account for this exit. There are additional questions about how the waste products can enter the perivascular space through aquaporin 4 channels. The resistance through and around the astrocytic endfeet might be low enough for water movements, but not for movements of larger molecules like Aβ. An alternative mechanism might be an outflow of waste-loaded fluid into the CSF at the level of the ependymal cell layer. The ependymal cells have no tight junctions and constitute a panglial syncytium with the underlying astrocytes. This might represent an alternate exit through this intraglial pathway into the CSF. As the ECS expands during sleep, the diffusional resistance through the ependymal cells into the ventricles could also be reduced. In any event, there is a mixing between CSF and glymphatic outflow either at the ventricular and/or subarachnoid level. Some CSF also ends up in the lymphatic system, especially when the CSF pressure is lower than venous sinus pressure (see above). Thus the meningeal lymphatic system has an important role in the drainage and clearance of

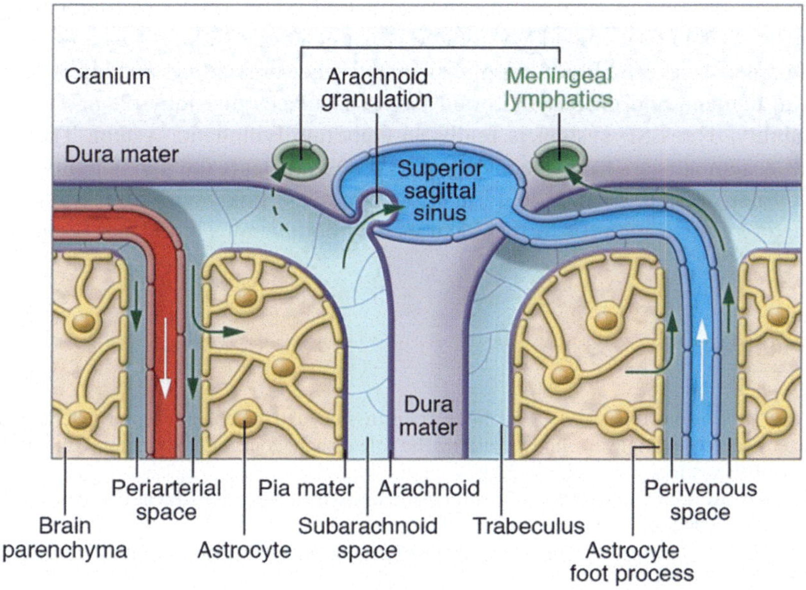

Fig. 2.3 Close-up view of ISF and CSF circulation. The perivascular glymphatic drainage system transports CSF and solutes into the brain via a periarterial pathway, whereas ISF and solutes exit the brain via the perivenous glymphatic pathway. CSF can enter the venous system via arachnoid granulations, and CSF macromolecules and immune cells are transported mainly along the dural lymphatic vessels into the lymph nodes and extracranial systemic circulation. (From Ref. [76] with permission. Copyright American Association of Clinical Investigation)

the brain. Glymphatic efflux and CSF drainage merge under certain circumstances in the meningeal lymphatic system [72] (see also Fig. 2.3). This hypothesized glymphatic pathway is very attractive and would account for one function of sleep and the variable ECS. Its dysfunction might be one underlying pathophysiological mechanism for neurological and psychiatric diseases with exciting therapeutic potential [73]. However, the so far poorly explained exit pathway warrants further investigation.

2.8 Conclusions

The brain has similar support systems as other organs of the body. Yet, there is a difference. Neurons not only have demands concerning supply and waste product removal, but they also need a very stable extracellular microenvironment. This is due to their specialized signal function with ion currents highly sensitive to electrochemical driving forces. An additional complication is that neuronal synaptic and action potentials themselves change this stable environment. Neurons therefore are in a very close interactive relationship with their microenvironment, the composition of which will respond to changes in neuronal activity. This microenvironment of the parenchyma is in turn embedded in three larger elaborate support systems that promote homeostasis. The blood supply, as in any other organ, has specializations that react to demand and protect the brain from substances that may change neuronal excitability. The CSF system is really an elaborate lymphatic system. The glymphatic system appears to be a convection system that clears the parenchyma of substances detrimental to neuronal function and then links with the brain's own circulatory CSF system for further clearance by both venous blood and lymphatics.

References

1. Fuxe K, Borroto-Escuela DO. Volume transmission and receptor-receptor interactions in heteroreceptor complexes: understanding the role of new concepts for brain communication. Neural Regen Res. 2016;11(8):1220–3.
2. Ransom BR. Glial modulation of neural excitability mediated by extracellular pH: a hypothesis. Prog Brain Res. 1992;94:37–46.
3. Walz W. Role of astrocytes in the clearance of excess extracellular potassium. Neurochem Int. 2000;36(4–5):291–300.
4. Raichle ME. The restless brain: how intrinsic activity organizes brain function. Philos Trans R Soc Lond Ser B Biol Sci. 2015;370(1668):20140172.
5. DiNuzzo M, Nedergaard M. Brain energetics during the sleep-wake cycle. Curr Opin Neurobiol. 2017;47:65–72.
6. Attwell D, Laughlin SB. An energy budget for signaling in the grey matter of the brain. J Cereb Blood Flow Metab. 2001;21(10):1133–45.
7. Hyder F, Rothman DL, Bennett MR. Cortical energy demands of signaling and nonsignaling components in brain are conserved across mammalian species and activity levels. Proc Natl Acad Sci U S A. 2013;110(9):3549–54.
8. Alle H, Roth A, Geiger JR. Energy-efficient action potentials in hippocampal mossy fibers. Science (New York, NY). 2009;325(5946):1405–8.
9. Somjen GG. Ion regulation in the brain: implications for pathophysiology. Neuroscientist. 2002;8(3):254–67.
10. Balestrino M, Aitken PG, Somjen GG. The effects of moderate changes of extracellular K+ and Ca2+ on synaptic and neural function in the CA1 region of the hippocampal slice. Brain Res. 1986;377(2):229–39.
11. Torres A, Wang F, Xu Q, Fujita T, Dobrowolski R, Willecke K, et al. Extracellular Ca(2)(+) acts as a mediator of communication from neurons to glia. Sci Signal. 2012;5(208):ra8.
12. Ruusuvuori E, Kaila K. Carbonic anhydrases and brain pH in the control of neuronal excitability. Subcell Biochem. 2014;75:271–90.
13. Murphy-Royal C, Dupuis J, Groc L, Oliet SHR. Astroglial glutamate transporters in the brain: regulating neurotransmitter homeostasis and synaptic transmission. J Neurosci Res. 2017;95(11):2140–51.
14. Aukland K, Reed RK. Interstitial-lymphatic mechanisms in the control of extracellular fluid volume. Physiol Rev. 1993;73(1):1–78.
15. Benias PC, Wells RG, Sackey-Aboagye B, Klavan H, Reidy J, Buonocore D, et al. Structure and distribution of an unrecognized Interstitium in human tissues. Sci Rep. 2018;8(1):4947.
16. Xie L, Kang H, Xu Q, Chen MJ, Liao Y, Thiyagarajan M, et al. Sleep drives metabolite clearance from the adult brain. Science (New York, NY). 2013;342(6156):373–7.
17. Ito H, Kanno I, Kato C, Sasaki T, Ishii K, Ouchi Y, et al. Database of normal human cerebral blood flow, cerebral blood volume, cerebral oxygen extraction fraction and cerebral metabolic

rate of oxygen measured by positron emission tomography with 15O-labelled carbon dioxide or water, carbon monoxide and oxygen: a multicentre study in Japan. Eur J Nucl Med Mol Imaging. 2004;31(5):635–43.
18. Xiao F, Nicholson C, Hrabe J, Hrabetova S. Diffusion of flexible random-coil dextran polymers measured in anisotropic brain extracellular space by integrative optical imaging. Biophys J. 2008;95(3):1382–92.
19. Thorne RG, Nicholson C. In vivo diffusion analysis with quantum dots and dextrans predicts the width of brain extracellular space. Proc Natl Acad Sci U S A. 2006;103(14):5567–72.
20. Hrabe J, Hrabetova S, Segeth K. A model of effective diffusion and tortuosity in the extracellular space of the brain. Biophys J. 2004;87(3):1606–17.
21. Nicholson C, Hrabetova S. Brain extracellular space: the final frontier of neuroscience. Biophys J. 2017;113(10):2133–42.
22. Dietzel I, Heinemann U. Dynamic variations of the brain cell microenvironment in relation to neuronal hyperactivity. Ann N Y Acad Sci. 1986;481:72–86.
23. Zoremba N, Homola A, Slais K, Vorisek I, Rossaint R, Lehmenkuhler A, et al. Extracellular diffusion parameters in the rat somatosensory cortex during recovery from transient global ischemia/hypoxia. J Cereb Blood Flow Metab. 2008;28(10):1665–73.
24. Iozzo RV, Gubbiotti MA. Extracellular matrix: the driving force of mammalian diseases. Matrix Biol. 2018;71–72:1–9.
25. George N, Geller HM. Extracellular matrix and traumatic brain injury. J Neurosci Res. 2018;96(4):573–88.
26. Song I, Dityatev A. Crosstalk between glia, extracellular matrix and neurons. Brain Res Bull. 2018;136:101–8.
27. Almutairi MM, Gong C, Xu YG, Chang Y, Shi H. Factors controlling permeability of the blood-brain barrier. Cell Mol Life Sci. 2016;73(1):57–77.
28. Testa D, Prochiantz A, Di Nardo AA. Perineuronal nets in brain physiology and disease. Semin Cell Dev Biol. 2019;89:125–35.
29. Rapoport SI, Ohno K, Pettigrew KD. Drug entry into the brain. Brain Res. 1979;172(2):354–9.
30. Hudetz AG. Blood flow in the cerebral capillary network: a review emphasizing observations with intravital microscopy. Microcirculation (New York, NY: 1994). 1997;4(2):233–52.
31. Zlokovic BV. Neurovascular mechanisms of Alzheimer's neurodegeneration. Trends Neurosci. 2005;28(4):202–8.
32. Smith KJ, Ainslie PN. Regulation of cerebral blood flow and metabolism during exercise. Exp Physiol. 2017;102(11):1356–71.
33. Sweeney MD, Zhao Z, Montagne A, Nelson AR, Zlokovic BV. Blood-brain barrier: from physiology to disease and back. Physiol Rev. 2019;99(1):21–78.
34. Kolinko Y, Kralickova M, Tonar Z. The impact of pericytes on the brain and approaches for their morphological analysis. J Chem Neuroanat. 2018;91:35–45.
35. Marques F, Sousa JC, Brito MA, Pahnke J, Santos C, Correia-Neves M, et al. The choroid plexus in health and in disease: dialogues into and out of the brain. Neurobiol Dis. 2017;107:32–40.
36. Pelvig DP, Pakkenberg H, Stark AK, Pakkenberg B. Neocortical glial cell numbers in human brains. Neurobiol Aging. 2008;29(11):1754–62.
37. Bribian A, Figueres-Onate M, Martin-Lopez E, Lopez-Mascaraque L. Decoding astrocyte heterogeneity: new tools for clonal analysis. Neuroscience. 2016;323:10–9.
38. Cai D, Cohen KB, Luo T, Lichtman JW, Sanes JR. Improved tools for the Brainbow toolbox. Nat Methods. 2013;10(6):540–7.
39. Martin-Lopez E, Garcia-Marques J, Nunez-Llaves R, Lopez-Mascaraque L. Clonal astrocytic response to cortical injury. PLoS One. 2013;8(9):e74039.
40. Oberheim NA, Goldman SA, Nedergaard M. Heterogeneity of astrocytic form and function. Methods Mol Biol. 2012;814:23–45.
41. Oberheim NA, Takano T, Han X, He W, Lin JH, Wang F, et al. Uniquely hominid features of adult human astrocytes. J Neurosci Off J Soc Neurosci. 2009;29(10):3276–87.
42. Kimelberg HK. The problem of astrocyte identity. Neurochem Int. 2004;45(2–3):191–202.

43. Emsley JG, Macklis JD. Astroglial heterogeneity closely reflects the neuronal-defined anatomy of the adult murine CNS. Neuron Glia Biol. 2006;2(3):175–86.
44. Kiyoshi CM, Du Y, Zhong S, Wang W, Taylor AT, Xiong B, et al. Syncytial isopotentiality: a system-wide electrical feature of astrocytic networks in the brain. Glia. 2018;66(12):2756–69.
45. Bayraktar OA, Fuentealba LC, Alvarez-Buylla A, Rowitch DH. Astrocyte development and heterogeneity. Cold Spring Harb Perspect Biol. 2014;7(1):a020362.
46. Oberheim NA, Wang X, Goldman S, Nedergaard M. Astrocytic complexity distinguishes the human brain. Trends Neurosci. 2006;29(10):547–53.
47. Pino A, Fumagalli G, Bifari F, Decimo I. New neurons in adult brain: distribution, molecular mechanisms and therapies. Biochem Pharmacol. 2017;141:4–22.
48. Mizrak D, Levitin HM, Delgado AC, Crotet V, Yuan J, Chaker Z, et al. Single-cell analysis of regional differences in adult V-SVZ neural stem cell lineages. Cell Rep. 2019;26(2):394–406.e5.
49. Chan-Palay V, Palay SL. The form of velate astrocytes in the cerebellar cortex of monkey and rat: high voltage electron microscopy of rapid Golgi preparations. Z Anat Entwicklungsgesch. 1972;138(1):1–19.
50. Nishiyama A, Boshans L, Goncalves CM, Wegrzyn J, Patel KD. Lineage, fate, and fate potential of NG2-glia. Brain Res. 2016;1638(Pt B):116–28.
51. Vigano F, Mobius W, Gotz M, Dimou L. Transplantation reveals regional differences in oligodendrocyte differentiation in the adult brain. Nat Neurosci. 2013;16(10):1370–2.
52. Eugenin-von Bernhardi J, Dimou L. NG2-glia, more than progenitor cells. Adv Exp Med Biol. 2016;949:27–45.
53. Almeida RG, Lyons DA. On myelinated axon plasticity and neuronal circuit formation and function. J Neurosci Off J Soc Neurosci. 2017;37(42):10023–34.
54. Monier A, Adle-Biassette H, Delezoide AL, Evrard P, Gressens P, Verney C. Entry and distribution of microglial cells in human embryonic and fetal cerebral cortex. J Neuropathol Exp Neurol. 2007;66(5):372–82.
55. Lloyd AF, Davies CL, Miron VE. Microglia: origins, homeostasis, and roles in myelin repair. Curr Opin Neurobiol. 2017;47:113–20.
56. Hristovska I, Pascual O. Deciphering resting microglial morphology and process motility from a synaptic prospect. Front Integr Neurosci. 2015;9:73.
57. Kataoka H, Kim SW, Plesnila N. Leukocyte-endothelium interactions during permanent focal cerebral ischemia in mice. J Cereb Blood Flow Metab. 2004;24(6):668–76.
58. Kulik T, Kusano Y, Aronhime S, Sandler AL, Winn HR. Regulation of cerebral vasculature in normal and ischemic brain. Neuropharmacology. 2008;55(3):281–8.
59. Spector R, Keep RF, Robert Snodgrass S, Smith QR, Johanson CE. A balanced view of choroid plexus structure and function: focus on adult humans. Exp Neurol. 2015;267:78–86.
60. Rash JE, Duffy HS, Dudek FE, Bilhartz BL, Whalen LR, Yasumura T. Grid-mapped freeze-fracture analysis of gap junctions in gray and white matter of adult rat central nervous system, with evidence for a "panglial syncytium" that is not coupled to neurons. J Comp Neurol. 1997;388(2):265–92.
61. Jimenez AJ, Dominguez-Pinos MD, Guerra MM, Fernandez-Llebrez P, Perez-Figares JM. Structure and function of the ependymal barrier and diseases associated with ependyma disruption. Tissue Barriers. 2014;2:e28426.
62. Rodriguez EM, Blazquez JL, Pastor FE, Pelaez B, Pena P, Peruzzo B, et al. Hypothalamic tanycytes: a key component of brain-endocrine interaction. Int Rev Cytol. 2005;247:89–164.
63. Garcia-Caceres C, Balland E, Prevot V, Luquet S, Woods SC, Koch M, et al. Role of astrocytes, microglia, and tanycytes in brain control of systemic metabolism. Nat Neurosci. 2019;22(1):7–14.
64. Suarez J, Romero-Zerbo SY, Rivera P, Bermudez-Silva FJ, Perez J, De Fonseca FR, et al. Endocannabinoid system in the adult rat circumventricular areas: an immunohistochemical study. J Comp Neurol. 2010;518(15):3065–85.
65. Tabata H. Diverse subtypes of astrocytes and their development during corticogenesis. Front Neurosci. 2015;9:114.

66. Liu X, Zhang Z, Guo W, Burnstock G, He C, Xiang Z. The superficial glia limitans of mouse and monkey brain and spinal cord. Anat Rec (Hoboken, NJ: 2007). 2013;296(7):995–1007.
67. Ma Q, Ineichen BV, Detmar M, Proulx ST. Outflow of cerebrospinal fluid is predominantly through lymphatic vessels and is reduced in aged mice. Nat Commun. 2017;8(1):1434.
68. Abbott NJ, Pizzo ME, Preston JE, Janigro D, Thorne RG. The role of brain barriers in fluid movement in the CNS: is there a 'glymphatic' system? Acta Neuropathol. 2018;135(3):387–407.
69. Legros C, Chesneau D, Boutin JA, Barc C, Malpaux B. Melatonin from cerebrospinal fluid but not from blood reaches sheep cerebral tissues under physiological conditions. J Neuroendocrinol. 2014;26(3):151–63.
70. Plog BA, Nedergaard M. The glymphatic system in central nervous system health and disease: past, present, and future. Annu Rev Pathol. 2018;13:379–94.
71. Faghih MM, Sharp MK. Is bulk flow plausible in perivascular, paravascular and paravenous channels? Fluids Barriers CNS. 2018;15(1):17.
72. Da Mesquita S, Fu Z, Kipnis J. The meningeal lymphatic system: a new player in neurophysiology. Neuron. 2018;100(2):375–88.
73. Rasmussen MK, Mestre H, Nedergaard M. The glymphatic pathway in neurological disorders. Lancet Neurol. 2018;17(11):1016–24.
74. Lei Y, Han H, Yuan F, Javeed A, Zhao Y. The brain interstitial system: anatomy, modeling, in vivo measurement, and applications. Prog Neurobiol. 2017;157:230–46.
75. Sweeney MD, Sagare AP, Zlokovic BV. Blood-brain barrier breakdown in Alzheimer disease and other neurodegenerative disorders. Nat Rev Neurol. 2018;14(3):133–50.
76. Louveau A, Plog BA, Antila S, Alitalo K, Nedergaard M, Kipnis J. Understanding the functions and relationships of the glymphatic system and meningeal lymphatics. J Clin Invest. 2017;127(9):3210–9.

Chapter 3
Glial Cells During the Life Cycle

Abstract Astrocytes, oligodendrocytes, and oligodendrocyte precursor cells have a lifelong close relationship with each other and with neurons. Radial glia evolve from neuroepithelial cells and serve as multipotent progenitor cells creating neural precursor cells and later oligodendrocyte precursor cells and even later astrocytes. Some radial glia survive in special niches into adulthood. Radial glia also form an important organizational scaffold for cortical structure and layers. This close relationship continues into adulthood. Astrocytes form a close spatial and functional relationship with neurons. This involves various homeostatic processes, for example, in directing local blood flow to areas with increased neuronal activity. It extends to transmitter, ion, and energy homeostasis among others. In addition, the astrocytic syncytium forms a parallel and complementary signal system, involved in the synchronization of neuronal networks. It is in fact difficult to separate clearly all the homeostatic and signaling processes as many serve both functions at the same time. Oligodendrocyte precursors are at the receiving end of synapses with neurons and are therefore under close control as they are the most proliferative cell type evolving from the neuroepithelium under normal conditions. Oligodendrocytes too have a homeostatic role (providing insulation and support for axonal conduction) and react with adaptive myelination to increased action potential traffic. Microglia are cells of the immune system which invade the brain when the blood–brain barrier is still permeable and then are trapped in the brain, exhibiting self-renewal. While these cells are regularly spaced apart and become activated when there is damage, this does not mean they are in between pathological events dormant. During development, they participate in synaptic stripping, removing underused and aberrant synapses. In adults, they participate in neuronal processing with interactions like those of astrocytes. In the aged organism, shredding of myelin fragments with associated activation of glial cells is a major event as it leads to the start of an inflammatory positive feedback cycle. There is also microglial priming which indicates that inflammatory responses to stress or disturbances have disproportional reaction amplitudes. All these processes can cause chronic low-key inflammation in the aging organism.

Keywords Astrocytes · Brain energy metabolism · Gliogenesis · Ion homeostasis · Microglia · Myelination · Neurogenesis · Neurotrophic factors · Neurovascular unit

W. Walz, *The Gliocentric Brain*, https://doi.org/10.1007/978-3-031-48105-5_3

· Oligodendrocytes · Oligodendrocyte precursor cells · Potassium siphoning · Saltatory conduction · Synaptic plasticity · Transmitter homeostasis · Tripartite synapse

3.1 Introduction

This chapter gives an overview of glial cell function in the healthy body to set the stage for further deliberations on their role during brain injury in later chapters. It describes the role of cell lineage and neural development with its implications for cell renewal in the adult organism, the role of glial cells in the normal, adult brain, and finally the changes in the aging brain without underlying pathologies.

3.2 Neuro- and Gliogenesis

The developing ectoderm consists of three parts, one of which is the neural tube. The neural tube is formed by invagination of the neural plate. Consisting of neuro-epithelial cells, the neural tube expands by self-renewal. Over time, the cells down-regulate their epithelial features and become more glia-like by containing glial fibrillary acidic protein (GFAP), glycogen, and glutamate aspartate transporter (GLAST), all exclusive glial markers in the central nervous system (CNS). They are now known as radial glia. However, there is increasing consensus that this cell type may not actually be a glial cell but a distinct neural cell type [1]: it functions as a neural progenitor cell and not as a bona fide glial cell type. The role of the radial glial cell is complicated. It has a bipolar morphology. The cell body is situated close to the ventricular surface of the neural tube and one single long process per radial glia extends all the way through the neural tube to the pial surface. Radial glia exhibit asymmetric cell division and serve as multipotent progenitor cells. These multipotent progenitor cells generate early on neural precursor cells. At a later stage, they produce oligodendrocyte precursor cells, and even later, astrocytes. By the end of this process, most of the remaining radial glia differentiate directly into mature astrocytes, a process that is irreversible. Some radial glia, however, become specialized glial cells that keep their radial structure in select areas of the CNS: tanycytes (circumventricular organs, floor of ventricles, hypothalamus), Muller cells (retina), and Bergmann glia (cerebellar cortex). Another population of radial glia seems to survive into adulthood to function as neural stem cells. As pointed out in the previous chapter, the subventricular zone (SVZ) has neural stem cells that continue to supply neuroblasts for migration to the olfactory bulb to replace mature neurons. These neural stem cells seem to be remnants of the original radial glia [2]. The same seems to be true of the second major site of adult neurogenesis, the sub-granular zone of the hippocampus [3].

The characteristics of developmental neurogenesis are best described for the cortex, but the general principles are similar for other areas. Radial glia undergo asymmetric cell division for self-renewal and neural progenitor generation. These progenitor cells initially divide to create more progenitor cells, and finally, toward the end of the neurogenesis phase, they deplete the progenitor pool by differentiating into neural precursors [4]. Once they have exited the cell cycle, these neuronal precursors migrate along the single long processes of each radial glia toward the pial surface. Many will accumulate there, creating a mantle zone (cortical plate in the cerebral cortex). This mantle zone expands and becomes layered. Neuronal identity is determined by the pattern of transcription factors at the time of its creation by radial glia division [5]. The precursor cells then migrate along the radial process to specified sites. The youngest neurons are the ones that are coming to a rest at the most outward layers. Intrinsic factors at the time of the cell division seem to determine the neuronal subtype. However, environmental factors at the target sites have an influence on the identity of the future neuronal subtype. Furthermore, the influence of these environmental factors seems to be restricted to narrow time windows [6]. Tangential migrations of interneurons with the involvement of microglia also occur [7].

After the neurons are in their final position, a phase with extensive synaptogenesis occurs. This phase coincides with the genesis of astrocytes by the radial glia [8]. In rodents, it occurs around the time of birth when radial glia switch from neurogenesis to gliogenesis. This switch is initiated by the activation of Notch1 and Jak/STAT (Janus kinase/signal transducer and activator of transcription) signaling pathways in the radial glia [9]. The astrocytes migrate without guidance by radial glia and continue to divide locally after migration is completed. In rodents, they organize into non-overlapping domains within a few weeks. Experimental evidence shows that neurons need exposure to soluble astrocytic factors and contact with astrocytic adhesion molecules to gain competency for synaptogenesis. Therefore, continued exposure to astrocytic factors is necessary for the maturing of chemical synapses [10]. Neuronal circuit activity feeds back on astrocytic properties to strengthen this circuit. It has been shown for glutamatergic synapses that the higher the neuronal activity, the closer the scaffolding by astrocytic processes around the synapses [11]. There are also indications that astrocytes might develop circuit-specific properties and as such form a close relationship with this circuit ("astromere" [12]). Thrombospondins and glypicans are examples of soluble astrocytic factors, whereas neuroligins and ephrins are important adhesion molecules [10]. The factors secreted by the astrocytes to support synaptogenesis and its maintenance have some specificity for the type of synapse (GABAergic, glutamatergic, cholinergic, and dopaminergic). Astrocytes are also involved in synapse elimination. Removal of unwanted synapses during development is controlled by astrocytes in two ways. The targeted synapses are either phagocytosed by the astrocytes themselves, or astrocytes release transforming growth factor beta (TGF-β) which induces complement protein expression on the selected synapse for removal by microglia [13]. Thus, there is from the very beginning of circuit development a close interaction between neuronal circuits and neighboring astrocytes. This early relationship foreshadows the very close homeostatic and signal-processing interactions between the two compartments in the adult organism.

Oligodendrocyte precursor cells (OPCs) arise from neuroepithelial cells around ventricles and from radial glial cells in several waves. The OPCs migrate outward and in the event of a disturbance, surviving OPCs from a previous wave can substitute [14]. Their migration route follows mainly blood vessels into the parenchyma. The migration is also controlled by soluble factors (e.g., morphogenetic protein and sonic hedgehog), extracellular matrix components (laminin, fibronectin), and neural activity [15]. Once at their destination, OPCs proliferate under the influence of signals (PDGF or platelet-derived growth factor seems to be the most important one). Finally, after the establishment of functional circuits, some OPCs downregulate inhibitory factors that suppress differentiation (Notch, Wnt, or wingless and int-1) and develop into mature, myelinating oligodendrocytes. Some OPCs (or NG2 or neuron-glia antigen 2 cells) remain as progenitors in the parenchyma under synaptic control of neurons (see previous chapter). They make up at least 8% of the cells in the adult parenchyma. These OPCs have a turnover rate and are recruited to replace myelinating oligodendrocytes in the adult organism. There is evidence of increasing heterogeneity in the NG2/OPC family after the development stage [16]. In humans, myelination of axons starts before birth (30 weeks gestational age) but largely occurs after birth until about five years of age, with most occurring in the first year after birth [17].

As pointed out in the previous chapter, microglia are cells of the immune system from erythromyeloid progenitor cells that migrate into the brain in two waves (in humans), where they proliferate and disperse throughout the parenchyma. Once the brain barrier is closed, the microglia are "trapped" in the brain and are no longer replenished by their circulating precursor cells. They develop in the brain under the influence of soluble factors (colony-stimulating factor 1 and TGF-β among others), adopting a new identity, different from their bone marrow origin. There is evidence of heterogeneity across brain regions as with all other cells termed "glia." They self-renew and in the adult brain constitute about 5% of the cells [18]. During development, they participate in neuronal circuit establishment by modulating it. Their major role is to eliminate neurons that underwent apoptosis. In close interaction with astrocytes, they respond to circuit activity by strengthening active circuits and weakening those circuits with less of a functional impact. These processes include synaptic pruning and promotion of post-synaptic spine formation. The microglia interact with all cell types in the microenvironment to accomplish their tasks in modifying circuit development. Some examples are layer V pyramidal neuron density which is regulated by insulin-like growth factor release by microglia. A subset of microglia expressing integrin is involved in stimulating myelination [19]. Astrocytic secretion of IL-33 (interleukin 33) controls the capability of microglia to phagocytose dendritic spines [20].

This overview shows that the precision and sophistication of neuronal circuit signaling need the involvement of all glial elements in the parenchyma during development. Neurons and astrocytes work together in establishing meaningful circuits, assisted by microglia which use their origins as immune cells to aggressively eliminate apoptotic neurons or detrimental synapses. They are also important relay stations for coordinating signals. Finally, oligodendrocytes and their precursors modify the speed of certain circuits. They have a close signaling relationship with neurons, just like the microglia. The strategies used to establish meaningful circuits in development by the full use of interaction with all glial elements is not restricted to the developmental phase. They continue for maintenance, repair, and response to behavioral challenges in the adult organism.

3.3 Glial Cell Function in the Adult Brain

This part is a review of the roles of astrocytes, oligodendrocytes, NG2 cells, and microglia in the adult non-pathological brain and focuses on functional concepts. It concentrates on in vivo findings rather than cell culture studies and currently proven, functionally significant contributions of glial cells to normal brain function.

3.3.1 Astrocytes

In light of the demands by neurons from their macro- and microenvironment, it is tempting to separate the astrocytic functions into homeostatic and signal roles. This cannot be done easily as astrocytes integrate both functions, using the same molecular mechanisms to support neuronal function as well as be part of the signal processing circuit. As noted in the previous chapter: astrocytes have a domain organization, are extensively connected by gap junctions to form a syncytium, have different functions in different brain regions (astromere concept) and their processes wrap around synapses and form endfeet which cover all blood vessels.

3.3.1.1 Neurovascular Unit

The concept of a neurovascular unit is now very much accepted. The brain lacks significant energy stores, despite its large energy demand. Its only stores are those of glycogen in astrocytes and they do not reach anywhere the energy storage capacity of muscle or liver [21]. The brain is therefore reliant on a continuous supply of blood flow as its energy source. Subsequently, blood flow to the brain and within the different brain regions must react quickly and decisively to energy demands. Flexible blood flow also plays a role in regulating brain temperature and in preventing overheating. Such a feedback system linking metabolic state and blood flow is known from all other organs. In the brain adenosine, carbon dioxide, hydrogen, and lactate accumulation play a role in this feedback system as signal links between the metabolic status of the brain parenchyma and blood flow [22]. This is a feedback homeostatic system similar to other organs. In the brain, however, there is a second system, which acts as a feed-forward homeostatic process which is more powerful than the feedback metabolic control. It links increased blood flow to increased signaling activity of neurons irrespective of the metabolic status of the region. This is accomplished by a neurovascular unit, which consists of neurons, astrocytes, endothelial cells, and contractile elements. Astrocytes are a critical link in this unit as their processes are close to synapses and their endfeet wrap around endothelial cells. It is now acknowledged that neurovascular coupling is based on different mechanisms in capillaries and arterioles. For capillaries, there is reasonable agreement of astrocytic involvement from in vivo experiments [23]. The following mechanism links neuronal activity with the diameter of capillaries in this neurovascular unit. Synaptic activity results in the release of adenosine triphosphate (ATP) from synaptic endings. The ATP molecules diffuse to neighboring astrocytic processes and activate ionotropic $P2X_1$ purinergic receptors. This activation leads to calcium influx. The increase in astrocytic calcium triggers a signal cascade involving phospholipase D2, diacylglycerol lipase, cyclooxygenase 1, and prostaglandin E synthase to produce prostaglandin E2. Prostaglandin E2 diffuses out of the astrocytes and interacts with the prostaglandin EP_4 receptor on pericytes, the contractile elements around capillary endothelial cells. Activation of this receptor leads to dilation of the capillaries. This cascade of effects links regions of increased synaptic activity to the relaxation of capillaries and increased blood flow [23]. Arterioles also react to increased synaptic activity with dilation; however, the role of astrocytes is considered controversial. In vivo experiments showed that nitric oxide (NO) is released by postsynaptic endings diffusing directly to vascular smooth muscle cells, the contractile elements of arterioles, and thereby dilating the arterioles. This results in a coupling of synaptic activity and blood flow through arterioles without using astrocytes as an intermediary element [23]. However, other investigators, using mainly brain slices, find that the astrocytic calcium level is correlated with arteriole diameter [24]. Thus, coupling mechanisms of arterioles and capillaries exist, but the role of astrocytes in the diameter control of arterioles remains controversial [25]. In the brain parenchyma, capillaries represent the highest resistance element in blood circulation, and the neurovascular unit acting through astrocytes on these capillaries is therefore functionally very significant [26].

3.3.1.2 Energy Metabolism

As pointed out in the previous chapter, the major site of energy consumption in the brain is the synapse. In neurons with myelinated axons, 80% of the energy is used for synaptic transmission [27]. In most, but not all neurons the presynaptic endings contain a high density of mitochondria [28]. However, the synapses are far removed from the cellular energy machinery in the cytoplasm, which limits flexible ATP supply. Astrocytes are in a strategic position to address this problem. They offer the shortest and most direct connection between capillaries and synapses via their endfeet around vessels and processes very close to the synapses. This intriguing association has been the focus of extensive research in the past but remains a highly controversial issue. The most accepted in vivo finding is the "astrocyte-neuron lactate shuttle" hypothesis [29], which postulates a glutamate- and potassium-dependent mechanism that is tightly connected to synaptic activity. The hypothesis contends that aerobic glycolysis in astrocytes can be upregulated and is responsive to factors indicative of increased activity. The key signal is the release of glutamate by excitatory presynaptic endings and to a lesser extent the increase of the potassium concentration in the extracellular space (ECS) by action and synaptic potentials. The uptake of glutamate with sodium via cotransport into astrocytes activates glucose uptake. The source of the glucose taken up by the astrocytic endfeet is the endothelial cells. These cells release glucose taken up from the blood. In astrocytes, the glucose turns into lactate via aerobic glycolysis [30]. As previously mentioned, astrocytes are capable of upregulating their aerobic glycolysis if stimulated by glutamate and potassium uptake [31]. The resulting lactate diffuses with the help of passive carriers into the neurons as there is a concentration gradient from astrocytes to neurons [32]. In the neurons, the lactate is converted to pyruvate which is used by neuronal mitochondria to oxidatively produce ATP [33]. In astrocytes, the sole glycogen store of the brain, additional glucose molecules can be made available due to glycogenolysis. This process is stimulated by various neuronal signals, namely, noradrenaline, vasoactive intestinal peptide, and the purinergic agonists adenosine and ATP. High glycolysis rates mean increased production of advanced glycation end products (AGEs). These AGEs are implicated as some of the causal agents of neurodegeneration [34]. Enzymes of the glyoxalase system, which are far more abundant in astrocytes than in neurons, detoxify these end products. Neurons therefore have less of this protective mechanism and are therefore more susceptible to AGEs than astrocytes. Neurons have limited capacity to upregulate glycolysis. However, they accumulate glucose and use it in the pentose phosphate pathway to produce nicotinamide adenine dinucleotide phosphate (NADPH). NADPH is important for reducing glutathione and is therefore available as a scavenger for reactive oxygen radicals (ROS). ROS are created due to the high oxidative stress in the neuronal mitochondria [35]. The astrocyte–neuron lactate shuttle puts astrocytes in a strategic position to couple synaptic activity and energy metabolism. The main mediator of the activity-dependent regulation of this shuttle is glutamate. Most in vivo investigations find support for the shuttle hypothesis; however, there have been some findings that are not consistent with the hypothesis. This concerns mainly the finding of upregulation of neuronal glucose uptake during activity [36].

3.3.1.3 The Tripartite Synapse

Astrocytes exhibit receptors for many neurotransmitters, such as glutamate, acetylcholine, ATP, gamma-aminobutyric acid (GABA), and endocannabinoids. Most of these receptors are metabotropic and their activation results in calcium signals across gap junctions into neighboring astrocytes. Astrocytes release neuroactive substances as a response, so-called gliotransmitters. Confirmed gliotransmitters are glutamate, D-serine, ATP, adenosine, and GABA [37]. This gliotransmitter cargo is mainly released in vesicles. This vesicular release is calcium and SNARE (soluble-dependent *N*-ethylmaleimide-sensitive factor protein receptor) dependent [38]. The action of gliotransmitters on neuronal activity can be either inhibitory or excitatory, depending on the circuit. This concept of astrocytes receiving information from the presynaptic ending, modulating it, and then in turn releasing signals which in the end influence the activity at the postsynaptic ending is known as the tripartite synapse. Implicit in this concept is the active participation of astrocytes in brain signal transmission [39]. This active participation is clearly different from its homeostatic function as a passive support cell with regard to blood flow or energy metabolism. Both roles, homeostatic and signaling, can combine to function in an integrative way (see further below). This tripartite synapse concept is now generally acknowledged [40], despite arguments that the concept only applies to extreme, pharmacological and not to physiological conditions [41]. Nevertheless, Panatier et al. [42] found that in the hippocampus, astrocytes detect and respond to a single-synaptic stimulation. In the following paragraph are selected examples of how the tripartite synapse concept works within functional circuits. D-serine is now acknowledged as a gliotransmitter [43]. The release of purinergic gliotransmitters modulates excitation and inhibition in neuronal circuits. ATP uses the P2Y receptors to modulate neuronal calcium currents and causes a reduction in excitation and heterosynaptic depression [44]. Activation of P2Y1 receptors increases interneuron activity and therefore GABA release which in turn mediates inhibition. Adenosine decreases (A1 receptors) or increases (A2A receptors) neuronal activity [45].

In the mouse dorsal striatum are two subtypes of medium spiny neurons. They are functionally distinct but spatially intermingled, representing the basal ganglia's direct and indirect pathways. When active, the synapses of the direct pathway (which use D1 dopamine receptors) release endocannabinoid which acts on receptors of dedicated adjacent astrocytic processes [46]. This interaction increases the intracellular free calcium in these dedicated astrocytes which in turn release glutamate at specific D1 synapses. This glutamate release enhances synaptic transmission in the direct pathway synapses. The same is true for the separate synapse–astrocyte–synapse arrangement of the indirect pathway (which uses D2 dopamine receptors). Thus, although they are spatially intermingled, astrocytes functionally couple the separate functional pathways and enhance their signaling. Despite being part of a syncytium, astrocytes are predetermined in their dedication to either the D1 or D2 pathway. They show no plasticity in switching their dedication between D1 and D2 pathways. The astrocytes act like an excitatory interneuron between synapses of the same functional pathway (D1 or D2) at least within the context of this mouse model (see Fig. 3.1).

Another example concerns the amygdala. The amygdala is involved in the fear response of the organism. In the medial portion of the central amygdala, stimulation of a neuron increases the calcium of adjacent astrocytes via the CB1R cannabinoid receptors. In response to the resulting calcium increase, the astrocytes release ATP and adenosine, which in turn increases the likelihood of GABA release in presynaptic endings of inhibitory neurons of the lateral portion of the amygdala via A2AR purinergic receptors [47]. At the same time, the adenosine interacts with A1R

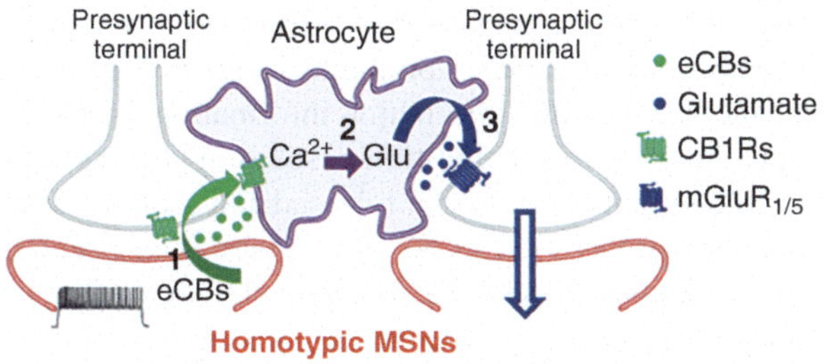

Fig. 3.1 Schematic representation of the signaling mechanism underlying heteroneuronal potentiation. Activation of medium spiny neurons (MSNs) stimulates the release of endocannabinoids (eCBs) that activate CB1R endocannabinoid receptors in astrocytes (1) and elevate intracellular calcium, which then stimulate the release of glutamate (2) that enhances synaptic transmission (3) selectively in homotypic MSNs through activation of presynaptic mGluR glutamate receptors. (From Ref. [46] with permission. Copyright American Association of the Advancement of Science)

purinergic receptors of glutamatergic synaptic endings of excitatory neurons of the basolateral amygdala, decreasing glutamate release. Both presynaptic endings target output neurons of the medial portion of the central amygdala (CeM). Thus, astrocytic activation decreases the firing rate of these CeM neurons through two different mechanisms. This astrocytic activity has functional consequences as it will reduce the fear expression [47].

A third example involves the mouse hippocampus. Several reports [48, 49] showed that astrocytic calcium increases due to neuronal release of GABA and cannabinoids. In response to the calcium increase, astrocytes release glutamate and ATP/adenosine. The time course and intensity of the release of the glutamatergic and purinergic agonists vary with the firing frequency and duration of the neuron which inputs on the astrocyte. The resulting complicated release pattern of these gliotransmitters causes a profound modulation of the activity of a neuronal circuit. In this case, it is the CA3-CA1 (*Cornu Ammonis* 3 and 1) hippocampal synapse which as the result of the calcium signals and subsequent gliotransmitter release of a single astrocyte changes its activity [49].

The pattern that seems to emerge from these examples of the functional impact of the tripartite synapse is a modulatory role for astrocytes in connecting two circuits that are not directly coupled by synapses. The astrocytic role seems to be predetermined, resembling an interneuron with modulatory effects. The role of gap junctions and calcium waves between astrocytes that are part of different astrocytic pathways is not yet clear.

3.3.1.4 Transmitter Homeostasis

Transmitter substances are the main signal of chemical synapses. The signal function of a chemical substance depends on various properties and conditions. Significantly, the substance, once released due to a stimulus, must not linger too long in the synaptic cleft, but be removed or inactivated as fast as possible. A quickly inactivated signal substance is a major prerequisite for neuronal processing, as in the axon to dendrite transmission the major information is coded in the action potential frequency. This requires a high-time resolution process at the postsynaptic ending, made possible only if the pulses of released and available transmitters in the cleft per action potential are as short as possible. There is clear and undisputed evidence for astrocytic processes close to the synaptic cleft being involved in transmitter inactivation processes [50].

Glutamate is the most abundant excitatory transmitter, and astrocytes are directly involved in its control. There are indications that glutamate, released at the presynaptic ending, is more efficiently taken up by astrocytes than by neurons [51]. In neurons, the taken-up glutamate is recycled into synaptic vesicles. Astrocytes use the high-affinity glutamate transporters EAAT-1 and EAAT-2 (excitatory amino acid transporters 1 and 2), which are preferentially expressed in these cells. In astrocytes, the majority of accumulated glutamate is turned into glutamine via the astrocyte-specific enzyme glutamine synthetase [52]. However, some of the accumulated glutamate is used in the oxidative metabolism [53]. A small portion of the accumulated glutamate might be released via vesicles as a gliotransmitter. There is also de novo synthesis of glutamine in the astrocytes [54]. There are passive transporters for glutamine in neurons and astrocytes as well as a concentration gradient from astrocytes to neurons. This results in a net flux of glutamine from astrocytes to neurons, where the glutamine is turned into glutamate to be available for packaging into synaptic vesicles. The glutamine release by astrocytes does not result in any change in neuronal excitability, as glutamine has no receptors connected to signaling systems. This mechanism is known as the glutamate–glutamine cycle.

EAAT-1 and EAAT-2 are part of the solute carrier 1 (SLC1) family. These carriers are sodium/potassium-dependent transporters. In addition, astrocytes express transporters of the SLC6 family, which are sodium/chloride dependent. These transporters shuttle dopamine, serotonin, noradrenaline, GABA, and glycine [55]. There is good evidence for significant uptake of the inhibitory transmitters GABA and glycine into astrocytes during synaptic transmission, although there are regional differences [52]. Neurons are more effective in the reuptake of GABA than of glutamate released by the presynaptic endings [56]. Most of the GABA released during synaptic transmission is thereby taken back up into presynaptic endings. The remaining GABA is taken up by astrocytes. The concept of the glutamate–glutamine cycle must be modified to reflect such a glutamate/GABA-glutamine cycle in neurons where GABA is synthesized from glutamate by glutamic acid decarboxylase. A large part of the glutamate, acting as a precursor for GABA in neurons, arrives via glutamine from the astrocytes. In this way, part of the astrocytic glutamine is transferred to neurons, turned into glutamate, and then into GABA to be packaged into vesicles. The portion of the released GABA that is taken up by astrocytes is mainly metabolized by GABA-transaminase. During depolarization, the accumulated GABA from the astrocytes can be released through reverse transport by the SLC6 transporters and in addition by outward diffusion through anion channels. In this way, GABA could act as a gliotransmitter. Astrocytes, however, seem not to be able to de novo synthesize GABA [57]. Uptake of monoamine neurotransmitters into astrocytes has been demonstrated in a variety of preparations. Given that these neurotransmitters may play an important role not only in synaptic, but also in volume transmission, the role of astrocytes may be more complicated than initially assumed. A key question is whether this uptake plays a functional role in vivo [58]. Key functional enzymes for monoamine metabolism like monoamine oxidase B and catecholamine O methyl transferase are strongly expressed in astrocytes [59], and recently evidence accumulated for a functional role of astrocytes in these non-amino acid transmitters (see further below).

3.3.1.5 Ion Homeostasis

As pointed out in the previous chapter, neuronal activity changes some of the ion concentrations in the narrow extracellular space (ECS) of the brain. These ion changes can alter ion channel kinetics, which in turn impact neuronal excitability. Astrocytes are equipped to counteract these extracellular changes.

Functionally, the most important astrocytic ion homeostatic mechanism is related to the potassium ion. Two homeostatic mechanisms work simultaneously in astrocytes. A combination of Na/K-ATPase and the neutral ion carrier K/Na/2Cl leads to an accumulation of KCl in astrocytes with the sodium ion cycling through the two mechanisms [60]. This KCl accumulation is rapid and can change the astrocytic potassium content by a large margin [61, 62]. This uptake, however, has its limits as the KCl accumulation leads to concomitant water uptake and therefore cell swelling. When the extracellular potassium concentration decreases again, the accumulated KCl and water are released with a similar time course. The release mechanism seems to involve potassium inward rectifier channels [63]. An alternative mechanism that avoids the net water movements is the spatial buffering of potassium. Accumulation of potassium in a restricted part of the ECS by neuronal activity leads to a strong depolarization of the adjacent cell membrane of the astrocytic syncytium. This strong depolarization of the astrocytic membrane is due to the high potassium permeability, which is a typical astrocytic property not found in neurons. As a result, these parts of the astrocytic syncytium, exposed to high external potassium concentrations, experience a strong depolarization, whereas the remainder of the syncytium is at close to normal resting membrane potential. Under these conditions, the syncytium assumes the duality like opposing poles of a battery. The part of the syncytium exposed to the high potassium concentration is depolarized and mirrors the part of the positive pole. The remainder of the syncytium at normal resting potential mirrors the negative pole as this part is more negative than the membrane depolarized by high potassium. The high potassium permeability of the astrocytic cell membrane, together with the low resistance coupling via gap junctions, leads to ion currents that equalize the potential difference with potassium the only charge carrier able to cross the astrocytic cell membrane [64]. As a result, potassium ions enter at the site of high potassium and leave at the sites with a close-to-normal membrane potential [65]. There is no energy needed, no significant volume changes are involved, and the mechanism is extremely flexible as it is solely dependent on potassium-induced depolarization and not on anatomical specializations. The potassium is redistributed to areas with close to normal extracellular potassium [66, 67]. The relative contribution of these two homeostatic potassium clearance mechanisms under different situations and brain regions is a matter of dispute [63, 68]. A strong case has been made with in vivo experiments [69] that extracellular potassium changes are responsible for brain activity state changes on a longer time scale, such as rhythmic locomotor activity in the spinal cord, sleep-wake cycle, and improvement of motor performance in the cerebral cortex. In these cases, the homeostatic potassium mechanisms also serve as a slowly acting signal in neuronal state transition.

Neuronal excitability is highly sensitive to small changes in intra- as well as extracellular pH ("potential of hydrogen" or acidity). Acidosis depresses neuronal excitability, whereas alkalosis enhances this excitability [70]. Neuronal activity results in an alkaline shift in the extracellular space as well as in astrocytes, whereas the intracellular pH of neurons shows an acid shift. The amplitudes depend on the intensity of the neuronal activity [71]. The shifts are mainly caused by H^+ (hydrogen ion) and HCO_3^- (bicarbonate ion) movements through ion channels and GABA as well as glutamate ionotropic channels. This is a similar situation to the potassium movements during neuronal activity, where neuronal signal processing by itself alters the information flow. In addition, metabolic and respiratory acidosis can impose limits on neuronal functioning. Astrocytes are equipped to buffer these changes in the neuronal environment. They express electrogenic sodium bicarbonate cotransporter NBCe1 as well as carbonic anhydrase II. NBCe1 acts as a bicarbonate sensor of the ECS and shuttles bicarbonate across the astrocytic membrane irrespective of the H^+ concentration (pH). In most situations, this will result in bicarbonate secretion by astrocytes and protect the neurons from intra- and extracellular acidification [72]. NBCe1 works in concert with carbonic anhydrase, which will deliver the necessary bicarbonate from CO_2 for export into the ECS by the bicarbonate shuttle [73]. Thus, to protect neuronal signaling from the detrimental impact of pH changes, the astrocytes take an acid load. The implication is that astrocytes are more resistant to internal pH changes than neurons.

3.3.1.6 Synaptic Plasticity

Dendritic spines are the postsynaptic endings of glutamatergic synapses. Experience-dependent changes in the morphology of these spines are indications of the strength of synaptic transmission. These changes reflect partly the synaptic plasticity that occurs during a learning experience [74]. Peridendritic astrocyte processes (PDAPs) are highly motile during synaptic plasticity which occurs during learning (potentiation) processes [75]. These astrocytic processes cover a larger spine area in stimulated synapses, and they phagocytose weak synapses [76]. The trigger for astrocytic plasticity seems to be the activation of astrocytic metabotropic glutamate receptors (mGluRs) by presynaptically released glutamate. The mGluR activation causes calcium increases in the astrocytes, which are a prerequisite for astrocytic plasticity.

These processes could just be a simple homeostatic adjustment of the astrocytes to neuronal activity, for example, to take up excess glutamate. However, recent studies showed that for neuronal potentiation and enhanced memory to occur in hippocampal synapses, astrocytic activation is a prerequisite at least in the hippocampus [77]. Increased neuronal activity without astrocytic activation did not lead to potentiation, but memory impairment. Astrocytic activation needs a calcium increase, which is normally accomplished by neuronally released glutamate and mGluR activation. If in experiments, where neurons are continuously activated, astrocytes are also stimulated experimentally by optogenetic or chemogenetic means, the neurons react with initiation of long-term potentiation. It seems that if astrocytic calcium is increased during neuronal activity, synaptic strength is increased. This seems to indicate that the plasticity of peridendritic astrocyte processes during learning is necessary for neuronal memory enhancement in addition to homeostatic events that are caused by glutamate interaction with astrocytes (glutamate–glutamine shuttle and lactate shuttle, see above). It is to be concluded that astrocytes are actively involved in at least some of the learning and memory processes in the brain.

3.3.1.7 Synchronization of Neuronal Network Activity

About 80% of the synapses that release acetylcholine (basal forebrain), dopamine (ventral segmental area/substantia nigra), histamine (tuberomammillary nucleus), norepinephrine (locus coeruleus), and serotonin (raphe nuclei) are not part of conventional synapses. Rather, they form axon varicosities and participate in volume transmission [78]. These transmitter substances therefore have a role more compatible with modulators as their release affects structures other than postsynaptic endings, namely, blood vessels, other axons, microglia, and astrocytes. Interestingly, these are all transmitter substances that are involved in changing brain states, such as sleep, wakefulness, and different states of alertness and attention. This is accomplished by synchronization of the activity of a large and defined population of neurons [44]. A single astrocytic domain can reach up to 140,000 synapses and is directly coupled through gap junctions with another seven to nine astrocytic domains, creating a giant syncytium. Transmitter substances released by these varicosities and acting on adjacent astrocytic receptors could therefore induce rapid and wide-spread changes in brain state and behavior through mechanisms using the astrocytic syncytium. Therefore, the role of the astrocytes in the synchronization of brain activity and behavior has been proposed [44]. Clear evidence points now to a decisive role of directed calcium waves across the glial syncytium as participants in these state transitions [79]. There are two main mechanisms by which astrocytes could serve as a conduit of these widespread effects. One is by modulating the potassium uptake and release sites and dynamics and therefore changing the neuronal activity by modulating the potassium activity in the ECS and subsequently ion channel and synaptic dynamics. Mild reduction of extracellular potassium increases the signal-to-noise ratio [80]. Mild accumulation of extracellular potassium increases neuronal high-frequency oscillations [69]. It has been shown that serotonin, norepinephrine, and acetylcholine have different effects on the extracellular potassium baseline. These manipulations of the potassium baseline cause modulation of the somatosensory adaptation and amplitude [81]. Furthermore, the typical EEG of sleeping mice can be transitioned to the EEG reflective of wakefulness by manipulating the extracellular potassium concentration to amounts induced by perfusion of these transmitter substances [82]. The second mechanism that enables astrocytes to widespread modulate brain states is by gliotransmitter release. Manipulation of cortical inhibition (release of GABA by interneurons) interferes with the filtering of incoming sensory information. This manipulation therefore changes the activity of cortical neurons to various inputs with consequences for the brain alertness states. Acetylcholine, serotonin, and norepinephrine all modulate this cortical inhibition. There is now good evidence that at least one of these, serotonin, exerts its effect through astrocytes. Serotonin effects on cortical inhibition are mediated exclusively through modification of P2Y and A2A receptors, as inhibition of these receptors abolishes these serotonin effects. Furthermore, disruption of astrocyte metabolism impaired these serotonin effects on evoked inhibition [83, 84].

3.3.1.8 Supply of Neurotrophic Factors

Astrocytes secrete peptides that act on mature neurons to ensure their survival and prevent neurodegeneration. There are large regional differences in astrocytic capability to secrete these factors [85]. This is even though neurons by themselves are releasing additional or even some of the same factors to act in a paracrine fashion on their own survival. The most important of the factors released by astrocytes are the following: brain-derived neurotrophic factor (BDNF), glial cell line-derived neurotrophic factor (GDNF), ciliary neurotrophic factor (CNTF), cerebral dopamine neurotrophic factor (CDNF), nerve growth factor (NGF), and neurotrophin-3 (NT-3). While there is strong evidence from cell cultures for astrocytic neurotrophic factor release to impact neuronal survival and synapse formation, information from in vivo systems is sparser and more complicated. BDNF and GDNF seem to interact to assist learning-based synapse formation in the hippocampus [86]. The other neurotrophic factors seem to play a more important role in development and after neuronal injury, as well as in neurodegenerative diseases. These issues will be discussed in a later chapter.

3.3.1.9 Control of Systemic Metabolism

In addition to the above functions, astrocytes are involved in the central nervous system control of systemic parameters. Astrocytes seem to play the role as sensors of these parameters and subsequently to transfer this information for neuronal processing and execution in a feedback loop [87, 88]. This includes the hypoxic ventilatory and cardiovascular response where brainstem astrocytes release ATP in response to hypoxia. This response can be induced by optogenetic stimulation of the brainstem astrocytes alone [89]. This set of astrocytes is sensitive to PO_2 (partial pressure of oxygen) and via purinergic signaling communicates this information to C1 brainstem neurons. The C1 neurons are catecholaminergic and glutamatergic neurons located in the rostral ventrolateral medulla. In addition, other brainstem astrocytes act as glucose sensors via their glucose transporter type 2 (GLUT2) and then transmit this information to neighboring neurons. Related to this function is the ability of hypothalamic astrocytes to measure the insulin concentration and in response to this signal interact with proopiomelanocortin (POMC) neurons to influence systemic glucose availability. In addition, astrocytes seem to modulate glucose transfer across the blood–brain barrier in response to insulin [90]. Regulating breathing in relation to the pCO_2 (partial pressure of carbon monoxide)/pH status of the brain is the function of neurons in the retrotrapezoid nucleus (RTN) at the ventral surface of the brainstem. There is now considerable evidence that astrocytes adjacent to the RTN neurons respond to pH decreases by calcium elevations. This triggers ATP release which in turn activates the RTN neurons [91]. In addition, ventral brainstem astrocytes sense CO_2 directly via CO_2-sensitive connexin-26 hemichannels and as a response transmit signals via ATP release to "CO_2-sensitive" neurons [92]. While direct oxygen and glucose sensing seems to be a property of all astrocytes, CO_2/pH sensing is a function unique to astrocytes in the ventral brainstem.

3.3.1.10 Integration of Astrocytic Functions

Most of these astrocytic functions are executed simultaneously. These multiple functions are especially intensive when neighboring neurons are highly active. Astrocytic functions cannot easily be separated into homeostatic and signal roles as there is no clear separation line. The role of glutamate can illustrate this. Glutamate is released by neurons in an activity-dependent manner. In astrocytes, glutamate is involved in blood flow regulation, energy metabolism, and astrocyte–neuron signaling, control of transmitter availability, and long-term control of synaptic plasticity. All of which are executed simultaneously. In addition, astrocytes release glutamate to modulate neuronal functions, including signal processing.

3.3.2 Oligodendrocytes

3.3.2.1 Saltatory Conduction

The key role of oligodendrocytes is the electrical isolation of axons to increase the speed of action potential conduction. To fulfill this function, there are as many as several thousand myelin sheaths wrapped around axons (see Fig. 3.2). These myelin sheaths decrease the electrical capacity of the axonal membrane and therefore speed up the time course of the voltage change evoked by currents. This is the main

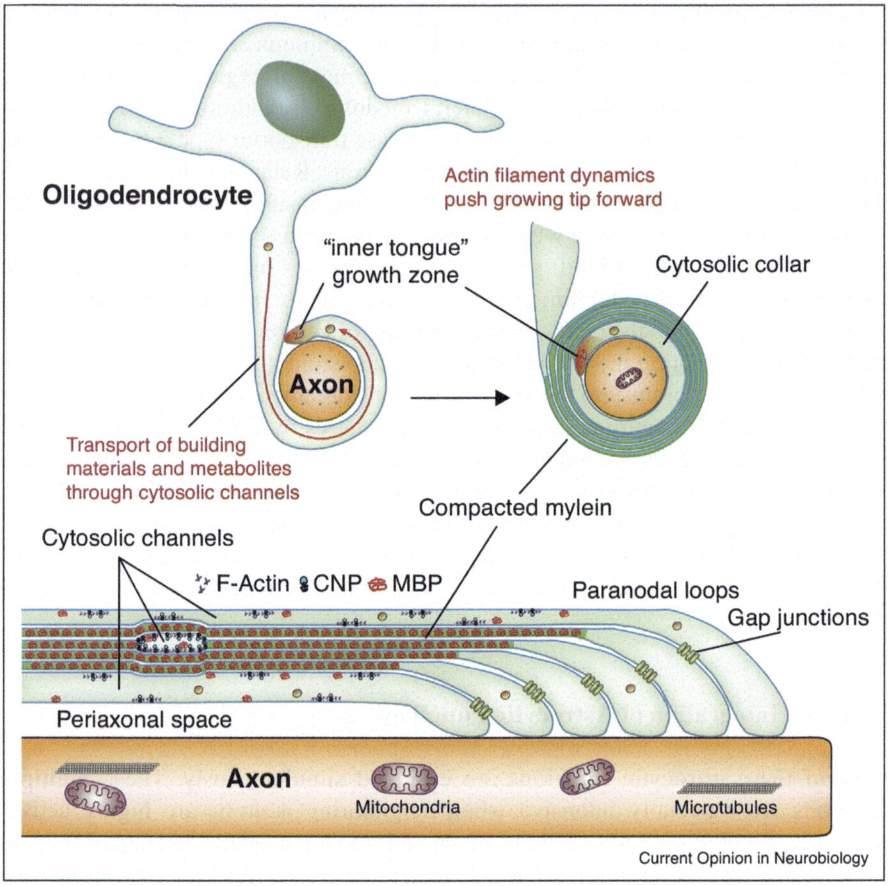

Fig. 3.2 Model of myelin growth dynamics beginning with an oligodendrocyte process that initiates wrapping of an axonal segment. Building materials and metabolites for membrane production and myelin growth are transported through a cytosolic channel system toward the growing tip, the inner tongue of the myelin wrap. The force that pushes the inner tongue forward relies on actin filament assembly–disassembly cycles. Stacked layers of plasma membranes are subsequently compacted in a zipper-like fashion and myelin basic protein (MBP) is essential for membrane compaction and cytoplasm extrusion. The membrane-associated 2′,3′-cyclic nucleotide 3′-phosphodiesterase (CNP) interacts with the actin cytoskeleton and counteracts the compacting force of MBP. This helps maintain non-compacted cytosolic channels in mature sheaths, which are critical for metabolite diffusion and the continued delivery of adaxonal membrane proteins (such as MCT1 and NMDA receptors), and thus axonal metabolic support. (From Ref. [142] with permission. Copyright Elsevier)

mechanism with which conduction speed is increased [93]. There are small regularly appearing gaps (nodes of Ranvier) in the myelin wrapping where the axon is unmyelinated. These nodes serve as "booster" stations to bring the action potential amplitude back to its original height. In addition to increasing the action potential speed, this type of conduction also reduces the energy requirements of the axons for the conduction of action potentials, as almost all the conduction mode is passive and does not involve transmembrane ion movements. However, this saltatory conduction is not the only function of oligodendrocytes and their myelin sheath. These functions are discussed below.

3.3.2.2 Potassium Siphoning

The anatomy of the myelin sheath is complicated. There is a periaxonal space between the innermost sheath and the axon surface with dimensions comparable with the synaptic cleft. There are cytoplasmic channels between the myelin sheaths and enlarged paranodal loops. The two latter structures are connected to astrocytic processes via gap junctions. In addition, astrocytic gap junctions connect to the oligodendrocyte cell body. This architecture contributes to the panglial syncytium, which functionally connects oligodendrocytes/myelin, astrocytes, and ependymal cells (see previous chapter). While sodium channels are restricted to the nodal membrane of the axon, potassium channels are in the axonal membrane facing the periaxonal space. During action potential propagation, potassium accumulates in the periaxonal space without interfering with sodium channel operation. The periaxonal space is highly depolarized during the passing of an action potential. The panglial syncytium connects the cytoplasmic fingers reaching the innermost myelin and paranodal extension with astrocytes. The astrocytic resting membrane potential is around −90 mV. Thus, there is a huge driving force for potassium currents carried through inward rectifier $K_{IR}4.1$ channels from myelin to astrocytes and into the ECS away from the active axon [94]. This is known as the potassium-siphoning mechanism of the panglial syncytium.

3.3.2.3 Oligodendrocyte-Lactate Shuttle

Although saltatory conduction is more energy efficient than conventional action potential propagation at the unmyelinated axon, the axons are normally far away from the neuronal cell body, making energy supply an issue. During saltatory conduction, the axons release glutamate into the periaxonal space. The glutamate release increases with the action potential frequency. The intermodal/paranodal myelin membranes contain NMDARs (*N*-methyl-D-aspartic acid receptors) to which the glutamate molecules bind. The NMDAR activation results in lactate release by the myelin via monocarboxylate MCT1 transporters. This lactate is taken up by axons to be used as an energy substrate. The panglial syncytium might involve astrocytes in this oligodendrocyte-lactate shuttle, especially if the axonal activity phase lasts longer [95].

3.3.2.4 Adaptive Myelination

When axons increase their activity over time in vivo, the thickness of their surrounding myelin sheets increases [96]. However, it is not clear if this effect is glutamate-mediated, or even specific to glutamatergic neurons. Indeed, it seems that adult oligodendrocytes react to the changes in the activity pattern of axons by changing the myelination pattern. This process is called adaptive myelination [97]. It is not clear if this is a homeostatic mechanism supporting potassium and energy homeostasis or a process that increases the axonal signal efficiency, or both.

3.3.3 Oligodendrocyte Precursor Cells

These cells are also known as NG2 cells (see the previous chapter). In the adult brain, they are in all areas and make up 2–9% of all cells [98]. As the name implies, in the adult, non-injured brain, they can generate new myelinating oligodendrocytes, but no other cell types. They are an interesting cell type as in the adult brain, they connect to neurons via glutamatergic and GABAergic synapses with the OPCs constituting the postsynaptic part of these synapses. They are therefore at the receiving end of a conventional signal pathway originating in neurons. Although the synapses are functional, their role in the adult CNS is somewhat of a quagmire. The synapses receive signals from neurons, whose cell body can be a long distance away. Local glutamatergic neurons do not connect to OPCs via synapses [97], but local GABAergic interneurons do [99]. This might be not such a big surprise as glutamatergic neurons make long-distance connections, whereas GABAergic neurons are mostly local interneurons. In the adult, OPCs are continuously and highly physically active. They send out processes to probe their environment, and they move around without intruding into other OPCs' territory. They turn into mature myelinating oligodendrocytes at a low rate in the healthy adult brain. This is done by differentiation, but it is not an asymmetric division. OPCs that differentiate into oligodendrocytes or die by apoptosis are replaced by division of existing OPCs and migration into the space vacated by the original OPC [100]. This also means that the synapses of these highly motile OPCs are of a transient nature and are constantly remodeled. Increased neuronal activity increases the OPC dynamics. They proliferate and differentiate into oligodendrocytes at a higher rate. This in turn increases myelination (see above) and seems to improve motor performance [101]. There is no agreement if this is due to neuronal signals at the neuron-OPC synapse [102]. However, in the corpus callosum in vivo, it was shown that the frequency and duration of neuronal activity at neuron-OPC synapses determined OPCs' self-renewal and differentiation into oligodendrocytes [103].

3.3.4 Microglia

Microglia are macrophages that migrate into the brain early in development. When the blood–brain barrier is sealed, microglia are "trapped" and replenished by self-renewal (see previous chapter). They play a role in development (see above, this chapter), and in brain injury and degeneration (see following chapters). In the healthy adult brain, microglia are evenly spaced and actively survey the adjacent area for damage. A less well-known fact is their involvement in neuronal signaling and homeostasis. The small cell body of these microglia stays in place, but its fine processes constantly extend and move around a defined area exhibiting no overlap with neighboring microglia. The processes seem to regularly probe the vicinity of pre- and postsynaptic endings [104], sensing neural activity at the synaptic level and regulating it in a similar way to astrocytes. Depletion of finely ramified microglia leads to synaptic dysfunction and deficits in learning and cognition [105, 106]. Microglia monitor firing frequency at synapses as they possess ionotropic (GluR) and metabotropic (mGluR) glutamate receptors as well as purinergic receptors for ATP and its breakdown products adenosine diphosphate (ADP) and adenosine. There are microglial receptors for all other transmitter substances [106]. As a response to neural activity, microglia modulate synaptic transmission. Their processes dampen overactive neurons and are involved in long-term potentiation (LTP). They also change the morphology of dendritic spines [107–109]. They are therefore intimately involved in synaptic plasticity by increasing meaningful activity and protecting neurons from hyperactivity. The mechanisms with which microglia accomplish this modulation of synaptic activity are cell–cell contact and the release of extracellular vesicles, tumor necrosis factor-alpha (TNF-α), interleukin-1β, brain-derived neurotrophic factor (BDNF), and prostaglandin E2 [106]. This modulatory function of neuronal signaling by microglia is less prominent in the literature than microglia activation and their macrophage-like actions during brain inflammation. The involvement of astrocytic processes in this interaction has not been studied in detail as well. It may well be that in the future, one has to rename the concept of the tripartite synapse into a quadripartite one.

However, so far, there no coherent overall functional concept emerged that would explain the interactions between neurons and microglia in the normal brain. This is reminiscent of the situation of astrocytic–neuronal signaling about 15 years ago. However, a few concepts emerge now. Direct contact of a normal, ramified microglia with a synapse results in increased synaptic activity at this particular synapse. Lipopolysaccharide (LPS) activation of microglia leads to a retraction of this process and termination of the stimulating effect of microglia. If a normal, healthy mouse was treated with LPS, microglia in the cortex retracted their processes and became activated. This led to a desynchronization of neuronal activity. The same happened if the animal was not treated with LPS, but the microglia were selectively ablated. It was concluded that in the healthy brain, microglia are involved in the local synchronization of neuronal activity and therefore plasticity. If there is an immune challenge, the microglia switch roles and retreat from their role as signal modulators and transfer to their role as immune cells [110]. There have been many attempts to categorize microglia subtypes in the past: phagocytic and surveilling, inflammatory or not (M1 and M2), amoeboid and ramified [111]. In addition, there are changes in the brain region, gender, and aging [111]. It seems to be more realistic to view microglia as a very plastic cell type that is the product of many environmental factors with inherent flexibility as these factors change. This explains the regional variability as well as differences in transcriptional profiles. It has been suggested [111] that microglia might act as "transformers" whose role is to integrate the status of their individual microenvironments and as a result impact changes on their surrounding other cell types including neurons. In this way, there would be no rigid distinction between normal and pathological status but a smooth adaptation along a sliding scale in both directions as the need arises. Pharmacologically or genetically disabled microglia result in learning deficits, memory loss, and reduced social interactions in rodents [112–115]. The changes were reverted when the brain areas were again repopulated with functional microglia.

To recapitulate, "resting" microglia cell bodies are normally stationary, but their processes are highly flexible and move in defined territories and specifically seek out synapses and monitor their activity. Neuronal signals seem to communicate with microglia on several levels. Microglia increase the number of their processes and the contact time of these processes with synapses when neuronal activity increases. They exhibit hyper-ramification. These changes are caused by hippocampal long-term potentiation [109]. They are also induced by increased glutamatergic synaptic traffic, whereas GABAergic neurotransmission has an inhibitory effect on these changes. Interestingly, the changes by glutamate and GABA are not caused by receptors for these transmitters on the microglial cell membrane but by modification of ATP release from neurons and/or astrocytes under the control of glutamate or GABA. ATP directly applied to microglia can cause these phenotypical changes through P2X and P2Y receptors on the microglia cell membrane [116]. On another level, other factors, which are secreted from healthy neurons, act directly on microglial receptors to keep them in their normal non-activated state. The two most prominent are fractalkine (a chemokine) and cluster of differentiation 200 (CD200; a glycoprotein) [117]. Fractalkine has a more complex role in pathological states [118]. Microglia are not only adapting to the degree of normal activity in non-pathological situations in a graded fashion; a prolonged decrease in neuronal activity shifts microglia gradually to a more "reactive" phenotype. This phenotype is more pronounced as neurons not only decrease their normal activity but also release specific "trouble" signals (see the following chapters).

The implication is therefore that a reduction of neuronal activity caused by pathological processes results in a shift of the microglial phenotype to a less ramified appearance, somewhat closer to the reactive phenotype [119].

It should be pointed out that the involvement of the microglia in physiological processes seems not unique to the brain. Slowly reports emerge that show other immune cells are involved in the normal function of their organ and interact with other cells. These recent examples are macrophages and enteric neurons in the gut during peristalsis [120] and prevention of tissue damage by regulatory T cells in lung tissue [121] and in skeletal muscle [122].

3.4 Glial Cell Changes During Normal, Non-pathological Aging

Aging is a risk factor for many pathological processes. It is time-dependent, in that the more advanced the age, the higher the disease risk. While this is true for all organs, the brain is especially susceptible due to the high likelihood of neurodegenerative processes in older age. Even in the absence of any pathology, aging causes a time-dependent deterioration of neural processes with some compensatory mechanisms. The question is whether the aging process is not only a risk but also a contributing cause when it comes to neurodegenerative pathologies. The following explores the changes in the normal aging brain without underlying pathology. Not surprisingly, given its frequent association with chronic inflammation [123], myelin is often at the center of processes that can lead to deterioration of brain function over a lifespan. In older mice, myelin fragments are constantly shed off the sheaths and these myelin pieces appear in the ECS [124]. This process leads to a thinning of the myelin sheaths and a decrease in the average myelin fiber length. This myelin reduction is a main contributing factor to the shrinkage of the brain volume in old age [125, 126]. The reason for myelin decrease seems to be twofold. On the one hand, aging oligodendrocytes are not as effective in producing myelin as younger ones, perhaps due to oxidative stress [127]. On the other hand, the time phase for differentiation of aging OPCs is substantially longer than in younger cells. In the adult body, OPCs produce almost exclusively oligodendrocytes. With aging, this production shifts from oligodendrocytes, which are produced less, to an increased production of astrocytes [128]. The result is a thinned-out myelin sheath. The appearance of myelin fragments around the sheaths and in the ECS leads to its phagocytosis by microglia cells [129]. Over time, this leads to myelin overload and the development of lysosomal inclusions in the microglia of aging organisms. This activation of microglia also increases their number during aging [130]. The microglia cell bodies are also enlarged due to the non-degraded debris that is accumulated over the entire life span not just myelin fragments from aging myelin sheaths [131]. As will be shown in the following chapters, activated microglia release factors that shift astrocytes into reactive types. One of the consequences of this age-induced microglia activation is the release of inflammatory cytokines [132]. Normal aging also causes the transformation of many astrocytes into reactive astrocytes with the toxic reactive type being the dominant one. The latter is known to be upregulated by microglial-secreted inflammatory cytokines [133]. During aging, individual astrocytes increase their domains in a way that they are now overlapping to an extent never seen in younger animals [134]. The cytotoxic reactive astrocytes are damaging to both neurons and oligodendrocytes. Thus, there is a strong possibility for a vicious positive feedback cycle, starting with reduced oligodendrocyte function → myelin shedding → microglia activation by myelin fragments → cytokine release by activated microglia → transformation of normal astrocytes into toxic reactive astrocytes by these inflammatory cytokines → release of toxic substances by the toxic reactive astrocytes → reduction in oligodendrocytes/OPC function. The deterioration of the myelin sheath as well as the release of inflammatory and toxic substances impacts the neuronal function negatively over time and creates subtle age-related deficits in neural processing, which increases over time as this cycle continues. Another process affecting both, astrocytes and microglia, is that their calcium responses are disrupted, stressing their relationships with each other and with neurons [135, 136].

This slow transformation of normal astrocytes to toxic astrocytes under the influence of pro-inflammatory cytokines secreted from activated microglia is subtle at first, then increases in intensity. Interestingly, the brain areas most susceptible to neurodegeneration with age, such as the hippocampus and striatum, see more inflammatory reactive genes upregulated than other areas like the cortex [133]. This positive feedback cycle with myelin in the center might not be the only contributor to the subtle change of astrocytes to the reactive toxic subtype. For example, even in non-Alzheimer cohorts, misfolded proteins are accumulating with age [137]. With normal aging, more CD8 T leukocytes invade the brain parenchyma and modify microglial properties without any apparent accompanying pathology [138]. It has been postulated [139] that, in the aging hippocampus neurons express complement C3. This expression leads to increased phagocytosis of synapses, but not neuronal cell bodies, by activated microglia and astrocytes. The resulting loss of synapses is a contributing factor to cognitive decline. Microglial priming, another microglial-based process during normal aging, creates problems for neuronal signaling. This priming of aging microglia does not affect their inflammatory reaction at baseline. However, small and subtle inflammatory stimuli will lead to a disproportionate and overblown inflammatory response of these microglia [140]. The appearance of toxic reactive astrocytes in the aging brain has other consequences, involving the differential expression of genes whose products favor synapse elimination and therefore could cause cognitive decline [141]. Expression of genes that are important for astrocytic homeostatic functions is – in contrast – not affected. The major consequence of this differential expression is the release of cytokines, such as CXCL10 and CXCL5, which attract and activate immune cells, including microglia. Other changes involve complement system factors, secretory factors such as neurotrophins, peptidase inhibitors, and cholesterol synthesis. These changes interfere with the homeostatic and protective properties of astrocytes.

The question is why the glial cells change to a less efficient and more toxic phenotype during aging rather than to a more protective phenotype. For example, why is the toxic reactive astrocyte type rather than the protective reactive phenotype upregulated in aging? One answer may be that after the reproductive years of an organism have passed, evolutionary pressure for proper brain function and therefore survival is less than during the earlier lifespan when the organism is involved in reproduction and nurturing of offspring.

3.5 Conclusions

There is a close interaction between neurons and all glial cell types. These partnerships start from the very beginning in the neural tube during neuronal genesis and continue throughout life, even though the purpose and nature of the partnerships change over time. Brain signaling is based on the close cooperation of neurons with glial cells. Neurons are closely interacting with each other via chemical and electrical synapses, as well as exchanging information through the ECS with the panglial syncytium and microglia. Neurons also have chemical synapses with OPCs, and their synapses have close contacts with microglial processes if these microglia are in a non-activated state. All these communication avenues serve at the same time as information processors as well as support systems to balance neuronal demand and activity.

The combination of these functional brain compartments with different properties, tolerance limits, structures, and regulatory processes gives the brain signaling system flexibility, resilience, and plasticity. Since each compartment contributes different properties and regulatory processes, the compartments collectively work together as a functional unit with enough flexibility to shift emphasis depending on needs and conditions. This situation begs another question, which is whether this partnership has the resilience to withstand damage and to undertake rescue and repair processes after injury or during disease. The following chapters will investigate this question further.

References

1. Barry DS, Pakan JM, McDermott KW. Radial glial cells: key organisers in CNS development. Int J Biochem Cell Biol. 2014;46:76–9.
2. Falk S, Gotz M. Glial control of neurogenesis. Curr Opin Neurobiol. 2017;47:188–95.
3. Xu L, Tang X, Wang Y, Xu H, Fan X. Radial glia, the keystone of the development of the hippocampal dentate gyrus. Mol Neurobiol. 2015;51(1):131–41.
4. Gotz M, Huttner WB. The cell biology of neurogenesis. Nat Rev Mol Cell Biol. 2005;6(10):777–88.
5. Ohtsuka T, Kageyama R. Regulation of temporal properties of neural stem cells and transition timing of neurogenesis and gliogenesis during mammalian neocortical development. Semin Cell Dev Biol. 2019;95:4.
6. Shen Q, Wang Y, Dimos JT, Fasano CA, Phoenix TN, Lemischka IR, et al. The timing of cortical neurogenesis is encoded within lineages of individual progenitor cells. Nat Neurosci. 2006;9(6):743–51.
7. Squarzoni P, Thion MS, Garel S. Neuronal and microglial regulators of cortical wiring: usual and novel guideposts. Front Neurosci. 2015;9:248.
8. Allen NJ, Lyons DA. Glia as architects of central nervous system formation and function. Science (New York, NY). 2018;362(6411):181–5.
9. Kamakura S, Oishi K, Yoshimatsu T, Nakafuku M, Masuyama N, Gotoh Y. Hes binding to STAT3 mediates crosstalk between Notch and JAK-STAT signalling. Nat Cell Biol. 2004;6(6):547–54.

10. Allen NJ, Eroglu C. Cell biology of astrocyte-synapse interactions. Neuron. 2017;96(3):697–708.
11. Lopez-Hidalgo M, Schummers J. Cortical maps: a role for astrocytes? Curr Opin Neurobiol. 2014;24(1):176–89.
12. Bayraktar OA, Fuentealba LC, Alvarez-Buylla A, Rowitch DH. Astrocyte development and heterogeneity. Cold Spring Harb Perspect Biol. 2014;7(1):a020362.
13. Reemst K, Noctor SC, Lucassen PJ, Hol EM. The indispensable roles of microglia and astrocytes during brain development. Front Hum Neurosci. 2016;10:566.
14. Kessaris N, Fogarty M, Iannarelli P, Grist M, Wegner M, Richardson WD. Competing waves of oligodendrocytes in the forebrain and postnatal elimination of an embryonic lineage. Nat Neurosci. 2006;9(2):173–9.
15. Mangin JM, Li P, Scafidi J, Gallo V. Experience-dependent regulation of NG2 progenitors in the developing barrel cortex. Nat Neurosci. 2012;15(9):1192–4.
16. Spitzer SO, Sitnikov S, Kamen Y, Evans KA, Kronenberg-Versteeg D, Dietmann S, et al. Oligodendrocyte progenitor cells become regionally diverse and heterogeneous with age. Neuron. 2019;101(3):459–71.e5.
17. van Tilborg E, de Theije CGM, van Hal M, Wagenaar N, de Vries LS, Benders MJ, et al. Origin and dynamics of oligodendrocytes in the developing brain: implications for perinatal white matter injury. Glia. 2018;66(2):221–38.
18. Monier A, Adle-Biassette H, Delezoide AL, Evrard P, Gressens P, Verney C. Entry and distribution of microglial cells in human embryonic and fetal cerebral cortex. J Neuropathol Exp Neurol. 2007;66(5):372–82.
19. Thion MS, Ginhoux F, Garel S. Microglia and early brain development: an intimate journey. Science (New York, NY). 2018;362(6411):185–9.
20. Vainchtein ID, Chin G, Cho FS, Kelley KW, Miller JG, Chien EC, et al. Astrocyte-derived interleukin-33 promotes microglial synapse engulfment and neural circuit development. Science (New York, NY). 2018;359(6381):1269–73.
21. Bak LK, Walls AB, Schousboe A, Waagepetersen HS. Astrocytic glycogen metabolism in the healthy and diseased brain. J Biol Chem. 2018;293(19):7108–16.
22. Iadecola C. The neurovascular unit coming of age: a journey through neurovascular coupling in health and disease. Neuron. 2017;96(1):17–42.
23. Mishra A, Reynolds JP, Chen Y, Gourine AV, Rusakov DA, Attwell D. Astrocytes mediate neurovascular signaling to capillary pericytes but not to arterioles. Nat Neurosci. 2016;19(12):1619–27.
24. Mehina EMF, Murphy-Royal C, Gordon GR. Steady-state free Ca(2+) in astrocytes is decreased by experience and impacts arteriole tone. J Neurosci. 2017;37(34):8150–65.
25. Rosenegger DG, Tran CH, Wamsteeker Cusulin JI, Gordon GR. Tonic local brain blood flow control by astrocytes independent of phasic neurovascular coupling. J Neurosci. 2015;35(39):13463–74.
26. Blinder P, Tsai PS, Kaufhold JP, Knutsen PM, Suhl H, Kleinfeld D. The cortical angiome: an interconnected vascular network with noncolumnar patterns of blood flow. Nat Neurosci. 2013;16(7):889–97.
27. Rangaraju V, Calloway N, Ryan TA. Activity-driven local ATP synthesis is required for synaptic function. Cell. 2014;156(4):825–35.
28. Devine MJ, Kittler JT. Mitochondria at the neuronal presynapse in health and disease. Nat Rev Neurosci. 2018;19(2):63–80.
29. Magistretti PJ, Allaman I. A cellular perspective on brain energy metabolism and functional imaging. Neuron. 2015;86(4):883–901.
30. Chuquet J, Quilichini P, Nimchinsky EA, Buzsaki G. Predominant enhancement of glucose uptake in astrocytes versus neurons during activation of the somatosensory cortex. J Neurosci. 2010;30(45):15298–303.
31. Obel LF, Muller MS, Walls AB, Sickmann HM, Bak LK, Waagepetersen HS, et al. Brain glycogen-new perspectives on its metabolic function and regulation at the subcellular level. Front Neuroenerg. 2012;4:3.

32. Machler P, Wyss MT, Elsayed M, Stobart J, Gutierrez R, von Faber-Castell A, et al. In vivo evidence for a lactate gradient from astrocytes to neurons. Cell Metab. 2016;23(1):94–102.
33. Pellerin L, Magistretti PJ. Sweet sixteen for ANLS. J Cereb Blood Flow Metab. 2012;32(7):1152–66.
34. Munch G, Westcott B, Menini T, Gugliucci A. Advanced glycation endproducts and their pathogenic roles in neurological disorders. Amino Acids. 2012;42(4):1221–36.
35. Belanger M, Yang J, Petit JM, Laroche T, Magistretti PJ, Allaman I. Role of the glyoxalase system in astrocyte-mediated neuroprotection. J Neurosci. 2011;31(50):18338–52.
36. Lundgaard I, Li B, Xie L, Kang H, Sanggaard S, Haswell JD, et al. Direct neuronal glucose uptake heralds activity-dependent increases in cerebral metabolism. Nat Commun. 2015;6:6807.
37. Durkee CA, Araque A. Diversity and specificity of astrocyte-neuron communication. Neuroscience. 2019;396:73–8.
38. Bohmbach K, Schwarz MK, Schoch S, Henneberger C. The structural and functional evidence for vesicular release from astrocytes in situ. Brain Res Bull. 2018;136:65–75.
39. Araque A, Parpura V, Sanzgiri RP, Haydon PG. Tripartite synapses: glia, the unacknowledged partner. Trends Neurosci. 1999;22(5):208–15.
40. Savtchouk I, Volterra A. Gliotransmission: beyond black-and-white. J Neurosci. 2018;38(1):14–25.
41. Fiacco TA, McCarthy KD. Multiple lines of evidence indicate that Gliotransmission does not occur under physiological conditions. J Neurosci. 2018;38(1):3–13.
42. Panatier A, Vallee J, Haber M, Murai KK, Lacaille JC, Robitaille R. Astrocytes are endogenous regulators of basal transmission at central synapses. Cell. 2011;146(5):785–98.
43. Papouin T, Henneberger C, Rusakov DA, Oliet SHR. Astroglial versus neuronal D-Serine: fact checking. Trends Neurosci. 2017;40(9):517–20.
44. Pacholko AG, Wotton CA, Bekar LK. Astrocytes-the ultimate effectors of long-range neuromodulatory networks? Front Cell Neurosci. 2020;14:581075.
45. Panatier A, Vallée J, Haber M, Murai KK, Lacaille JC, Robitaille R. Astrocytes are endogenous regulators of basal transmission at central synapses. Cell. 2011;146(5):785–98.
46. Martin R, Bajo-Graneras R, Moratalla R, Perea G, Araque A. Circuit-specific signaling in astrocyte-neuron networks in basal ganglia pathways. Science (New York, NY). 2015;349(6249):730–4.
47. Martin-Fernandez M, Jamison S, Robin LM, Zhao Z, Martin ED, Aguilar J, et al. Synapse-specific astrocyte gating of amygdala-related behavior. Nat Neurosci. 2017;20(11):1540–8.
48. Matos M, Bosson A, Riebe I, Reynell C, Vallee J, Laplante I, et al. Astrocytes detect and upregulate transmission at inhibitory synapses of somatostatin interneurons onto pyramidal cells. Nat Commun. 2018;9(1):4254.
49. Covelo A, Araque A. Neuronal activity determines distinct gliotransmitter release from a single astrocyte. elife. 2018;7:e32237.
50. Octeau JC, Chai H, Jiang R, Bonanno SL, Martin KC, Khakh BS. An optical neuron-astrocyte proximity assay at synaptic distance scales. Neuron. 2018;98(1):49–66.e9.
51. Rothstein JD, Dykes-Hoberg M, Pardo CA, Bristol LA, Jin L, Kuncl RW, et al. Knockout of glutamate transporters reveals a major role for astroglial transport in excitotoxicity and clearance of glutamate. Neuron. 1996;16(3):675–86.
52. Schousboe A. Metabolic signaling in the brain and the role of astrocytes in control of glutamate and GABA neurotransmission. Neurosci Lett. 2019;689:11–3.
53. McKenna MC, Stridh MH, McNair LF, Sonnewald U, Waagepetersen HS, Schousboe A. Glutamate oxidation in astrocytes: roles of glutamate dehydrogenase and aminotransferases. J Neurosci Res. 2016;94(12):1561–71.
54. Sonnewald U. Glutamate synthesis has to be matched by its degradation - where do all the carbons go? J Neurochem. 2014;131(4):399–406.
55. Torres GE, Amara SG. Glutamate and monoamine transporters: new visions of form and function. Curr Opin Neurobiol. 2007;17(3):304–12.

56. Schousboe A, Wellendorph P, Frolund B, Clausen RP, Krogsgaard-Larsen P. Astrocytic GABA transporters: pharmacological properties and targets for antiepileptic drugs. Adv Neurobiol. 2017;16:283–96.
57. Hertz L, Rothman DL. Glucose, lactate, beta-Hydroxybutyrate, acetate, GABA, and succinate as substrates for synthesis of glutamate and GABA in the glutamine-glutamate/GABA cycle. Adv Neurobiol. 2016;13:9–42.
58. Kimelberg HK. Occurrence and functional significance of serotonin and catecholamine uptake by astrocytes. Biochem Pharmacol. 1986;35(14):2273–81.
59. Petrelli F, Dallerac G, Pucci L, Cali C, Zehnder T, Sultan S, et al. Dysfunction of homeostatic control of dopamine by astrocytes in the developing prefrontal cortex leads to cognitive impairments. Mol Psychiatry. 2018;25:732.
60. MacVicar BA, Feighan D, Brown A, Ransom B. Intrinsic optical signals in the rat optic nerve: role for K(+) uptake via NKCC1 and swelling of astrocytes. Glia. 2002;37(2):114–23.
61. Ballanyi K, Grafe P, ten Bruggencate G. Ion activities and potassium uptake mechanisms of glial cells in Guinea-pig olfactory cortex slices. J Physiol. 1987;382:159–74.
62. Dufour S, Dufour P, Chever O, Vallee R, Amzica F. In vivo simultaneous intra- and extracellular potassium recordings using a micro-optrode. J Neurosci Methods. 2011;194(2):206–17.
63. D'Ambrosio R, Gordon DS, Winn HR. Differential role of KIR channel and Na(+)/K(+)-pump in the regulation of extracellular K(+) in rat hippocampus. J Neurophysiol. 2002;87(1):87–102.
64. Kiyoshi CM, Du Y, Zhong S, Wang W, Taylor AT, Xiong B, et al. Syncytial isopotentiality: a system-wide electrical feature of astrocytic networks in the brain. Glia. 2018;66(12):2756–69.
65. Hounsgaard J, Nicholson C. Potassium accumulation around individual purkinje cells in cerebellar slices from the Guinea-pig. J Physiol. 1983;340:359–88.
66. Chen KC, Nicholson C. Spatial buffering of potassium ions in brain extracellular space. Biophys J. 2000;78(6):2776–97.
67. D'Ambrosio R, Maris DO, Grady MS, Winn HR, Janigro D. Impaired K(+) homeostasis and altered electrophysiological properties of post-traumatic hippocampal glia. J Neurosci. 1999;19(18):8152–62.
68. Macaulay N, Zeuthen T. Glial K(+) clearance and cell swelling: key roles for cotransporters and pumps. Neurochem Res. 2012;37(11):2299–309.
69. Rasmussen R, Nicholas E, Petersen NC, Dietz AG, Xu Q, Sun Q, et al. Cortex-wide changes in extracellular potassium ions parallel brain state transitions in awake behaving mice. Cell Rep. 2019;28(5):1182–94.e4.
70. Ruusuvuori E, Kaila K. Carbonic anhydrases and brain pH in the control of neuronal excitability. Subcell Biochem. 2014;75:271–90.
71. Chesler M, Kraig RP. Intracellular pH transients of mammalian astrocytes. J Neurosci. 1989;9(6):2011–9.
72. Theparambil SM, Naoshin Z, Defren S, Schmaelzle J, Weber T, Schneider HP, et al. Bicarbonate sensing in mouse cortical astrocytes during extracellular acid/base disturbances. J Physiol. 2017;595(8):2569–85.
73. Fern R. Astrocytes snipH out bicarbonate to regulate acid-base. J Physiol. 2017;595(8):2417.
74. Wilbrecht L, Holtmaat A, Wright N, Fox K, Svoboda K. Structural plasticity underlies experience-dependent functional plasticity of cortical circuits. J Neurosci. 2010;30(14):4927–32.
75. Perez-Alvarez A, Navarrete M, Covelo A, Martin ED, Araque A. Structural and functional plasticity of astrocyte processes and dendritic spine interactions. J Neurosci. 2014;34(38):12738–44.
76. Lushnikova I, Skibo G, Muller D, Nikonenko I. Synaptic potentiation induces increased glial coverage of excitatory synapses in CA1 hippocampus. Hippocampus. 2009;19(8):753–62.
77. Adamsky A, Kol A, Kreisel T, Doron A, Ozeri-Engelhard N, Melcer T, et al. Astrocytic activation generates De Novo neuronal potentiation and memory enhancement. Cell. 2018;174(1):59–71.e14.

78. Descarries L, Mechawar N. Ultrastructural evidence for diffuse transmission by monoamine and acetylcholine neurons of the central nervous system. Prog Brain Res. 2000;125:27–47.
79. Lim EY, Ye L, Paukert M. Potential and realized impact of Astroglia Ca(2 +) dynamics on circuit function and behavior. Front Cell Neurosci. 2021;15:682888.
80. Wang F, Smith NA, Xu Q, Fujita T, Baba A, Matsuda T, et al. Astrocytes modulate neural network activity by Ca^2+–dependent uptake of extracellular K+. Sci Signal. 2012;5(218):ra26.
81. Wotton CA, Cross CD, Bekar LK. Serotonin, norepinephrine, and acetylcholine differentially affect astrocytic potassium clearance to modulate somatosensory signaling in male mice. J Neurosci Res. 2020;98(5):964–77.
82. Ding F, O'Donnell J, Xu Q, Kang N, Goldman N, Nedergaard M. Changes in the composition of brain interstitial ions control the sleep-wake cycle. Science (New York, NY). 2016;352(6285):550–5.
83. Wotton CA, Quon EF, Palmer AC, Bekar LK. Corticosterone and serotonin similarly influence GABAergic and purinergic pathways to affect cortical inhibitory networks. J Neuroendocrinol. 2018;30(4):e12592.
84. Quon EF, Wotton CA, Bekar LK. Evidence for astrocyte purinergic signaling in cortical sensory adaptation and serotonin-mediated neuromodulation. Mol Cell Neurosci. 2018;88:53–61.
85. Poyhonen S, Er S, Domanskyi A, Airavaara M. Effects of neurotrophic factors in glial cells in the central nervous system: expression and properties in neurodegeneration and injury. Front Physiol. 2019;10:486.
86. Ledda F, Paratcha G, Sandoval-Guzman T, Ibanez CF. GDNF and GFRalpha1 promote formation of neuronal synapses by ligand-induced cell adhesion. Nat Neurosci. 2007;10(3):293–300.
87. Marina N, Turovsky E, Christie IN, Hosford PS, Hadjihambi A, Korsak A, et al. Brain metabolic sensing and metabolic signaling at the level of an astrocyte. Glia. 2018;66(6):1185–99.
88. Garcia-Caceres C, Balland E, Prevot V, Luquet S, Woods SC, Koch M, et al. Role of astrocytes, microglia, and tanycytes in brain control of systemic metabolism. Nat Neurosci. 2019;22(1):7–14.
89. Marina N, Tang F, Figueiredo M, Mastitskaya S, Kasimov V, Mohamed-Ali V, et al. Purinergic signalling in the rostral ventro-lateral medulla controls sympathetic drive and contributes to the progression of heart failure following myocardial infarction in rats. Basic Res Cardiol. 2013;108(1):317.
90. Garcia-Caceres C, Quarta C, Varela L, Gao Y, Gruber T, Legutko B, et al. Astrocytic insulin Signaling couples brain glucose uptake with nutrient availability. Cell. 2016;166(4):867–80.
91. Gourine AV, Kasymov V, Marina N, Tang F, Figueiredo MF, Lane S, et al. Astrocytes control breathing through pH-dependent release of ATP. Science (New York, NY). 2010;329(5991):571–5.
92. Meigh L, Greenhalgh SA, Rodgers TL, Cann MJ, Roper DI, Dale N. CO(2)directly modulates connexin 26 by formation of carbamate bridges between subunits. elife. 2013;2:e01213.
93. Stadelmann C, Timmler S, Barrantes-Freer A, Simons M. Myelin in the central nervous system: structure, function, and pathology. Physiol Rev. 2019;99(3):1381–431.
94. Battefeld A, Klooster J, Kole MH. Myelinating satellite oligodendrocytes are integrated in a glial syncytium constraining neuronal high-frequency activity. Nat Commun. 2016;7:11298.
95. Saab AS, Tzvetavona ID, Trevisiol A, Baltan S, Dibaj P, Kusch K, et al. Oligodendroglial NMDA receptors regulate glucose import and axonal energy metabolism. Neuron. 2016;91(1):119–32.
96. Mitew S, Gobius I, Fenlon LR, McDougall SJ, Hawkes D, Xing YL, et al. Pharmacogenetic stimulation of neuronal activity increases myelination in an axon-specific manner. Nat Commun. 2018;9(1):306.
97. Habermacher C, Angulo MC, Benamer N. Glutamate versus GABA in neuron-oligodendroglia communication. Glia. 2019;67:2092.
98. Dawson MR, Polito A, Levine JM, Reynolds R. NG2-expressing glial progenitor cells: an abundant and widespread population of cycling cells in the adult rat CNS. Mol Cell Neurosci. 2003;24(2):476–88.

99. Rudy B, Fishell G, Lee S, Hjerling-Leffler J. Three groups of interneurons account for nearly 100% of neocortical GABAergic neurons. Dev Neurobiol. 2011;71(1):45–61.
100. Hughes EG, Kang SH, Fukaya M, Bergles DE. Oligodendrocyte progenitors balance growth with self-repulsion to achieve homeostasis in the adult brain. Nat Neurosci. 2013;16(6):668–76.
101. Gibson EM, Purger D, Mount CW, Goldstein AK, Lin GL, Wood LS, et al. Neuronal activity promotes oligodendrogenesis and adaptive myelination in the mammalian brain. Science (New York, NY). 2014;344(6183):1252304.
102. Bechler ME, Swire M, Ffrench-Constant C. Intrinsic and adaptive myelination-A sequential mechanism for smart wiring in the brain. Dev Neurobiol. 2018;78(2):68–79.
103. Nagy B, Hovhannisyan A, Barzan R, Chen TJ, Kukley M. Different patterns of neuronal activity trigger distinct responses of oligodendrocyte precursor cells in the corpus callosum. PLoS Biol. 2017;15(8):e2001993.
104. Wake H, Moorhouse AJ, Jinno S, Kohsaka S, Nabekura J. Resting microglia directly monitor the functional state of synapses in vivo and determine the fate of ischemic terminals. J Neurosci. 2009;29(13):3974–80.
105. Ikegami A, Haruwaka K, Wake H. Microglia: lifelong modulator of neural circuits. Neuropathology. 2019;39(3):173–80.
106. York EM, Bernier LP, MacVicar BA. Microglial modulation of neuronal activity in the healthy brain. Dev Neurobiol. 2018;78(6):593–603.
107. Ji K, Akgul G, Wollmuth LP, Tsirka SE. Microglia actively regulate the number of functional synapses. PLoS One. 2013;8(2):e56293.
108. Li Y, Du XF, Liu CS, Wen ZL, Du JL. Reciprocal regulation between resting microglial dynamics and neuronal activity in vivo. Dev Cell. 2012;23(6):1189–202.
109. Pfeiffer T, Avignone E, Nagerl UV. Induction of hippocampal long-term potentiation increases the morphological dynamics of microglial processes and prolongs their contacts with dendritic spines. Sci Rep. 2016;6:32422.
110. Akiyoshi R, Wake H, Kato D, Horiuchi H, Ono R, Ikegami A, et al. Microglia enhance synapse activity to promote local network synchronization. eNeuro. 2018;5(5):ENEURO.0088-18.2018.
111. Wright-Jin EC, Gutmann DH. Microglia as dynamic cellular mediators of brain function. Trends Mol Med. 2019;25(11):967–79.
112. Torres L, Danver J, Ji K, Miyauchi JT, Chen D, Anderson ME, et al. Dynamic microglial modulation of spatial learning and social behavior. Brain Behav Immun. 2016;55:6–16.
113. Rogers JT, Morganti JM, Bachstetter AD, Hudson CE, Peters MM, Grimmig BA, et al. CX3CR1 deficiency leads to impairment of hippocampal cognitive function and synaptic plasticity. J Neurosci. 2011;31(45):16241–50.
114. Zhan Y, Paolicelli RC, Sforazzini F, Weinhard L, Bolasco G, Pagani F, et al. Deficient neuron-microglia signaling results in impaired functional brain connectivity and social behavior. Nat Neurosci. 2014;17(3):400–6.
115. Parkhurst CN, Yang G, Ninan I, Savas JN, Yates JR 3rd, Lafaille JJ, et al. Microglia promote learning-dependent synapse formation through brain-derived neurotrophic factor. Cell. 2013;155(7):1596–609.
116. Fontainhas AM, Wang M, Liang KJ, Chen S, Mettu P, Damani M, et al. Microglial morphology and dynamic behavior is regulated by ionotropic glutamatergic and GABAergic neurotransmission. PLoS One. 2011;6(1):e15973.
117. Biber K, Neumann H, Inoue K, Boddeke HW. Neuronal 'On' and 'Off' signals control microglia. Trends Neurosci. 2007;30(11):596–602.
118. Pawelec P, Ziemka-Nalecz M, Sypecka J, Zalewska T. The impact of the CX3CL1/CX3CR1 Axis in neurological disorders. Cell. 2020;9(10):2277.
119. Szepesi Z, Manouchehrian O, Bachiller S, Deierborg T. Bidirectional microglia-neuron communication in health and disease. Front Cell Neurosci. 2018;12:323.
120. Gabanyi I, Muller PA, Feighery L, Oliveira TY, Costa-Pinto FA, Mucida D. Neuro-immune interactions drive tissue programming in intestinal macrophages. Cell. 2016;164(3):378–91.

121. Arpaia N, Green JA, Moltedo B, Arvey A, Hemmers S, Yuan S, et al. A distinct function of regulatory T cells in tissue protection. Cell. 2015;162(5):1078–89.
122. Villalta SA, Rosenthal W, Martinez L, Kaur A, Sparwasser T, Tidball JG, et al. Regulatory T cells suppress muscle inflammation and injury in muscular dystrophy. Sci Transl Med. 2014;6(258):258ra142.
123. Kopper TJ, Gensel JC. Myelin as an inflammatory mediator: myelin interactions with complement, macrophages, and microglia in spinal cord injury. J Neurosci Res. 2018;96(6):969–77.
124. Safaiyan S, Kannaiyan N, Snaidero N, Brioschi S, Biber K, Yona S, et al. Age-related myelin degradation burdens the clearance function of microglia during aging. Nat Neurosci. 2016;19(8):995–8.
125. Bartzokis G, Cummings JL, Sultzer D, Henderson VW, Nuechterlein KH, Mintz J. White matter structural integrity in healthy aging adults and patients with Alzheimer disease: a magnetic resonance imaging study. Arch Neurol. 2003;60(3):393–8.
126. Peters R. Ageing and the brain. Postgrad Med J. 2006;82(964):84–8.
127. Sowell ER, Peterson BS, Thompson PM, Welcome SE, Henkenius AL, Toga AW. Mapping cortical change across the human life span. Nat Neurosci. 2003;6(3):309–15.
128. Doucette JR, Jiao R, Nazarali AJ. Age-related and cuprizone-induced changes in myelin and transcription factor gene expression and in oligodendrocyte cell densities in the rostral corpus callosum of mice. Cell Mol Neurobiol. 2010;30(4):607–29.
129. Hill RA, Li AM, Grutzendler J. Lifelong cortical myelin plasticity and age-related degeneration in the live mammalian brain. Nat Neurosci. 2018;21(5):683–95.
130. Mosher KI, Wyss-Coray T. Microglial dysfunction in brain aging and Alzheimer's disease. Biochem Pharmacol. 2014;88(4):594–604.
131. Streit WJ, Xue QS, Tischer J, Bechmann I. Microglial pathology. Acta Neuropathol Commun. 2014;2:142.
132. Sierra A, Gottfried-Blackmore AC, McEwen BS, Bulloch K. Microglia derived from aging mice exhibit an altered inflammatory profile. Glia. 2007;55(4):412–24.
133. Clarke LE, Liddelow SA, Chakraborty C, Munch AE, Heiman M, Barres BA. Normal aging induces A1-like astrocyte reactivity. Proc Natl Acad Sci U S A. 2018;115(8):E1896–e905.
134. Grosche A, Grosche J, Tackenberg M, Scheller D, Gerstner G, Gumprecht A, et al. Versatile and simple approach to determine astrocyte territories in mouse neocortex and hippocampus. PLoS One. 2013;8(7):e69143.
135. Guerra-Gomes S, Viana JF, Nascimento DSM, Correia JS, Sardinha VM, Caetano I, et al. The role of astrocytic calcium signaling in the aged prefrontal cortex. Front Cell Neurosci. 2018;12:379.
136. Olmedillas Del Moral M, Asavapanumas N, Uzcátegui NL, Garaschuk O. Healthy brain aging modifies microglial calcium signaling in vivo. Int J Mol Sci. 2019;20(3):589.
137. Yu L, De Jager PL, Yang J, Trojanowski JQ, Bennett DA, Schneider JA. The TMEM106B locus and TDP-43 pathology in older persons without FTLD. Neurology. 2015;84(9):927–34.
138. Ritzel RM, Crapser J, Patel AR, Verma R, Grenier JM, Chauhan A, et al. Age-associated resident memory CD8 T cells in the central nervous system are primed to potentiate inflammation after ischemic brain injury. J Immunol. 2016;196(8):3318–30.
139. Shi Q, Colodner KJ, Matousek SB, Merry K, Hong S, Kenison JE, et al. Complement C3-deficient mice fail to display age-related hippocampal decline. J Neurosci. 2015;35(38):13029–42.
140. Niraula A, Sheridan JF, Godbout JP. Microglia priming with aging and stress. Neuropsychopharmacology. 2017;42(1):318–33.
141. Boisvert MM, Erikson GA, Shokhirev MN, Allen NJ. The aging astrocyte transcriptome from multiple regions of the mouse brain. Cell Rep. 2018v22(1):269–85.
142. Saab AS, Nave KA. Myelin dynamics: protecting and shaping neuronal functions. Curr Opin Neurobiol. 2017;47:104–12.

Chapter 4
Reactive Microglia and Astrocyte Phenotype Transitions: A Framework

Abstract This chapter gives an overview of the reactions of microglia and astrocytes to pathological events. In the focus are the microglia as they are normally the first cell type to sense and react to these detrimental events. Astrocytes become only partially reactive without the previous reaction of microglia. Thus, microglia are seen as "transformers" and the "sensome" to pathological events, whereas astrocytes act as a "reserve force" with a graded response in relation to the pathological impact. If the disturbance is severe enough, microglia give up their homeostatic and signal role and revert to phenotypes with features specific to the challenges. Contemporary reports no longer distinguish specific microglial phenotype endpoints as a response to these different challenges. Rather, microglia are seen as cells with reactions and phenotypes depending on the context of the disturbances. In this respect, two extreme phenotypes are pro-inflammatory and toxic microglia on one hand as well as anti-inflammatory and protective microglia on the other hand. There are large differences within each of these two groups. Astrocytes are more reluctant to give up their homeostatic roles as long as there are still viable neurons in their vicinity. They are not as likely as microglia to proliferate and migrate. However, they are flexible enough to take on functions that are known from immune cells, if necessary.

Keywords Astrocytes · Brain cell proliferation · Brain injury · Microglia · Neuroinflammation · Reactive microgliosis · Reactive astrogliosis · Sensome

The previous chapters described a close and flexible partnership between neurons and all types of glial cells covering most aspects of the normal operation of the central nervous system (CNS). The question is now if during injury, disease, or degeneration, these different partners exhibit enough versatility to rescue as much as possible of the CNS function. The answer is clearly that all glial cells can switch into altered phenotypes to address the disturbance. Almost all adult neurons are, due to their integration into circuits, not as flexible to switch states, phenotypes, or even migrate and proliferate. This role of changing phenotype and therefore major functions is taken over by the glial cells. The resulting change in phenotypes of the glial cells is called reactive gliosis as it is a reaction to a disturbance and as the phenotypes of the glial participants are visibly changing. Major factors are the interactions

W. Walz, *The Gliocentric Brain*, https://doi.org/10.1007/978-3-031-48105-5_4

between astrocytes and microglia and to a lesser extent interactions with components of the blood–brain barrier and invading cells from the immune system. Both astrocytes and microglia constantly receive clues about the health and activity of neurons. They also have receptors for signals that are reflective of problems and dysfunction. Thus, the reactions of these two cell types are context-dependent and only as good as the dynamic picture that these environmental clues provide. Faulty readings, overreactions, and inactions are frequent. The location of the glial cells within a brain region and close to a specific neuronal circuit will also play a role in the plasticity of the responses. These context-dependent reactions of both cell types of course are contrasted in most publications, which still report the existence of distinct endpoint categories of disease phenotypes. Rather, both cell types react flexibly to the signals they receive from the environment and from each other. This is also the reason why these glial cells show different phenotypes during an injury and its aftermath. Especially, microglia can function at first as a toxic cell type, then as a protective type, and – with chronic conditions – switch back to a detrimental type. The underlying reason is a misreading of environmental clues and therefore a misjudgment about the state of health of the brain region. Each situation is "judged" by its own merit, and the resulting phenotype shift represents an attempt to optimize the responses for protection and repair. Of course, similar situations provoke similar response patterns. This is the reason that many investigators attempted to categorize the phenotypes with pathological situations. This is reflected in categorizing microglia reactions into categories akin to those of other immune cells. Most publications still distinguish only M1 (pro-inflammatory and toxic) and M2 (anti-inflammatory and protective) types. This categorizing was subsequently extended to astrocytic A1 and A2 types. There is no denying that detrimental or toxic as well as protective and restorative phenotypes of both cell types exist. They are, however, extreme cases and may only exist for certain time periods or extreme situations [1].

A key process is inflammation. If this occurs, toxic phenotypes appear to dominate, depending on the degree of the inflammation. It is important that the inflammation phase is as short as possible. Chronic inflammation is very detrimental as it keeps microglia in a permanent toxic or semi-toxic state and prevents coordination between microglia and astrocytic phenotypes. Here are some basic facts regarding gliosis.

1. Gliosis is caused by any disturbance to normal function. It is even present during normal aging (see the previous chapter).
2. Gliosis is a graded process and not an all-or-nothing event.
3. Neuronal damage, inflammation, and time courses modify the process.
4. Microglia are an immune cell type and highly flexible in giving up their homeostatic and signal roles in favor of becoming a classical macrophage-like cell. It is now acknowledged as the brain cell type with the most plastic behavior [2]. Microglia are normally the cell type reacting first to abnormal events, thanks to a multitude of surface receptors. This multitude of receptors serves to sense disturbances in the normal function of the CNS and is the reason why some researchers call microglia a "sensome" [3]. There are "ON" and "OFF" recep-

tors monitoring neuronal health, pathogen recognition receptors, and damage/danger recognition receptors.

5. The astrocytes are always involved, and they enlarge (hypertrophy) and might proliferate (hyperplasia). The increase in glial fibrillary acidic protein (GFAP) density in the astrocytic cytoplasm is a hallmark of reactive astrocytes and indicates the degree of hypertrophy. Yet, reactive astrocytes keep their domain- and (normally) their syncytium-like organization.
6. An important hallmark is the interactions between microglia and astrocytes. However, oligodendrocytes and oligodendrocyte progenitor cells are often involved.
7. Inflammation involves microglia and often, but not always, other invading cells of the immune system.
8. Regarding neuronal survival and regeneration, reactive gliosis is a two-edged sword. Protective and supportive elements of the response are present together with damaging and outright toxic elements. The degree of inflammation is key in this regard. The more intense the inflammatory response, the more detrimental to the health of neurons. Chronic inflammation is close to a worst-case scenario.
9. If the intensity of damage and reactive gliosis is strong enough, a completely remodeled tissue is the result: the glial scar tissue.
10. The foremost priority of a gliotic response is to counteract any breach in the blood–brain barrier and to seal off any leak from the blood plasma.
11. Neurons are as much as possible shielded. For example, microglia and invading immune cells rigorously eliminate cells deemed infected by viruses. However, if neurons are infected their cytolysis is not a first option. This is to protect neuronal circuits as much as possible.
12. Gliosis has been classified as isomorphic or anisomorphic, depending on the fact if the reactive gliotic response leaves the basic structure of the affected region intact (isomorphic) or if the affected tissue is completely reorganized by a glial scar (anisomorphic). This is an older definition based on morphological criteria with no pertaining to functional aspects, although it is, of course, an indication of the intensity of the response.

One should not forget that microglia are involved in normal signaling with close interactions with neurons as well as in homeostatic roles [4]. This is not unlike most other tissue macrophages [5]. For example, there are indications that both microglia and astrocytes are involved in the synchronization of neuronal activities in the healthy brain [4, 6]. These microglial and astrocytic activities must be coordinated. However, so far, no mechanisms are known with which these two cell types communicate and coordinate their pattern generation.

Once recruited for activities to resolve damage or external challenges, microglia abandon all signal and homeostatic functions as they detach from their close relationships with neurons, change shape, and migrate. In the past, several microglia types have been identified after pathological events. Examples are disease-associated microglia, microglial neurodegenerative phenotype, activated response microglia, interferon-responsive microglia, human Alzheimer's disease microglia, microglia inflamed in multiple sclerosis, lipid-droplet-accumulating microglia, glioma-associated microglia, amyotrophic lateral sclerosis-associated microglia, Parkinson disease-associated microglia, white matter-associated microglia, axon tract-associated microglia, and proliferative-region-associated microglia. These have been reviewed by Paolicelli et al. [2] and are now – apart from disease-associated microglia – seen as context-dependent microglia states rather than endpoints for definite subcellular phenotypes. The problem is that sole dependence on various single-cell analytical methods such as transcriptomics or mass cytometry does not do justice to complex functional states [7]. Thus, microglia are clearly the most versatile and dynamic cell type in the brain. Most astrocytes do not abandon at least some homeostatic role in their reactive state as long as there are still viable neurons in their vicinity. An exception might be the barrier astrocyte, which seems to be the only reactive astrocyte phenotype that proliferates and does not exist in the vicinity of functional neurons.

What triggers a gliotic reaction? Is it solely the state of neuronal health or are there other indicators of tissue trouble, like blood flow or myelin integrity? Is there a cell type that is responsible for coordinating the gliotic response? Is there more than one trigger? And if so, do different triggers cause different types of gliosis? These are crucial questions, and I will attempt to deal with possible answers in the chapters that are dedicated to specific disease groups.

General answers to these questions are complicated:

1. Deterioration of neuronal health is an important trigger, but it is not the only one.
2. The kind of trigger has an important influence on the course of reactive gliosis.
3. The microglia seem to be the focus of the initiation and coordination of the response of the surrounding tissue to injury and disease.
4. There is no "all-or-none" threshold for microglia activation by neuronal signals. It seems rather that microglia show a graded response, during which they show increased interaction with synapses, morphological changes, and then various intensities of an activation pattern. Thus, there normally is no uncontrolled "run-away" microglial activation.
5. The mode of activation (the nature of the neuronal and other signals) seems to determine the activation phenotype of the microglia. Therefore, microglia activation is not a stereotypical event but rather a process that reacts flexibly and with endpoints tailored to the situation. Transition to a reactive state is subtle, smooth, incremental, and flexibly aligned with the cause and degree of the damage.

In the normal, adult brain microglia are intimately connected to neurons, surveying synapses, remodeling dysfunctional synapses, and removing apoptotic neurons. Microglia are involved in normal synaptic transmission, can phagocytose debris, can promote inflammation, and can resurrect their nature as an immune cell. Microglial cells seem even to support adult neurogenesis [8]. All these mechanisms are expressed according to the need and in response to a combination of environmental factors. So far, no coherent overall functional concept emerged which would explain the interactions between neurons and microglia in the normal brain. This is reminiscent of the situation of astrocytic–neuronal signaling about 10 years ago. There have been many attempts to categorize microglia subtypes in the past: phagocytic and surveilling, inflammatory or not (M1 and M2), amoeboid and ramified [9]. In addition, there are changes in the brain region, gender, and aging [9]. It seems to be more realistic to view microglia as a very plastic cell type that is the product of many environmental factors with inherent flexibility as these factors change. This explains the regional variability as well as differences in transcriptome profiles. It has been suggested [9] that microglia might act as "transformers" whose role is to integrate the status of their individual microenvironments and as a result impact changes on their surrounding other cell types including neurons. In this way, there would be no rigid distinction between normal and pathological status but a smooth adaptation along a sliding scale in both directions as the need arises. Pharmacologically or genetically disabled microglia result in learning deficits, memory loss, and reduced social interactions in rodents [10–13]. The changes were reverted when the brain areas were again repopulated with functional microglia.

To recapitulate from the previous chapter, "resting" microglia cell bodies are normally stationary, but their processes are highly flexible and move in defined territories and specifically seek out synapses and monitor their activity. Neuronal signals seem to communicate with microglia on several levels. Microglia increase the number of their processes and the contact time of these processes with synapses when neuronal activity increases. They exhibit hyper-ramification. These changes can be caused by hippocampal long-term potentiation [14]. They are also induced by increased glutamatergic synaptic traffic, whereas GABAergic (gamma-aminobutyric acid) neurotransmission has an inhibitory effect on these changes. Interestingly, the changes by glutamate and GABA are not caused by receptors for these transmitters on the microglial cell membrane, but by modification of adenosine triphosphate (ATP) release from neurons and/or astrocytes under the control of glutamate or GABA. ATP directly applied on microglia can cause these phenotypical changes through P2X and P2Y purinergic receptors on the microglia cell membrane [15]. On another level, other factors, which are secreted from healthy neurons, act directly on microglial receptors to keep them in their normal non-activated state. The two most prominent are fractalkine (a chemokine) and CD200 (cluster of differentiation 200, a glycoprotein) [16]. Microglia are not only adapting to the degree of normal activity in non-pathological situations in a graded fashion, a prolonged decrease in neuronal activity shifts microglia gradually to a more "reactive" phenotype. This phenotype is more pronounced as neurons not only decrease their normal activity but also release specific "trouble" signals. The implication is therefore that a reduction of neuronal activity caused by pathological processes results in a shift of the microglial phenotype to a less ramified appearance, somewhat closer to the reactive phenotype [17].

In general, microglia are more sensitive to pathological changes than astrocytes. Microglia react first and in turn induce reactive astrogliosis, in keeping with the microglial role as "sensome." The signals from microglia to astrocytes to indicate pathological changes vary, but the most important are cytokines [18]. Indeed, microglia are the major source of cytokines in the CNS and therefore drivers of neuroinflammation [19]. Moreover, astrocytic responses to pathogens without microglia involvement seem to be slow and delayed. It seems that, in the absence of microglia, astrocytes do not become fully activated [20]. Due to their organization into a syncytium and processes close to the blood–brain barrier and synapses, however, astrocytes are an important target for microglia to amplify and propagate the reactive gliotic response, especially if it involves a large inflammatory component. The astrocytic syncytium has therefore also been called a "reserve force" complementing the microglial "sensome" [18]. An exception is the disturbances originating due to blood–brain barrier disruption and blood flow abnormalities [21]. In these cases, the astrocytic endfeet near the endothelial cell layer have a major sensor and initiator function. To coordinate responses, astrocytes feedback on the microglial reaction with various signals. Most important are chemokines [22].

The microglial, astrocytic, and inflammatory responses to pathological events are discussed in more detail in subsequent separate chapters as are the specific considerations applying to the various disease groups.

References

1. Devanney NA, Stewart AN, Gensel JC. Microglia and macrophage metabolism in CNS injury and disease: the role of immunometabolism in neurodegeneration and neurotrauma. Exp Neurol. 2020;329:113310.
2. Paolicelli RC, Sierra A, Stevens B, Tremblay ME, Aguzzi A, Ajami B, et al. Microglia states and nomenclature: a field at its crossroads. Neuron. 2022;110(21):3458–83.
3. Hickman SE, Kingery ND, Ohsumi TK, Borowsky ML, Wang LC, Means TK, et al. The microglial sensome revealed by direct RNA sequencing. Nat Neurosci. 2013;16(12):1896–905.
4. Akiyoshi R, Wake H, Kato D, Horiuchi H, Ono R, Ikegami A, et al. Microglia enhance synapse activity to promote local network synchronization. eNeuro. 2018;5(5):ENEURO.0088.
5. Gordon S, Martinez-Pomares L. Physiological roles of macrophages. Pflugers Archiv Eur J Physiol. 2017;469(3–4):365–74.
6. Pacholko AG, Wotton CA, Bekar LK. Astrocytes-the ultimate effectors of long-range neuromodulatory networks? Front Cell Neurosci. 2020;14:581075.
7. Böttcher C, Schlickeiser S, Sneeboer MAM, Kunkel D, Knop A, Paza E, et al. Human microglia regional heterogeneity and phenotypes determined by multiplexed single-cell mass cytometry. Nat Neurosci. 2019;22(1):78–90.
8. Kreisel T, Wolf B, Keshet E, Licht T. Unique role for dentate gyrus microglia in neuroblast survival and in VEGF-induced activation. Glia. 2019;67(4):594–618.
9. Wright-Jin EC, Gutmann DH. Microglia as dynamic cellular mediators of brain function. Trends Mol Med. 2019;25(11):967–79.
10. Torres L, Danver J, Ji K, Miyauchi JT, Chen D, Anderson ME, et al. Dynamic microglial modulation of spatial learning and social behavior. Brain Behav Immun. 2016;55:6–16.
11. Rogers JT, Morganti JM, Bachstetter AD, Hudson CE, Peters MM, Grimmig BA, et al. CX3CR1 deficiency leads to impairment of hippocampal cognitive function and synaptic plasticity. J Neurosci. 2011;31(45):16241–50.
12. Zhan Y, Paolicelli RC, Sforazzini F, Weinhard L, Bolasco G, Pagani F, et al. Deficient neuron-microglia signaling results in impaired functional brain connectivity and social behavior. Nat Neurosci. 2014;17(3):400–6.

13. Parkhurst CN, Yang G, Ninan I, Savas JN, Yates JR 3rd, Lafaille JJ, et al. Microglia promote learning-dependent synapse formation through brain-derived neurotrophic factor. Cell. 2013;155(7):1596–609.
14. Pfeiffer T, Avignone E, Nagerl UV. Induction of hippocampal long-term potentiation increases the morphological dynamics of microglial processes and prolongs their contacts with dendritic spines. Sci Rep. 2016;6:32422.
15. Fontainhas AM, Wang M, Liang KJ, Chen S, Mettu P, Damani M, et al. Microglial morphology and dynamic behavior is regulated by ionotropic glutamatergic and GABAergic neurotransmission. PLoS One. 2011;6(1):e15973.
16. Biber K, Neumann H, Inoue K, Boddeke HW. Neuronal 'on' and 'off' signals control microglia. Trends Neurosci. 2007;30(11):596–602.
17. Szepesi Z, Manouchehrian O, Bachiller S, Deierborg T. Bidirectional microglia-neuron communication in health and disease. Front Cell Neurosci. 2018;12:323.
18. Liu LR, Liu JC, Bao JS, Bai QQ, Wang GQ. Interaction of microglia and astrocytes in the neurovascular unit. Front Immunol. 2020;11:1024.
19. Buffo A, Rolando C, Ceruti S. Astrocytes in the damaged brain: molecular and cellular insights into their reactive response and healing potential. Biochem Pharmacol. 2010;79(2):77–89.
20. Chen SH, Oyarzabal EA, Sung YF, Chu CH, Wang Q, Chen SL, et al. Microglial regulation of immunological and neuroprotective functions of astroglia. Glia. 2015;63(1):118–31.
21. Argaw AT, Asp L, Zhang J, Navrazhina K, Pham T, Mariani JN, et al. Astrocyte-derived VEGF-A drives blood-brain barrier disruption in CNS inflammatory disease. J Clin Invest. 2012;122(7):2454–68.
22. Xu J, Dong H, Qian Q, Zhang X, Wang Y, Jin W, et al. Astrocyte-derived CCL2 participates in surgery-induced cognitive dysfunction and neuroinflammation via evoking microglia activation. Behav Brain Res. 2017;332:145–53.

Chapter 5
Transition of Microglia to Reactive States

Abstract Microglia are fully involved in the homeostatic and signaling function of the brain parenchyma. At the same time, they can quickly transition into a macrophage-like cell function. They dissociate themselves from the normal functions and change their morphology, and they can migrate and proliferate if needed. They receive extensive signals from their environment and integrate these signals to judge the severity, location, source, and nature of the disturbance as much as possible and react with high flexibility to the challenge. It is important to register that there are no fixed or pre-assigned microglial reactive types. Rather, the cells use the context of all signals to find the appropriate response. Microglia can also change their response pattern during a pathological event, for example, transitioning from a pro-inflammatory phenotype to an anti-inflammatory one or vice versa. The signals can be released due to damage to neurons or other cells. In this case, the strength and appearance of related factors are important. They can also be pathogen-associated signals, only apparent if there is a threat. The microglial phenotype in case of a malfunction ranges from phagocytosis, cytotoxic actions, as well as recruitment of other glial cells, including astrocytes and invading cells of the immune system. Due to their sensitivity, microglia are a kind of first responder and a coordinator of the brain tissue response to pathological challenges.

Keywords Cytokines · Microglia classification · Neuroinflammation · Neuronal OFF/ON factors · Nucleotide-binding oligomerization domain-like receptors · Pattern recognition receptors · Phagocytosis · Reactive microglia · Retinoic acid-inducible gene-1-like receptors · Toll-like receptors

5.1 Microglia Classification

In the healthy tissues, most resident macrophages are involved in homeostatic tasks [1]. As we have seen in previous chapters, the microglia of the brain are no exception. This is probably due to the organism optimizing its resources and avoiding the existence of a "resting" sentinel cell with no functional role for most of its lifetime. Microglia are mainly distinguished from macrophages by their origin (see previous

W. Walz, *The Gliocentric Brain*, https://doi.org/10.1007/978-3-031-48105-5_5

chapters). The properties of microglia as immune cells are coopted by the central nervous system (CNS) to assist in normal development and healthy adult brain in being involved in brain cell development, synaptic pruning, scavenging of surplus cells, and similar events. These microglia are actively involved in neuronal function and change their shape and processes continuously in response to neuronal activity (see previous chapters). Here we are dealing with microglia as immune cells after transitioning to a reactive state by pathological events.

In the normal adult brain, microglia are not a homogeneous group of cells. There are regional differences, differences in conjunction with the type of neuronal pathway in the vicinity, and differences depending on the type of neuronal messengers used. Examples, as presented by different gene expression [2], are as follows:

- The satellite microglia, which interact with the axon initial segment of neurons.
- The microglia supporting neurogenesis, which are involved in neuroblast survival in the adult neurogenic subventricular zone and further on in the rostral migratory stream.
- Hox8b (homeobox8) – microglia, which are involved in the function of the corticostriatal neuronal circuit.
- Dark microglia, which interact with blood vessels and synapses and accumulate during the aging process. Apart from these defined subtypes, the naming of the general microglia population in the healthy adult as "resting" is now considered misplaced [3]. As pointed out in previous chapters, the microglia processes are in a constant state of remodeling. It seems that microglia properties are defined by their microenvironment, namely, factors released from neurons and – to a lesser extent – from astrocytes and oligodendrocytes [4].

Microglial reactivity refers to the transformation of the "resting" homeostatic cell type into a different phenotype after exposure to disturbances in its environment. This reactivity transforms microglia into retracting its processes and acquiring an ameboid morphology. The cells migrate through the tissue, homing in on invaders or damaged tissue. They may also multiply. This is not an all-or-none process. It is progressively reacting to the severity of the situation, and it is reversible. The endpoint would be a fully immunocompetent macrophage called reactive microglia [3]. The characterization of the different stages of microglia activation is highly controversial. Analogous to the Th1/Th2 classification of T lymphocytes, similar activation states for microglia have been proposed [5]:

- M1 microglia express cytotoxic properties.
- M2a are involved in repair and regeneration.
- M2b microglia display immunoregulatory properties.
- M2c microglia is referred to as acquired deactivation phenotype.

However, it should be pointed out that this classification is being disputed and is an area of controversy [6]. In contrast, evidence is accumulating that microglia activation states have a high plasticity [3]. First, it depends on the microglia's "resting" state and its location and normal environment as well as its past activity pattern. Second, the combination of activation factors, their intensity, and their sequency or timing will have a massive effect. Whereas the M1/M2 classification and variations thereof suggest a choice of several stereotypical behavior patterns, the alternate view promotes a flexible course of action, depending on previous history, microenvironment, and impact of pathological factors.

5.2 Defining Reactive Microglia

Microglia are cells from the immune system, which invade the brain and serve as tissue-specific macrophages. Like other macrophages, they serve in the healthy tissue in other homeostatic functions. During distress, damage, or infection, they are further activated and resemble more their original purpose as macrophages. It should be kept in mind that in the normal and healthy brain, microglia are not "resting" as their processes are constantly probing the environment, although the cell body is stagnant (see previous chapters). It is the "ramified" state, which describes this stage the best. Another caveat is that microglia activation is a graded process depending on the history of the specific cell and the combined environmental factors impacting it. Upon reaching a more reactive state, microglia turn into ameboid cells, which consist of a round cell body and short, thick pseudopodia [7]. At this stage, they exhibit chemotaxis and move toward areas of damage or distress. Once in contact with a zone of distress or pathogen, the different microglia subtypes have several options of action. They can release cytotoxic substances, which kill pathogens, they can use phagocytosis to clear debris or pathogens, they release pro-inflammatory cytokines to communicate with other brain cells and attract blood-borne immune cells into the parenchyma (permeabilization of the blood–brain barrier), they can present antigen under certain conditions, and they can proliferate. Therefore, in areas of distress, the microglia density increases by migration and proliferation and the cells exhibit further activated phenotypes (reactive microgliosis). However, microglia must have the ability to terminate their responses. The M2 microglia phenotype (or variations thereof) is assumed to be involved in the inhibition of inflammation and restoration of homeostasis as well as of releasing protective factors. After injury or damage is resolved or stabilized, excess microglia are removed via apoptosis and revert slowly to a more normal state.

5.2.1 Cytokine Secretion

Below is a sample of representative cytokines released under different circumstances. The type of substance and the quantity that is released depend on the context of the disturbance [8]. CNS (central nervous system) infection causes the release of INF-γ (interferon gamma), which is a very potent upregulator of cytokine production, cytotoxicity, phagocytosis, and antigen presentation. It is also released by other cells and acts as an autocrine factor. Previous exposure to INF-γ seems to be necessary for them to express major histocompatibility complex (MHC) II (see next chapter). The main molecules released during physical brain injuries such as trauma and stroke are tumor necrosis factor-alpha (TNFα), interleukin 6 (IL-6), and IL-1β. The effects of these cytokines are dose- and time-dependent. During injury, the IL-6 release pattern exhibits a rapid spike and is then downregulated. In lower concentrations, TNFα has a protective effect on neurons; at higher levels, it damages neurons. IL-1β and TNFα are continuously released during neurodegeneration, and in this case (chronic inflammation), they seem to be heavily involved in this pathological process.

5.2.2 Phagocytosis and Cytotoxicity

Phagocytosis is a well-established process in organisms. In the human body, it is estimated that phagocytosis occurs over a billion times a day, mostly by macrophages [9]. Normal microglia in the healthy adult brain are known to phagocytose excess neuroblasts in the rostral migratory stream. This latent function is activated in pathological situations in some, but not all activated microglia. Astrocytes can also phagocytose in pathological situations, but they are far less efficient than microglia. CD68 (cluster of differentiation 68) is a transmembrane protein expressed in phagocytic microglia [10]. Apoptotic neurons, cell debris, and pathogens can be engulfed by microglia and broken down. If the phagocytosis involves a pathogen, some breakdown products can be presented as antigens on the cell surface. The phagocytic process involves first the formation of a phagocytic cup. The next step is the engulfing of the cup and the fusion with endosomes and lysosomes to form a phagolysosome [11]. The subsequent enzymatic breakdown involves a respiratory burst, which through superoxide anion, the highly acidic pH (power or potential of hydrogen), and hydrogen peroxide evolves in the production of reactive oxygen species (ROS). ROS release is a major mechanism by which a breakdown of the engulfed material is accomplished. ROS are completely restricted to the phagosome as they are very damaging to any structure. However, under certain conditions, a respiratory burst can be initiated by microglia without phagocytosis and result in the release of extracellular ROS. This is usually elicited by lipopolysaccharides (LPS) and involves the release of microglial glutamate [12] as well as a full pro-inflammatory response. The action of these externally released ROS is not at all as controlled as the release into the phagosome and leads to complications due to neuronal bystander damage. Interestingly, phagocytosis of brain material such as apoptotic cells and myelin is usually anti-inflammatory as it involves inhibition of the secretion of pro-inflammatory cytokines and release of transforming growth factor beta (TGFβ), which is the main anti-inflammatory cytokine [13]. Phagocytosis involving pathogens and initiated by Toll-like receptor (TLR) activation usually stimulates pro-inflammatory cytokine release.

5.2.3 Anti-inflammatory Factors

TGFβ is released as an anti-inflammatory molecule to counteract pro-inflammatory actions and to prevent harmful overreactions for the brain tissue. Other ant-inflammatory substances are interleukin 10 and arginase 1. To assist with repair processes and damage resolution, glucocorticoids and extracellular matrix proteins are secreted. Signal transducer and activator of transcription 3 (STAT3), a transcription factor, shifts into the nucleus of these microglia, and its action there inhibits the synthesis and release of most pro-inflammatory cytokines [14].

5.3 Control of Microglia Activation by Neuronal Factors

As mentioned before, microglia and neurons interact closely during development and normal adult nervous system function. Microglia can be activated by a multitude of factors, but there is a strong influence of neuronal activity on this process.

5.3.1 Neuronal Inhibitory OFF Factors

Inducing microglia reactivity is a complicated, multi-faceted process. The activation phenotype and intensity are very flexible and depend on the context. First, there are OFF factors. These are signals released from normally functioning neurons, which keep neighboring microglia in their normal, homeostatic, and non-reactive state. Very prominent among these signals are the chemokine fractalkine and the glycoproteins CD200 (Ox-2 membrane glycoprotein) and SIRPa (CD47 signal regulatory protein alpha) [15]. CD47 (cluster of differentiation 47) is a transmembrane protein on microglia, which, if activated by SIRPa, acts as a "don't eat me" signal to prevent upregulation of microglial phagocytosis. Reduced secretion of these factors by distressed neurons will make microglia more likely to move toward a more reactive phenotype.

5.3.2 Neuronal Activating ON Factors

Then there are neuronally released ON factors, which shift microglia into a more reactive stage. The two most prominent factors released by distressed neurons are adenosine triphosphate (ATP) and potassium. ATP leaks out of cells with damaged cell membranes and accumulates in the extracellular space, but only around damaged cells, as it is broken down quickly by ectonucleotidases. The appearance of increased amounts of extracellular ATP is therefore a good indicator of the site of cell damage. Another such damage signal is increased extracellular potassium. Its concentration is normally just below 3 mM and fluctuates slightly with neuronal activity. Higher increases (6–12 mM) are indicative of pathological processes. Both potassium and ATP work together in activating microglia and initiating chemotaxis to trouble spots. The activation of the microglial purinergic $P2Y_{12}$ receptor by extracellular ATP initiates chemotaxis toward the source of ATP and is therefore a key component in microglia motility, but not for retracting of its processes. The tonic activity of the two-pore domain potassium channel THIK1 (TWIK-related halothane-inhibited potassium channel) is potentiated by $P2Y_{12}$ signaling and contributes to depolarization when the extracellular potassium concentration increases to pathological levels. Such depolarization, mediated by THIK1, restricts microglial surveillance (probing its environment with its processes), but not chemotaxis [16]. It also initiates the release of cytokines, therefore supporting the inflammasome [16]. In combination, both $P2Y_{12}$ and THIK1 therefore tend to shift microglia from a normal homeostatic state with processes communicating with active neurons to an early-stage alert microglia, focused on and homing in on sites of trouble. The release of ATP and excess potassium from intracellular compartments seem to be potentiating damage signals, which result in a shift from a homeostatic cell toward an activated immune cell.

5.4 Pattern Recognition Receptor Activation

Pattern recognition receptors (PPRs) on microglia are a heterogeneous group that react to external signals with a shift toward activation [17]. These signals consist of two functional groups, with lots of overlap between them. There are pathogen-associated molecular patterns (PAMPs). These are signals released from external pathogens such as viruses and components of bacterial cell walls. Damage/danger-associated molecular patterns (DAMPs) are signals released from brain cells themselves and indicative of a pathological process, but one that is not necessarily caused by an infectious agent. These signals are key to understanding the shift toward a more activated microglial state. DAMPs upregulate the immune response of microglia during sterile conditions (no pathogens present). As mentioned, there is considerable overlap between PAMPs and DAMPs. The PRRs are therefore considered a group. There are four main PPRs on microglia. The most important are Toll-like receptors (TLRs). The others are NOD-like receptors (NLRs or nucleotide-binding oligomerization domain-like receptors) and RIG-I-like receptors (RLRs or retinoic acid-inducible gene-I-like receptors). The fourth group of receptors facilitates phagocytosis when activated and consists of Fc receptors (fragment crystallizable region receptor), complement receptors, and TREM2 (triggering receptor expressed on myeloid cells 2).

5.4.1 Toll-Like Receptors

There are 13 active TLRs in mice and 10 in humans, which are all expressed in microglia [18]. Microglia that are situated close to the meninges and circumventricular organs have a higher density of these receptors. The presence of pro-inflammatory stimuli increases TLR density. Various TLRs are also expressed on astrocytes and neurons, but not on oligodendrocytes [19]. Activation of members of this receptor family leads to an upregulation of the inflammatory responses. Six of the human TLRs are located on the cell membrane, and the others are located intracellularly in the membrane of the endolysosomal compartment in a way that the ligand-receptor binding site is inside the compartment [20]. The TLR receptors on the cell membrane are mainly activated by bacterial components, the most important are LPS (lipopolysaccharides) which are localized in the cell wall of Gram-negative bacteria (TLR4). Intracellular TLRs are more specialized for virus-specific nucleic acids. Single-stranded ribonucleic acid (ssRNA) is a ligand for TLR7 and TLR8 on the endolysosomal compartment. The rationale is that if ssRNA turns up in this compartment it is derived from a viral invader (non-self) and not a self-molecule. TLR3 is activated by double-stranded RNA (dsRNA), which is not part of the mammalian transcription machinery. TLR9 is activated by the appearance of unmethylated cytosine/guanine (CpG)-rich DNA (deoxyribonucleic acid) strands in this compartment. Again, such DNA is specific for bacteria (broken down in lysosomes after phagocytosis) and less common in mammalian cells. Upon activation, the TLRs release various adapter proteins, which induce the downstream signaling pathway. Almost all TLRs use MyD88 (myeloid differentiation 88) which is also called central adapter protein [21]. MyD88 release leads to the expression of genes involved in the immune response, like the production of pro-inflammatory cytokines TNFα, IL-1β, and IL-6. Another important adapter protein is TRIF (Toll interleukin 1 receptor-domain-containing adapter-inducing interferon-β). TRIF expression leads to upregulation of a different mixture of pro-inflammatory interleukins [22]. As can be seen from the above short review, TLR activation by PAMPs induces inflammatory responses to viruses and bacteria. However, one caveat is that DAMPs such as heat shock proteins, uric acid, aggregated proteins, and other endogenous signals can activate TLRs and cause chronic inflammation.

5.4.2 Nucleotide-Binding Oligomerization Domain-Like Receptors and Inflammasome Assembly

Nucleotide-binding oligomerization domain (NOD)-like receptors (NLRs) are a group of cytosolic receptors that are present in microglia and to a lesser extent in astrocytes and neurons. For the CNS, the best-described NLR is the NLRP3 (NLR family pyrin domain containing 3), and it will be highlighted here as an example of the role of the NLRs in microglia. NLRP3 expression in the cytoplasm is normally quite low. It needs a priming signal, which is usually the activation of TLRs as described above. This TLR activation acts through the transcription factor NF-κB (nuclear factor kappa-light-chain-enhancer of activated B cells) and among other results causes the expression of cytosolic NLRP3 (acting as a sensor), adaptor apoptosis-associated Speck-like protein (an adapter protein), and pro-caspase. This arrangement is passive until an activation step induces NLRP3 [23]. This activation can be caused by any combination of metabolic disturbances such as potassium outflow, extracellular ATP accumulation, crystalline material from various sources, and aggregational proteins in the cytoplasm. Lysosomal damage is another activator. These activation signals lead to the assembly of these three components. The final assembled product is called inflammasome, and it results in an active caspase-1. This caspase-1 will convert pro-IL1β and pro-IL18 into IL1β and IL18. The synthesis of these pro-interleukins was previously initiated by TLR activation. IL1β and IL18 are then released into the extracellular space and are strong pro-inflammatory signals. Thus, NLRP3 priming and activation steps combine and integrate various pathogenic and pathological features to contribute to a strong inflammatory response. It needs the combination (together with TLR interaction) of several stimuli. This strong inflammatory response is therefore not provoked by one stimulus alone, it needs a scenario of various factors. Most of the other microglial NLRs are activated by intracellular bacterial components [24].

5.4.3 Retinoic Acid-Inducible Gene-I-Like Receptors

The RIG-like receptor family (RLR) is in the cytoplasm of microglia and astrocytes. It recognizes virus-derived RNA. This family is also capable of detecting DNA viruses in the cytoplasm [25]. RLR activation upregulates type I interferon production, which will interfere with virus replication. In addition, it also upregulates the inflammatory response by promoting the expression of IL6 and TNFα.

5.4.4 Receptors Facilitating Phagocytosis

Triggering receptor expressed on myeloid cells 2 (TREM2) and its adapter protein DNAX-activating protein of 12 kDa (DAP12) are expressed at the cell membrane of microglia. TREM2 density is 300 times greater in microglia than in other cells [26]. A broad range of bacteria, various lipids, and nucleic acids from dying cells are ligands for TREM2. Its activation depresses TLR-induced pro-inflammatory cytokine production and facilitates phagocytosis and clearance of debris. It also enhances microglia proliferation and reduces microglia apoptosis. Microglia possess complement receptors (CR 1, 2, and 3) and Fc receptors (for antibodies like IgG – immunoglobulin G). These receptors support the opsonization of antibody-coated pathogens or other phagocytic objects [27]. The involvement of complement protein

C1q (complement component 1q) in phagocytosis usually results in the participating microglia initiating an anti-inflammatory response [28]. Heat shock protein 60 is expressed on apoptotic cells and directs the phagocytic microglial activity [29].

References

1. Timmerman R, Burm SM, Bajramovic JJ. Tissue-specific features of microglial innate immune responses. Neurochem Int. 2021;142:104924.
2. Stratoulias V, Venero JL, Tremblay M, Joseph B. Microglial subtypes: diversity within the microglial community. EMBO J. 2019;38(17):e101997.
3. Hirbec H, Rassendren F, Audinat E. Microglia reactivity: heterogeneous pathological phenotypes. Methods Mol Biol (Clifton, NJ). 2019;2034:41–55.
4. De Biase LM, Schuebel KE, Fusfeld ZH, Jair K, Hawes IA, Cimbro R, et al. Local cues establish and maintain region-specific phenotypes of basal ganglia microglia. Neuron. 2017;95(2):341–56.e6.
5. Mills CD, Kincaid K, Alt JM, Heilman MJ, Hill AM. Pillars article: M-1/M-2 macrophages and the Th1/Th2 paradigm. J Immunol. 2000;164:6166–73. J Immunol (Baltimore, MD: 1950). 2017;199(7):2194–201
6. Ransohoff RM. A polarizing question: do M1 and M2 microglia exist? Nat Neurosci. 2016;19(8):987–91.
7. Jurga AM, Paleczna M, Kuter KZ. Overview of general and discriminating markers of differential microglia phenotypes. Front Cell Neurosci. 2020;14:198.
8. Conti P, Lauritano D, Caraffa A, Gallenga CE, Kritas SK, Ronconi G, et al. Microglia and mast cells generate proinflammatory cytokines in the brain and worsen inflammatory state: suppressor effect of IL-37. Eur J Pharmacol. 2020;875:173035.
9. Arandjelovic S, Ravichandran KS. Phagocytosis of apoptotic cells in homeostasis. Nat Immunol. 2015;16(9):907–17.
10. Galloway DA, Phillips AEM, Owen DRJ, Moore CS. Phagocytosis in the brain: homeostasis and disease. Front Immunol. 2019;10:790.
11. Sierra A, Abiega O, Shahraz A, Neumann H. Janus-faced microglia: beneficial and detrimental consequences of microglial phagocytosis. Front Cell Neurosci. 2013;7:6.
12. Barger SW, Goodwin ME, Porter MM, Beggs ML. Glutamate release from activated microglia requires the oxidative burst and lipid peroxidation. J Neurochem. 2007;101(5):1205–13.
13. Lucas M, Stuart LM, Zhang A, Hodivala-Dilke K, Febbraio M, Silverstein R, et al. Requirements for apoptotic cell contact in regulation of macrophage responses. J Immunol (Baltimore, MD: 1950). 2006;177(6):4047–54.
14. Li Z, Song Y, He T, Wen R, Li Y, Chen T, et al. M2 microglial small extracellular vesicles reduce glial scar formation via the miR-124/STAT3 pathway after ischemic stroke in mice. Theranostics. 2021;11(3):1232–48.
15. Biber K, Neumann H, Inoue K, Boddeke HW. Neuronal 'On' and 'Off' signals control microglia. Trends Neurosci. 2007;30(11):596–602.
16. Madry C, Kyrargyri V, Arancibia-Cárcamo IL, Jolivet R, Kohsaka S, Bryan RM, et al. Microglial ramification, surveillance, and interleukin-1β release are regulated by the two-pore domain K(+) channel THIK-1. Neuron. 2018;97(2):299–312.e6.
17. Klegeris A. Regulation of neuroimmune processes by damage- and resolution-associated molecular patterns. Neural Regen Res. 2021;16(3):423–9.
18. Kumar V. Toll-like receptors in the pathogenesis of neuroinflammation. J Neuroimmunol. 2019;332:16–30.
19. Kouli A, Horne CB, Williams-Gray CH. Toll-like receptors and their therapeutic potential in Parkinson's disease and α-synucleinopathies. Brain Behav Immun. 2019;81:41–51.

20. Brencicova E, Diebold SS. Nucleic acids and endosomal pattern recognition: how to tell friend from foe? Front Cell Infect Microbiol. 2013;3:37.
21. Deguine J, Barton GM. MyD88: a central player in innate immune signaling. F1000prime Rep. 2014;6:97.
22. Kawai T, Akira S. Toll-like receptors and their crosstalk with other innate receptors in infection and immunity. Immunity. 2011;34(5):637–50.
23. Jo EK, Kim JK, Shin DM, Sasakawa C. Molecular mechanisms regulating NLRP3 inflammasome activation. Cell Mol Immunol. 2016;13(2):148–59.
24. Piancone F, La Rosa F, Marventano I, Saresella M, Clerici M. The role of the Inflammasome in neurodegenerative diseases. Molecules (Basel, Switzerland). 2021;26(4):953.
25. Crill EK, Furr-Rogers SR, Marriott I. RIG-I is required for VSV-induced cytokine production by murine glia and acts in combination with DAI to initiate responses to HSV-1. Glia. 2015;63(12):2168–80.
26. Painter MM, Atagi Y, Liu CC, Rademakers R, Xu H, Fryer JD, et al. TREM2 in CNS homeostasis and neurodegenerative disease. Mol Neurodegener. 2015;10:43.
27. Weinstein JR, Quan Y, Hanson JF, Colonna L, Iorga M, Honda S, et al. IgM-dependent phagocytosis in microglia is mediated by complement receptor 3, not Fcα/μ receptor. J Immunol (Baltimore, MD: 1950). 2015;195(11):5309–17.
28. Fraser DA, Pisalyaput K, Tenner AJ. C1q enhances microglial clearance of apoptotic neurons and neuronal blebs, and modulates subsequent inflammatory cytokine production. J Neurochem. 2010;112(3):733–43.
29. Goh YC, Yap CT, Huang BH, Cronshaw AD, Leung BP, Lai PB, et al. Heat-shock protein 60 translocates to the surface of apoptotic cells and differentiated megakaryocytes and stimulates phagocytosis. Cell Mol Life Sci. 2011;68(9):1581–92.

Chapter 6
Reactive Astrocytes

Abstract Similar to microglia, astrocytes can shift into reactive phenotypes. These phenotypes are – just as microglial ones – highly flexible and dependent on context. They are not preordained defined subtypes. In contrast to microglia, reactive astrocytes normally keep up homeostatic roles in the parenchyma. Astrogliosis is enabled by the integration of various signals from the environment, but the most important signals are cytokines from microglia. Special phenotypes of these reactive astrocytes are those surrounding perivascular cuffs with crucial functions involving leukocyte passage into the parenchyma and proliferative, border-forming reactive astrocytes.

Keywords Brain cell proliferation · Brain innate immunity · Microglia · Neuroinflammation · Proliferative · border-forming reactive astrocytes · Reactive astrocytes

6.1 Introduction

Astrocytes change their phenotype after disturbances in their environment like microglia. As will be shown, astrocytic reactivity has many similarities with microglia reactivity, but there are also fundamental differences. One similarity is that as with M1 (pro-inflammatory) and M2 (anti-inflammatory) microglia, A1 (toxic) and A2 (beneficial) astrocytes were proposed. However, this astrocytic classification has been, like the microglia one, phased out in favor of context-dependent, gradual, and flexible phenotypes [1]. Microglia are a cell type of the immune system, a tissue-specific macrophage, which in normal tissue serves a homeostatic role like other tissue macrophages. However, at the same time, microglia are surveying the tissue and changing their phenotype according to disturbances and "rediscovering" their macrophage purpose as a response to it. Astrocytes are central nervous system (CNS) satellite cells with homeostatic and signal involvement, which in need, due to disturbances, can adopt a different phenotype to adjust to the new challenges, including functions similar to an immune cell. There are close interactions between microglia and astrocytes during disturbances and their reactions are coordinated.

W. Walz, *The Gliocentric Brain*, https://doi.org/10.1007/978-3-031-48105-5_6

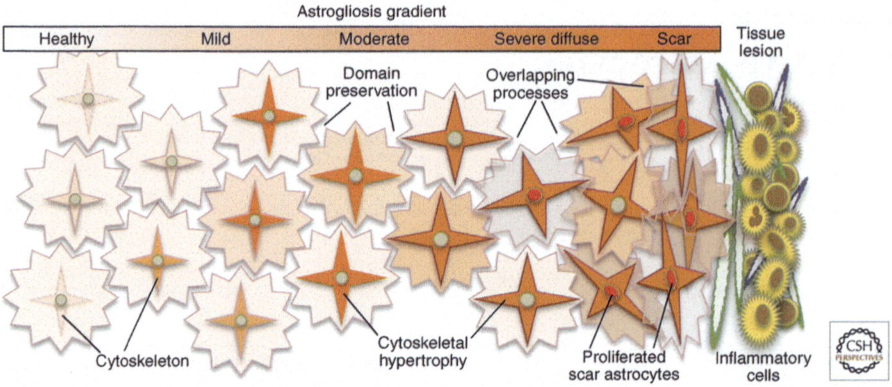

Fig. 6.1 Schema of astrogliosis gradient from mild to moderate to scar. In healthy CNS tissue, many astrocytes do not express detectable levels of the cytoskeletal protein, GFAP. In mild to moderate astrogliosis, most astrocytes upregulate GFAP and their cytoskeleton but preserve individual domains. In severe diffuse astrogliosis, there is also proliferation (depicted by red nuclei). Compact astroglial scars comprise newly proliferated astrocytes with densely overlapping processes that form borders to damaged tissue and inflammation. (From Ref. (16), Reproduced with permission from Cold Spring Harbor Laboratory Press)

Astrocytic reactivity (also called astrogliosis) is a reaction to disturbances of the brain microenvironment. The reaction cannot be seen in isolation from other cell types, especially neurons and microglia. In a similar way to microgliosis, reactive astrogliosis is graded and context-dependent. Its location and previous function as well as the intensity and nature of the injurious disturbance are important. Astrocytes react with hypertrophy, and the hallmark glial fibrillary acidic protein (GFAP) intensity increases to lighter impacts. They preserve their individual domains, and this situation is fully reversible once the situation is resolved (see Fig. 6.1). This stage of reactive astrocytes is also observed in focal injury in areas remote from the focus, as there is a clear gradient of reactivity in these injuries [2]. If the stimulus increases, the borders between domains can overlap and both hypertrophy and GFAP content will increase. This stage is not as easily reversed as the more diffuse non-overlapping astrocytic reactivity. There might be some proliferation of astrocytes. This is also the stage of astrocytes close to focal injuries. The third, most intensive stage of astrocytic reactivity is the border-forming reactive astrocyte (see below).

6.2 Signals that Shift Astrocytes Toward a Reactive State

Astrocytes shift into a reactive phenotype by more-or-less the same factors as microglia. Astrocytes react to damage-associated molecular patterns (DAMPs) and pathogen-associated molecular patterns (PAMPs) as well as most cytokines released from all brain cell types in the parenchyma. Invading cells of the immune system and pathogens can also evoke a transition to a reactive state. As such, astrocytes are considered part of the innate immune system [3]. Even molecules produced outside the parenchyma, which manage to diffuse into the brain, such as ammonium (in liver disease), serum proteins, or bacterial toxins (LPS or lipopolysaccharides), can cause this transition. Different signals involved in the genesis of astrocytic reactivity will shift the nature and type of the transformation in certain directions. One example is the up- or down-modulation of the astrocytic glutamate transporter GLT-1 in different pathological situations. This is an effective way to control the extracellular glutamate concentration around neurons [4]. In this way, astrogliosis is not a stereotypical event but has inherent flexibility to react within the context of the disturbed environment. More details are given in the following chapters where specific disease states are introduced.

6.3 Interactions with Microglia

During pathological processes, coordination of microglial and astrocytic reactivity is of prime importance. In many pathological situations, microglia are the first cell type to respond and induce reactions of the other cell types due to cytokine release [5]. However, there is feedback, especially by astrocytes, and therefore, there exists a two-way signaling mode between astrocytes and microglia. Virtually, all substances released by either cell type can cause a response in the other one. Examples are cytokines, chemokines, neurotransmitters, and neuromodulators as well as nitric oxide and reactive oxygen species. Astrocytic adenosine triphosphate (ATP) is a signal for neurons, but it will also react with P2Y12 and P2Y6 purinergic receptors in microglia [3]. One interesting pathway is the one using extracellular vesicles for long-range signaling and modulating gene expression. Among other substances, these vesicles can also contain messenger ribonucleic acid (mRNA) and miRNA (microRNA) for the modulation of gene expression of other cells. It has been demonstrated that the reactivity state of microglia influences the composition of proteins in these release vesicles to affect the response of astrocytes [6].

Cytokines released by microglia, such as Il-1β (interleukin 1beta), IL-6, and TNF-α (tumor necrosis factor-alpha), downregulate astrocytic P2Y1 receptor density. This transforms these astrocytes into a phenotype, which promotes reduced neuronal damage. Thus, microglia can be involved in the neuroprotective transformation of reactive astrocytes [7]. The opposite effect of microglia signaling has also been shown. In certain pathological situations, activated microglia can secrete IL-1α, TNF-α, and complement factor Cq1 to induce an astrocytic reactive state that is very neurotoxic and induces neuronal death [8].

6.4 Proliferation

In contrast to microglia, which proliferate extensively in almost all pathological situations, astrocyte proliferation is not common in the healthy as well as the diseased brain. There is less evidence of proliferation in mild to moderate astrogliosis. Severe diffuse astrogliosis has some evidence of proliferation of reactive astrocytes [9]. Compact scar formation with barrier astrocytes always involves the proliferation of the astrocytes (see below). This is in line with observations that neuronal loss does not cause astrocytic proliferation, but it seems to be that only invasive damage (like a stab wound) is accompanied by proliferating astrocytes. Diffuse injury does not involve astrocytic proliferation. Sonic Hedgehog is a major signal for induction of astrocytic proliferation [10].

6.5 Reactivity and Functions

What are the unique functions of the different stages of reactivity of these cells? As pointed out previously [11], reactive astrocytes are not necessarily non-homeostatic, as injury or disease are only selectively changing astrocytic functions. Their homeostatic role is normally present and mostly intact during pathological states, in contrast to reactive microglia. The one exception is the border astrocyte. Reactive astrocytes have many features in common with microglial reactivity, but there are also differences. One major difference is that astrocytes remain within their domain unless they die due to the insult. Unlike microglia, they do not migrate as a result of a focal injury, and they normally do not proliferate (see above).

6.6 Innate Immunity

An interesting feature of the involvement of reactive astrocytes in immune reactions is perivascular cuffs. During infections and autoimmune reactions, these cuffs form in all affected organs and the brain is no exception [12]. They form a kind of tertiary lymphoid organ. In the brain, blood-borne leukocytes spread between the parenchymal basement membrane and the retracted astrocytic endfeet in the perivascular space. The role of reactive astrocytes seems to attract these leukocytes. However, in a situation where such a cuff is formed, reactive astrocytes then in turn restrict the movement of these lymphocytes into the parenchyma and form tight junctions among themselves [13]. These astrocytes integrate various parenchymal signals, and under their instruction, they can ease the entrance of leukocytes into the parenchyma and therefore fulfill a pro-inflammatory role. Their alternate role is – if so instructed – to restrict leukocyte access to the parenchyma and therefore function in an anti-inflammatory role.

6.7 Proliferative, Border-Forming Reactive Astrocytes

This is a special astrocyte subtype, that involves proliferation and its appearance is not reversible [14]. It is permanent and normally involves a focal injury, neuronal death, and therefore tissue remodeling. It usually takes several weeks after neuronal death to reach a stable and final tissue architecture. The lesion core is no longer functionally part of the brain parenchyma, as there are no neural cells, and the core is isolated from functional tissue. The core is a fibrotic scar, which consists of extracellular matrix, crisscrossing blood vessels, endothelial cells, fibroblasts, and fibroblast-like cells. Initially, there are cells of the immune system present, which gradually disappear over time. Some of these cores can turn into fluid-filled cysts, which are not completely free of fibroblast-like cells and blood vessels [15]. Surrounding these core areas is a thin layer of elongated border-forming astrocytes. These astrocytes are created by proliferation after the insult and have a far higher density than normal or other reactive astrocyte types. They functionally seal the lesion core from the functional parenchyma, but it is not clear if this involves tight junctions. At least early in the formation of the lesion core, they prevent the spread of inflammation into functional or at least salvageable parenchyma tissue. Regenerating axons cannot cross the border astrocyte layer. The communication pathways across this border astrocyte layer are not known. It must be clearly stated that these astrocytes are border-forming astrocytes and not an astrocytic scar. They envelop and functionally separate the scar tissue, but they themselves are not scar tissue as astrocytes are neural parenchymal cells and reactive astrocytes after injury are not considered to form scar tissue. Scar tissue is defined as replacement of host organ parenchyma by mesenchymal (stromal) cells and fibrotic extracellular matrix [11]. Surrounding the border-forming astrocyte layer is a gradient of reactive astrocytes and microglia with diminishing reactivity and transitioning into healthy tissue as the distance from the border increases. Early after injury, just when the border is established, there will be intensive synapse formation, axon regeneration, and myelin remodeling taken place to reintegrate this area outside the scar into the functional neural parenchyma.

References

1. Escartin C, Galea E, Lakatos A, O'Callaghan JP, Petzold GC, Serrano-Pozo A, et al. Reactive astrocyte nomenclature, definitions, and future directions. Nat Neurosci. 2021;24(3):312–25.
2. Wang K, Walz W. Unusual topographical pattern of proximal astrogliosis around a cortical devascularizing lesion. J Neurosci Res. 2003;73(4):497–506.
3. Matejuk A, Ransohoff RM. Crosstalk between astrocytes and microglia: an overview. Front Immunol. 2020;11:1416.
4. Bianchi MG, Bardelli D, Chiu M, Bussolati O. Changes in the expression of the glutamate transporter EAAT3/EAAC1 in health and disease. Cell Mol Life Sci. 2014;71(11):2001–15.
5. Liberto CM, Albrecht PJ, Herx LM, Yong VW, Levison SW. Pro-regenerative properties of cytokine-activated astrocytes. J Neurochem. 2004;89(5):1092–100.

6. Drago F, Lombardi M, Prada I, Gabrielli M, Joshi P, Cojoc D, et al. ATP modifies the proteome of extracellular vesicles released by microglia and influences their action on astrocytes. Front Pharmacol. 2017;8:910.
7. Shinozaki Y, Shibata K, Yoshida K, Shigetomi E, Gachet C, Ikenaka K, et al. Transformation of astrocytes to a neuroprotective phenotype by microglia via P2Y(1) receptor downregulation. Cell Rep. 2017;19(6):1151–64.
8. Liddelow SA, Guttenplan KA, Clarke LE, Bennett FC, Bohlen CJ, Schirmer L, et al. Neurotoxic reactive astrocytes are induced by activated microglia. Nature. 2017;541(7638):481–7.
9. Sofroniew MV, Vinters HV. Astrocytes: biology and pathology. Acta Neuropathol. 2010;119(1):7–35.
10. Sirko S, Behrendt G, Johansson PA, Tripathi P, Costa M, Bek S, et al. Reactive glia in the injured brain acquire stem cell properties in response to sonic hedgehog. [corrected] Cell Stem Cell. 2013;12(4):426–39.
11. Sofroniew MV. Astrocyte reactivity: subtypes, states, and functions in CNS innate immunity. Trends Immunol. 2020;41(9):758–70.
12. Dahlgren MW, Molofsky AB. Adventitial cuffs: regional hubs for tissue immunity. Trends Immunol. 2019;40(10):877–87.
13. Sofroniew MV. Astrocyte barriers to neurotoxic inflammation. Nat Rev Neurosci. 2015;16(5):249–63.
14. Burda JE, Sofroniew MV. Reactive gliosis and the multicellular response to CNS damage and disease. Neuron. 2014;81(2):229–48.
15. Walz W, Cayabyab FS. Neutrophil infiltration and matrix metalloproteinase-9 in lacunar infarction. Neurochem Res. 2017;42(9):2560–5.
16. Sofroniew MV. Astrogliosis. Cold Spring Harb Perspect Biol. 2014;7(2):a020420.

Chapter 7
Neuroinflammation

Abstract Neuroinflammation is closely intertwined with the release of pro-inflammatory cytokines by microglia and astrocytes. These cytokines have a double function as they also serve as modulators during learning processes when they are released by these glial cells and modulate glutamatergic signaling as well as glutamatergic synaptic scaling. Another complication is that during stress, microglia use cytokines to recruit macrophages, which in turn release pro-inflammatory cytokines that interfere with cognitive processes. If microglial pattern recognition receptors are activated by pathogens or components of damaged cells, microglia release pro-inflammatory cytokines which orchestrate a repertoire of cellular behavior from microglia, astrocytes, and often also of recruited immune cells to target the causes of the aberrations. Ideally, this response is graded, localized, and timed to the challenge. There are also feedback processes to induce an anti-inflammatory response which will take over as the threat diminishes and activate repair processes. These processes require a lot of fine-tuning and often the inflammatory response is more damaging than the pathological cause. A huge problem is chronic inflammation which if long-lasting and severe can lead to demyelination. The chapter concludes with a comparison of neuroinflammation with peripheral inflammation outside the central nervous system.

Keywords Anti-inflammatory factors · Chronic neuroinflammation · Cytokine networks · Cytokine signaling · Microglia activation

7.1 Introduction

As in any other tissue in the body, the parenchyma can become inflamed to various degrees. Inflammation is part of the innate immune response and is targeted at invading foreign pathogens or detrimental products released by dying or damaged cells. It always involves the release of pro-inflammatory cytokines and chemokines, followed by changes to the phenotypes of parenchymal cells, but not necessarily the invasion of blood-borne immune cells. Inflammatory processes assist in the removal of a threat to the tissue. However, the challenge is to target and grade the

W. Walz, *The Gliocentric Brain*, https://doi.org/10.1007/978-3-031-48105-5_7

inflammatory response in line with the threat. Almost as important is to downgrade the response as the threat diminishes. This is the *key issue* in neuroinflammation, as otherwise the response can get more damaging than the threat. While tissue macrophages and dendritic cells initiate the inflammatory response in body tissues, in the central nervous system (CNS), the resident microglia are key in initiating, sustaining, and terminating the response.

The hallmark of the neuroinflammatory response is the timely and sequential release of various pro-inflammatory and anti-inflammatory cytokines and chemokines. In the healthy brain, some of these substances are used as messengers and modulators during normal function. Therefore, they have dual roles. Thus, neuroinflammation depends on cytokine and chemokine networks and is very context-dependent, as will be shown below (see Fig. 7.1).

7.2 Cytokine Signaling in Learning and Memory

Two forms of synaptic plasticity involve cytokine release from astrocytes and microglia in physiological conditions. The first one is long-term potentiation (LTP) and long-term depression. Here, I will focus on LTP [1]. It is a property of almost all excitatory synapses. If postsynaptic activation of AMPARs (α-amino-3-hyroxy-5--methyl-4-isoxazolepropionic acid receptors) by glutamate is enhanced beyond a certain degree, the resulting depolarization will cause the removal of the magnesium block of adjacent NMDARs (*N*-methyl-D-aspartate receptors). This removal will cause calcium influx through the NMDARs, which in turn will activate immediate early genes whose activity elicits structural changes to strengthen the synaptic efficacy. This mechanism is specific to the synapse in question and does not apply to the other synapses of the target neuron. Under physiological conditions, interleukin-1β (IL-1β), released from astrocytes and microglia, acts either directly or indirectly on NMDARs to potentiate the calcium conductance increase. IL-1β has therefore a role in the maintenance and may be induction of LTP. This is supported by experiments on IL-1R knockout mice, which exhibit reduced learning and spatial memory compared with the wild type [2].

Another cellular mechanism for learning and memory is synaptic scaling. In this case, AMPAR endocytosis of the whole target neuron, not just a particular synapse, is involved. Increases in AMPAR density make the neuron more responsive to glutamate input, and reductions have the opposite effect. Around relatively inactive hippocampal and cortical neurons, tumor necrosis factor α (TNFα) is released by astrocytes and microglia. This release increases the insertion rate of AMPARs, thus making the neuron more sensitive to glutamate input. TNFα therefore has a role in inactivity-induced synaptic scaling [3]. In the mouse dorsolateral striatum, TNFα release adapts to the need to either downregulate or upregulate glutamate signaling due to circumstances. A larger release of TNFα causes the removal of calcium-permeable AMPARs, and a reduced release has the opposite effect. Thus, under

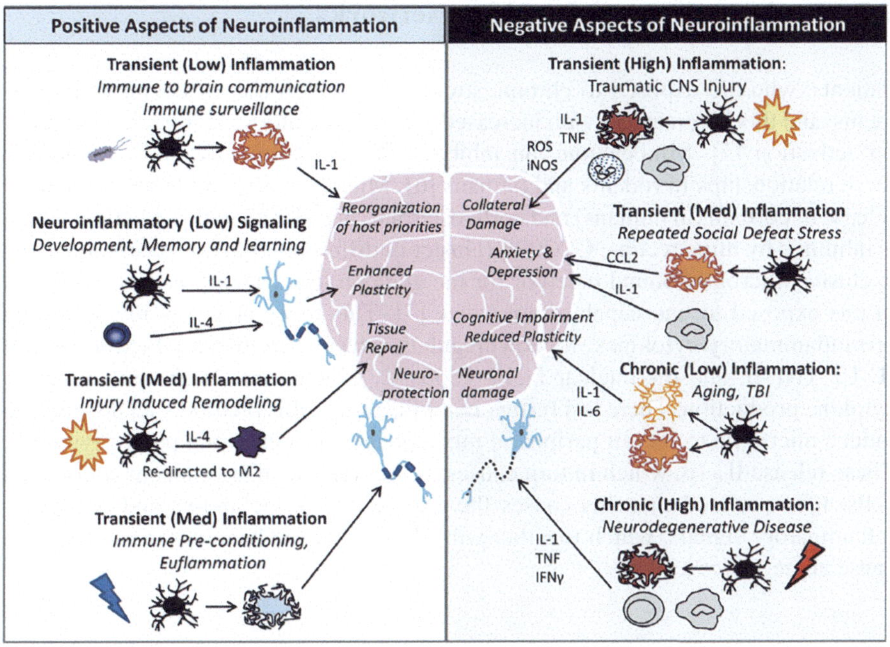

Fig. 7.1 Positive and negative aspects of neuroinflammation. The intensity and duration of inflammation account for much of whether immune signals are supportive or destructive to the central nervous system. On the left, we show examples of brief and controlled inflammatory responses that are generally considered beneficial to the host organism. For instance, immune-to-brain signals after infection lead to the subsequent reorganization of host priorities and induction of sickness behaviors. Additionally, there is an important maintenance role of IL-1 and IL-4 on learning and memory. Following traumatic CNS injury, IL-4-driven repolarization of macrophages (M2) has been proven to be highly effective in promoting recovery and axonal regrowth. Immune preconditioning, or euflammation, provides a method for training the innate immune system toward a more neuroprotective phenotype. Conversely, on the right, we demonstrate various maladaptive inflammatory responses. Chronic, uncontrolled inflammation is characterized by increased production of cytokines (IL-1 and TNF), reactive oxygen species (ROS), and other inflammatory mediators (inducible nitric oxide synthase). These markers are highly evident following trauma to the CNS and are accompanied by significant recruitment and trafficking of peripheral macrophages and neutrophils to the site of injury. The transient inflammation after repeated social defeat stress also leads to monocyte and macrophage recruitment and causes anxiety and depression. Additionally, a low-level and chronic inflammatory response driven by IL-1 and IL-6 is caused by aging, follows the acute phase of CNS trauma, and leads to reduced neuronal plasticity and cognitive impairments. A higher degree of chronic inflammation is greatly damaging to the nervous system and is characteristic of neurodegenerative diseases. (From Ref. (5). Copyright John Wiley and Sons. Reprinted with permission)

these circumstances, TNFα is a regulator of glutamatergic synaptic strength [4]. It is obvious from these examples that, in pathological situations, surges in cytokine release can interfere with the proper functioning of learning and memory processes.

7.3 Social Stress and Cytokine Networks

Patients, who are exposed to chronic stress, have a high risk of cognitive impairments, and this is consistent with increased pro-inflammatory cytokines and microglia activation [5]. Minocycline, an inhibitor of microglia activation, counteracts these relationships in rodents and humans [6]. Microglia respond to stress-induced release of cortisol (in humans) and corticosterone (in rodents) with activation, which is inhibited by minocycline. CD200R (cluster of differentiation 200 receptor and the exclusive microglia-bound receptor for the glycoprotein CD200) is downregulated in rats exposed to unescapable foot shock [7]. This receptor is key in controlling pro-inflammatory cytokines. Its downregulation results in increased expression of IL-1β, TNF-α, and the nuclear factor kappa B, which controls pro-inflammatory cytokine production. There is a further complication to this picture: social stress can induce microglia to recruit peripheral macrophages into the brain parenchyma [8]. These release IL-1β, which in turn causes further release of IL-1β from endothelial cells. It is assumed that this causes the release of prostaglandins and other pro-inflammatory signals, which together with IL-1β alter the behavior of neurons and cause anxiety [8, 9].

7.4 Cytokine-Mediated Sickness Behavior

This is a transient neuroinflammatory response not associated with pathological events in the brain parenchyma. There is no peripheral recruitment of immune cells, and the blood–brain barrier stays intact. Instead, a peripheral infection or insult uses the cytokine networks to induce coordinated behavioral adjustments through the modulation of neuronal circuits. The initiation starts in the peripheral immune system and involves the neurovascular unit, brain stem, and circumventricular organs of the brain [5], although the exact mechanisms are not yet known. This communication will activate microglia and cause the release of the cytokines IL-1β, TNFα, and IL-6. Subsequently, these cytokines modulate neuronal pathways in the hypothalamus and elsewhere to affect arousal, sleep–wake cycle, fever, food and water intake, and hyperalgesia [10, 11]. The neuronal response is the basis for a purposeful elicited coping behavior of the whole organism to an infection, which has not yet reached the brain parenchyma. Stimulation of this sickness behavior with subthreshold pathogens (LPS, lipopolysaccharides, a bacterial cell wall breakdown product) at various times is called immune preconditioning (euflammation). If after such a preconditioning period the organism is exposed to a full dose of LPS, the resulting sickness behavior is reduced and resolves faster [12].

7.5 Pathological Neuroinflammation

In this response, microglia are key, but astrocytes are also heavily involved. The response involves the recruitment of peripheral immune cells into the brain. The response is usually initiated by pathogen-associated molecular patterns (PAMPs) or damage-associated molecular patterns (DAMPs) on pattern recognition receptors (PRRs) on the microglia surface. This causes a shift of the nuclear factor kappa B into the nucleus with upregulation of pro-inflammatory cytokines (TNF-α, IL-6, IL-1α, and IL-1β), nitric oxide, chemokines, reactive oxygen species, proteolytic enzymes, and glutamate. Replication of intracellular pathogens will be blocked, and cell swelling and phagocytosis are induced. This repertoire can be adjusted to the circumstances. The response is attenuated and terminated by anti-inflammatory cytokine and neurotrophic factor release to initiate repair processes, but the potential for this response being uncontrolled for a long time is always present.

Part of this restraining feedback is the release of IL-10 by microglia to interact with astrocytes, which in turn release transforming growth factor β (TGF-β) to dampen microglia activation. If this fails, a reactive loop can develop turning neuroinflammation into full microgliosis (reactive and proliferative microglia) and astrogliosis, releasing more Clq (complement component 1q), a protein of the complement cascade, TNFα, and IL-1β. Finally, oligodendrocytes will become targeted by the large amounts of TNFα and can end up in activating demyelination. If the reactive loop reaches this state, synapses will be damaged. Prostaglandin D2 (PGD_2) is released from reactive microglia, increasing local blood flow and causing antigen presentation in astrocytes. Nitric oxide release by microglia causes reactive oxygen species (ROS) release by astrocytes. The resulting vasodilation supports the recruitment of monocytes by endothelial cells primed by these substances (IL-1β, IL-6, TNFα, nitric oxide, and ROS). This causes the downregulation of tight junctions. The resulting leaking blood–brain barrier makes it possible that components of the adaptive immune system can diffuse into the affected part of the parenchyma. This leads to the expression of specific receptors for T cells on the lumen side of endothelial cells and the secretion of chemokines by the endothelial cells. The invading T cells encounter inflammatory conditions in the affected part of the parenchyma and are activated to release pro-inflammatory cytokines into the environment. This in turn stimulates the microglia even more, leading to positive feedback of inflammatory microglia, astrocytes, and T cells.

7.6 Anti-inflammatory Actions and Repair

For obvious reasons, the inflammatory processes also need negative feedback and mechanisms that terminate the inflammation if the threat is resolved. Anti-inflammatory substances can be released with pro-inflammatory substances and work in the background till the threat starts to diminish [13]. In any event, the strength of the inflammatory response has also to be scaled down if the pathological conditions slowly decrease in intensity. Otherwise, the protective response (the inflammation) will become a bigger threat than the original damage. In fact, injury induces inflammatory responses and repair processes at the same time. Invading macrophages, not microglia, normally carry out the major portion of phagocytosis. This process clears dead cells and myelin debris. The myelin debris removal is the first necessary step in promoting axonal regeneration. One key mechanism seems to be mediated by IL-4. Over time during the inflammatory response, some microglia upregulate their IL-4 receptors. At the same time, invading $CD4^{+}$ T lymphocytes release IL-4. The IL-4 signal on these microglia induces an anti-inflammatory response that promotes angiogenesis, axon regeneration, and oligodendrocyte genesis for remyelination. Il-4 also acts as a neuroprotective signal on neurons. IL-10 and IL-13 are another group of anti-inflammatory cytokines. Manipulations that lead to increased impact of these cytokines are correlated with improved functional recovery after an insult.

7.7 Chronic Neuroinflammation

An inflammatory response can linger around with various intensities as chronic neuroinflammation. It is an inflammatory situation in the parenchyma, which never resolves. A major example is autoimmune responses against myelin. This situation will be dealt with in a separate chapter. It involves the infiltration of peripheral immune cells and the presence of auto-reactive T cells. The actions result in myelin loss with resulting axonal fragmentation. A similar framework is at play in the later stages of Alzheimer's disease. In this case, protein misfolding seems to be a trigger for the inflammatory responses of microglia, astrocytes, and infiltrating immune cells.

7.8 Comparison of Neuroinflammation with Peripheral Inflammation

In the tissues outside the CNS (central nervous system), PAMPs and DAMPS elicit an inflammatory response usually through resident dendritic cells. As there is no barrier to the vasculature, these cells can adapt to the seriousness of the threat and respond in a variable mode of action. This reaches from a nonspecific manner without outside engagement to a full-fledged response with antigen presentation, activation of the complement system, vasodilation, and extravasation of blood cells and components. In the brain, due to structural restraints, the situation is different. Dendritic cell density is very low (see later chapter), and macrophages are only present at distinct peripheral locations. The initiation of the inflammatory response rests primarily with microglia and – to a lesser degree – with astrocytes [13].

Another problem is the blood–brain barrier, which restricts communication of the innate immune cells (microglia and astrocytes) to a few restricted channels. The innate inflammatory response of the CNS is therefore not as efficient as in the periphery in recruiting outside help, such as complement cascades and leukocytes.

As shown above, another problem in the CNS is that pro-inflammatory cytokines play genuine roles in neuronal communication systems involving homeostatic microglia and astrocytes. These circuits are of functional importance in learning and sickness behavior. Thus, microglia (and astrocytes) play a context-dependent role that involves graded flexibility and feedback to tailor their inflammatory response and not to overreact. There is no place for a stereotypical all-or-nothing response.

References

1. Innes S, Pariante CM, Borsini A. Microglial-driven changes in synaptic plasticity: a possible role in major depressive disorder. Psychoneuroendocrinology. 2019;102:236–47.
2. Goshen I, Kreisel T, Ounallah-Saad H, Renbaum P, Zalzstein Y, Ben-Hur T, et al. A dual role for interleukin-1 in hippocampal-dependent memory processes. Psychoneuroendocrinology. 2007;32(8–10):1106–15.
3. Rizzo FR, Musella A, De Vito F, Fresegna D, Bullitta S, Vanni V, et al. Tumor necrosis factor and interleukin-1β modulate synaptic plasticity during neuroinflammation. Neural Plast. 2018;2018:8430123.
4. Lewitus GM, Pribiag H, Duseja R, St-Hilaire M, Stellwagen D. An adaptive role of TNFα in the regulation of striatal synapses. J Neurosci. 2014;34(18):6146–55.
5. DiSabato DJ, Quan N, Godbout JP. DNeuroinflammation: the devil is in the details. J Neurochem. 2016;139(Suppl 2):136–53.
6. Finnell JE, Wood SK. Putative inflammatory sensitive mechanisms underlying risk or resilience to social stress. Front Behav Neurosci. 2018;12:240.
7. Frank MG, Fonken LK, Annis JL, Watkins LR, Maier SF. Stress disinhibits microglia via down-regulation of CD200R: a mechanism of neuroinflammatory priming. Brain Behav Immun. 2018;69:62–73.
8. McKim DB, Weber MD, Niraula A, Sawicki CM, Liu X, Jarrett BL, et al. Microglial recruitment of IL-1β-producing monocytes to brain endothelium causes stress-induced anxiety. Mol Psychiatry. 2018;23(6):1421–31.
9. Quan N. In-depth conversation: spectrum and kinetics of neuroimmune afferent pathways. Brain Behav Immun. 2014;40:1–8.
10. Borniger JC, de Lecea L. Peripheral lipopolyssacharide rapidly silences REM-active LH(GABA) neurons. Front Behav Neurosci. 2021;15:649428.
11. Chen Q, Tarr AJ, Liu X, Wang Y, Reed NS, Demarsh CP, et al. Controlled progressive innate immune stimulation regimen prevents the induction of sickness behavior in the open field test. J Inflamm Res. 2013;6:91–8.
12. Tarr AJ, Liu X, Reed NS, Quan N. Kinetic characteristics of euflammation: the induction of controlled inflammation without overt sickness behavior. Brain Behav Immun. 2014;42:96–108.
13. Xanthos DN, Sandkühler J. Neurogenic neuroinflammation: inflammatory CNS reactions in response to neuronal activity. Nat Rev Neurosci. 2014;15(1):43–53.

Chapter 8
The Brain and the Immune System

Abstract The function and different layers of the immune system for the whole body are described. The notion of the brain as an "immune privileged site" is analyzed and found not accurate. The immune functions of the microglia (the brain macrophages) are suppressed by a functional brain environment, there are no antigen-presenting cells in the parenchyma, and the blood–brain barrier prevents the free exchange of immune components. However, these facts are somewhat misleading. The efflux of the brain fluids is sampled for antigens by dendritic cells, which if activated migrate to lymph nodes and activate brain-competent lymphocytes. These use blood circulation to arrive at the brain surface. If they then are reactivated by the dendritic cells and found brain competent – and only then – they will migrate into the brain to locations of existing inflammations. At these locations, lymphocytes and microglia interact and coordinate the immune response including subsequent anti-inflammatory processes by regulatory T lymphocytes. Microglia are not fully immunocompetent without inflammation and interactions with these lymphocytes to protect neurons from premature attacks by immune reactions. Thus, the brain is not immune privileged, but it is clearly different and has intrinsic mechanisms to protect neuronal functions from an overreaching immune reaction.

Keywords Adaptive immune response · Antigen-presenting cells · Brain immune privilege · Brain lymphatics · Cytotoxic T cells · Damage-associated molecular patterns · Immune system · Innate immune response · Lymph nodes · Lymphocytes · Major histocompatibility complex · Microglia · Pathogen-associated molecular patterns

8.1 Introduction

The brain function can be detrimentally affected by agents that are not normally present in the parenchyma but rather enter it from the outside and then multiply from within. These infectious agents are viruses, bacteria, protozoa, helminths, fungi, and (as a special case) prion proteins. All body organs are protected by the

W. Walz, *The Gliocentric Brain*, https://doi.org/10.1007/978-3-031-48105-5_8

immune system against such occurrences. The immune system has two components, both interacting with each other: the innate and the adaptive immune response.

Before delving into details, the influential concept of "immune privilege of the brain" has to be examined. This concept was created in the 1940s to address the fact that certain body organs do not reject implanted allografts (implanted foreign tissues). It was argued that this property would give these tissues not only certain privileges but also risks. The organs in question are the central nervous system (CNS), the eye, the testis, and the feto-maternal interface [1]. For the CNS, it is assumed that the reason is the protective environment of the brain. There is first and foremost the blood–brain barrier (BBB), but also the lack of antigen-presenting cells (APCs) circulating in the parenchyma or cerebrospinal fluid (CSF). There are also no lymph nodes within the brain. Such an immune privilege – if it indeed exists – can be the cause of excessive harm, as it may prevent complete clearance of infectious agents and may be detrimental to tumor suppression [2]. As with most issues, in view of recent findings, the concept of "CNS immune privilege" needs to be heavily modified, as pointed out later in this chapter.

At first, the organization of the body's immune defense will be briefly reviewed. This is necessary in order to understand the challenges the brain environment faces during an infection and during a challenge by malignant cells.

All nucleated cells, if infected by viruses or certain bacteria, release interferons, a group of signal proteins. Interferons signal other infected cells to undergo apoptosis and neighboring healthy cells to reduce protein synthesis and destroy double-stranded ribonucleic acid (RNA), which usually originates only from viruses. These interferons also interact with cells of the immune system to alert them to the site of infection.

The innate immune system is more specific and based on specialized cells. It is mainly based on macrophages and other monocytes. These cells react to pathogen-associated molecular patterns (PAMPs) like LPS (lipopolysaccharides, components of the outer membranes of Gram-negative bacteria) or double-stranded RNA (derived from viruses). PAMP receptor activation causes inflammatory responses including phagocytosis. A complication is the reaction of this system to damage-associated molecular patterns (DAMPs): these are signals that are released from the tissue during damage, irrespective if this damage is caused by pathogens or intrinsic factors. This process can trigger inflammation in tissue that is stressed but not invaded.

Adaptive, humoral immunity is focused on B lymphocytes. These cells reside in the spleen and peripheral lymph nodes. They can be activated against foreign peptides by interacting with T helper lymphocytes involving major histocompatibility complex (MHC) class II molecules on the surface. These specifically activated B lymphocytes expand their pool and release high-affinity antibodies against peptides into the extracellular fluid for distribution. They can also be activated by molecules other than peptides, for example, polysaccharides, deoxyribonucleic acid (DNA), and small molecules without the help of T lymphocytes and then release low-affinity antibodies against these substances into the body fluids.

All nucleated body cells also express MHC (major histocompatibility complex) class I molecules on their surface. They present peptides derived from cytosolic degradation to the interstitial space. Antigen-presenting cells (APCs), which are usually dendritic cells, but can also be macrophages and B lymphocytes, interact with these peptides presented by MHC I. The dendritic cells are strategically situated in barrier organs but also migrate to survey tissues. If they encounter a peptide presented by MHC I at the surface of a cell, they migrate to peripheral lymph nodes and activate CD8+ (cluster of differentiation 8) T lymphocytes specific for this antigen. Only when this peptide is derived from a pathogen (such as an invading virus), the cytotoxic T cells will activate and trigger an immune response. This pool of Class I-restricted cytotoxic T cells specific for this antigen (foreign peptide) will proliferate and invade the tissues. When this specific cytotoxic T lymphocyte encounters a body cell with the specific peptide attached to an MHC I molecule, it binds to it and responds threefold: first, to attack and destroy the infected cell (with a cocktail of cytotoxic perforins, granzymes, and granulysin granules), second, to multiply and therefore amplify the response to this antigen, and third, to signal to other immune cells that an infection has occurred.

Cell-mediated, adaptive immunity is based on CD4+ helper T lymphocytes and APCs. APCs phagocytose foreign particles or cells and present peptide fragments on their MHC class II molecules to the outside. These APCs then migrate to the lymph nodes and interact with CD4+ helper T lymphocytes. If the presented peptide antigen is recognized as foreign, CD4+ helper cells activate, expand their clone, and migrate out of the nodes into tissues. As these APCs also express MHC I, they will present peptide fragments with MHC I to CD8+ lymphocytes and strengthen the parallel response of these CD8+ cytotoxic lymphocytes. This parallel mechanism is called cross-presentation. Several subsets of CD4+ T helper cells exist. One set (known as Th1 or type 1 helper cell) interacts with B lymphocytes to potentiate the response for humoral immunity (see above). It also interacts with local macrophages, which present the same antigen on their surface via MHC II, derived from ingested bacteria or viruses. This recognition stimulates phagocytosis by the macrophages and the release of oxygen radicals to destroy foreign particles. Th2 cells activate when they encounter parasitic worm infections. These worms are usually too large to be phagocytosed; therefore, cytokines are released which lead to the coating of the helminths by antibodies. Another T helper lymphocyte subset, when interacting with bacteria specific to its antigen MHC II complex, induces tissue inflammation due to cytokine release. Due to additional chemokine release, this process also attracts neutrophils and macrophages into the infected tissue. This potentiates inflammation already present by the innate immune response (see above). The overall effect is the recruitment of leukocytes, stimulation of the production of antimicrobial substances, and promotion of repair processes. Memory T and B cells exist and preserve their specificity for certain antigens, even so the antigen is no longer present. If the same antigen appears again, these cells are activated and multiply in order to cut the response time short. There are central memory T and B cells in the lymph nodes and tissue-specific memory cells in certain organs, including the brain.

Due to the separation in MHC class I and II, the system can tailor its response to intra- and extracellular pathogenic challenges. A virus located in the extracellular body fluids is ingested by APCs and its breakdown peptides are presented on the APC surface together with MHC II. Only CD4+ T lymphocytes are activated and stimulate macrophage phagocytosis and cytotoxic processes to destroy the extracellular virus. The CD4+ T helper lymphocytes present peptide fragments from the virus to B lymphocytes. This activates mechanisms that result in these B lymphocytes releasing specific antibodies against components of the virus. These antibodies will only act extracellularly. In addition, B lymphocytes can release low-affinity antibodies against all kinds of circulating foreign extracellular molecules from breakdown products of bacterial cell walls to small toxic molecules. Once the virus invades body cells, peptide fragments will be presented by the invaded cells via MHC I to the outside. As a result of various processes, this will attract cytotoxic CD8+ lymphocytes, which will destroy infected cells and assist surrounding cells with their defense. Therefore, the immune system specifically can eliminate intra- and extracellular threats.

What exactly is an inflammation? Inflammation is a local phenomenon, which typically would be the result of an infection. The hallmark is the release by APCs, which encounter foreign particles, of pro-inflammatory cytokines and chemokines. This induces and directs the invasion of macrophages, neutrophils, and (to a lesser extent) T lymphocytes as well as the leakage of plasma proteins into the infected tissue. The process would be supported by increased blood flow supplying this area. Macrophages and neutrophils phagocytose and degrade the foreign particles. They also release enzymes and reactive oxygen species (ROS) to break down foreign substances. Thus, inflammation is a beneficial process that is directed against foreign invasion or substances. It is normally switched off after the threat subsides. Problems can arise if the intensity and duration of the inflammatory process are not adjusted to the severity of the threat and an overkill response results. Another problem occurs if the inflammation is in response not to a foreign challenge but due to the activation of the system by internal damage.

With this basic knowledge of the body's immune defense workings, one can see a few challenges for the brain environment due to its "immune privilege." There are no APCs migrating and surveying the brain parenchyma as microglia do not act as APCs to naïve lymphocytes, neither are naïve lymphocytes in the brain in any significant numbers. There are also no lymph nodes containing the various T and B lymphocytes within the brain area. The only "saving grace" would be the microglia cells, which are dormant cells of the immune system (macrophage-like), which are adapted to homeostatic and signal-processing tasks in the healthy brain. However, they are intrinsically involved in neuronal interaction and function and can easily react to disturbances in neuronal function. As such, they can act as macrophages and initiate and sustain an innate immune response with resulting inflammation. How they would initiate and sustain a more tailored immune response with the help and recruitment of components of the immune system involving MHC II class molecules from outside the brain is a key question.

8.2 Innate Immune Responses of the Brain

The innate immune system of the brain is based on two distinct cell types. For once, there are the microglia in the brain parenchyma, which were introduced in previous chapters. The microglia are the site of the basic innate immune defense of the brain. There are also border-associated macrophages at the brain boundaries. These consist of perivascular, subdural meningeal, and choroid plexus macrophages. These macrophages have a complex origin but are not related to the parenchymal microglia [3]. They are located in the perivascular fluid space (Virchow-Robin space) between the endothelial basement membrane and the endfeet of astrocytes [4]. Thus, the innate immune system is present in the brain. The adaptive immune system, which relies on antigen-presenting cells and circulating lymphocytes, in contrast, is not present in the brain parenchyma. However, as will be discussed below, there are cells representing this system, especially APCs associated with distinct brain compartments outside the parenchyma.

The innate immune response is normally based on sentinel cells that survey the tissue and respond to unusual molecules released by foreign pathogens or damaged cells due to the activation of their PAMPs. This process initiates and sustains an inflammatory response, deals with the abnormal situation, facilitates the adaptive immune response, and primes the tissue for the invasion of blood-borne immune cells. The primary sentinel cells responsible for this role in the brain are the microglia, although other cell types including neurons can feature some of these properties. Microglia initiate and sustain neuroinflammatory reactions to these threats. Neuroinflammation and its mechanisms were presented in a previous chapter. During neuroinflammatory conditions with invading pathogens (or aberrant molecules during neurodegeneration), microglia also present antigens via MHC I and II molecules on their surface. This presentation via MHC is an important link with the adaptive immune response of the brain (see below).

8.3 Adaptive Immune Response

Each threat by a pathogen is dealt with somewhat differently. In this chapter, a generalized account about the interactions between the brain and the immune system is given. Specific details for the different challenges are then given later in this monograph in subsequent chapters using the framework presented here.

In order to understand how the immune system reacts to antigens in the parenchyma, the anatomy of the brain including its barriers should be kept in mind. The fluid compartments of the brain have been introduced in a previous chapter. The presentation of MHC II by microglia in the non-inflamed brain is negligent. Furthermore, MHC II expression by microglia is prevented by neuronal activity [5], indicating that a functional brain environment keeps microglia from transitioning into an active immune cell. Microglia are therefore not capable of acting as efficient

APCs in the parenchyma nor can they migrate to lymph nodes and present the MHC II – antigen complex to T cells. The CSF and glymphatics contain antigens, which originate from pathogens in the parenchyma. At the brain surfaces, the most prominent APCs (cells capable of presenting antigens via MHC II to T and B lymphocytes) are dendritic cells [6]. These are not in the parenchyma, but at the borders where they are able to sample and survey the efflux from the brain into lymph and venous blood. They can also migrate to secondary lymph nodes, which no other brain cell type is capable of doing. The dendritic cells are in the choroid plexus, dura mater, leptomeninges, and perivascular spaces. They constantly sample possible antigens flowing out of the brain. In the healthy brain without accompanying neuroinflammation, the sampling of self-antigens by dendritic cells is a major mechanism for immune tolerance. Thus, all the fluid outflow of the brain (CSF, glymphatics) into lymph and venous sinuses is surveyed by dendritic cells. During conditions of neuroinflammation, the population of dendritic cells increases. Any foreign antigens are then presented together with MHC II molecules on the dendritic cell surface. Figure 8.1 shows the fluid movements exiting the brain containing foreign antigens if an infection is present in the brain. Drainage of CSF toward lymph nodes carries dendritic cells with the antigen/MHC II complex. These dendritic cells will flow from the CSF into the dural lymphatics and the nasal lymphatics (through the cibriform plate). Eventually, they will arrive at the cervical and lumbar lymph nodes (from CSF draining through the lumbar subarachnoid space). The perivascular pathways for fluid outflow from the brain are probably not large enough to allow dendritic cells to pass [7].

In the cervical and lumbar lymph nodes, the dendritic cells activate T and B lymphocytes by presenting the specific antigens. In addition, the dendritic cells imprint in these lymphocytes brain-specific-trafficking programs. This will make sure that these lymphocytes are destined for the brain. This is not unusual as similar mechanisms exist for skin and gut in other lymph nodes. The activated T lymphocytes enter the blood circulation. Those lymphocytes that are brain-competent leave the vessels upon reaching the leptomeninges. In the subarachnoid space, they interact with dendritic cells, thereby screening for dendritic cells that present the cognate antigen. If this confirmation happens, the T lymphocytes become reactivated. Only when this reactivation happens are the T lymphocytes able to breach the glia limitans and enter the brain parenchyma. This reactivation is therefore an important step in the interaction of the immune system and brain. Lymphocytes in the subarachnoid space, which have not been reactivated either undergo cell death or are transported out in the lymph, in a way that is similar to the antigen-activated dendritic cells (see Fig. 8.2).

Once these effector CD4+ T lymphocytes are in the brain parenchyma, their interactions with microglia are of utmost importance. The effector T lymphocytes are attracted to areas in the parenchyma where local inflammation increases the population of MHC II – antigen-expressing microglia. In this case of existing neuroinflammatory conditions, the activated sedentary microglia act as APCs for these invading T lymphocytes. The interaction of MHC II–antigen complexes of microglia and of CD4+ T lymphocytes is important for upregulating the brain immune

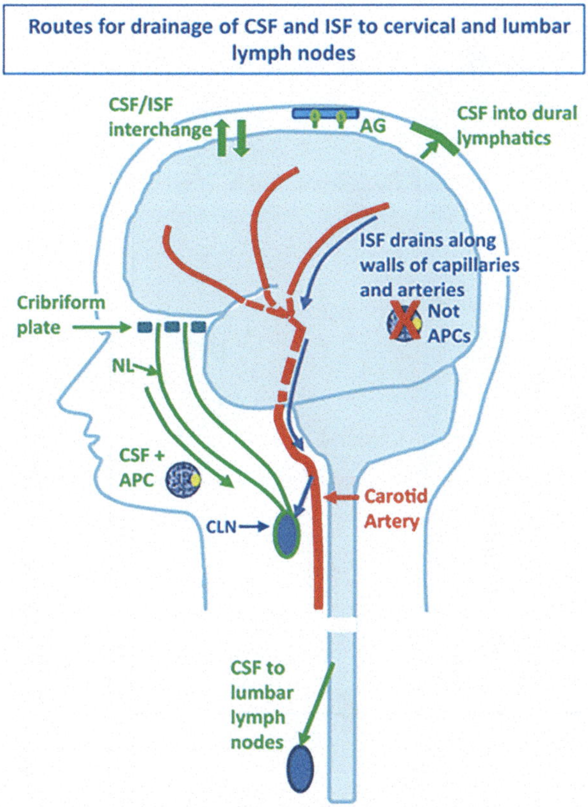

Fig. 8.1 Routes of drainage of cerebrospinal fluid and interstitial fluid to cervical and lumbar lymph nodes. Drainage pathways for CSF and interstitial fluid (ISF) to cervical lymph nodes. CSF and ISF drain to lymph nodes by different and distinct pathways. In humans, CSF drains into the blood of venous sinuses through well-developed arachnoid villi and granulations (AG). Lymphatic drainage of CSF occurs via nasal and dural lymphatics and along cranial and spinal nerve roots (outlined in *green*). Channels that pass from the subarachnoid space through the cribriform plate allow passage of CSF (*green line*) T cells and antigen-presenting cells (APCs) into nasal lymphatics (NL) and cervical lymph nodes (CLN). CSF from the lumbar subarachnoid space drains to lumbar lymph nodes. ISF from the brain parenchyma drains along basement membranes in the walls of cerebral capillaries and arteries (*blue arrows*) to cervical lymph nodes adjacent to the internal carotid artery just below the base of the skull. This narrow intramural perivascular drainage pathway does not allow the traffic of APC. There is an interchange between CSF and ISF (convective influx/glymphatic system), as CSF enters the surface of the brain alongside penetrating arteries. (From Ref. (6). Reprinted with permission from Springer Nature)

response. Both cell types change their operation once there is a physical contact of these complexes [8]. Interestingly, this interaction is often used to turn the CD4+ T cells into regulatory T cells, which act as a “brake” with anti-inflammatory properties. Under the direction of neuronal factors, microglia modulate the CD4+ T cell response heavily to target it specifically to identified locations and to prevent

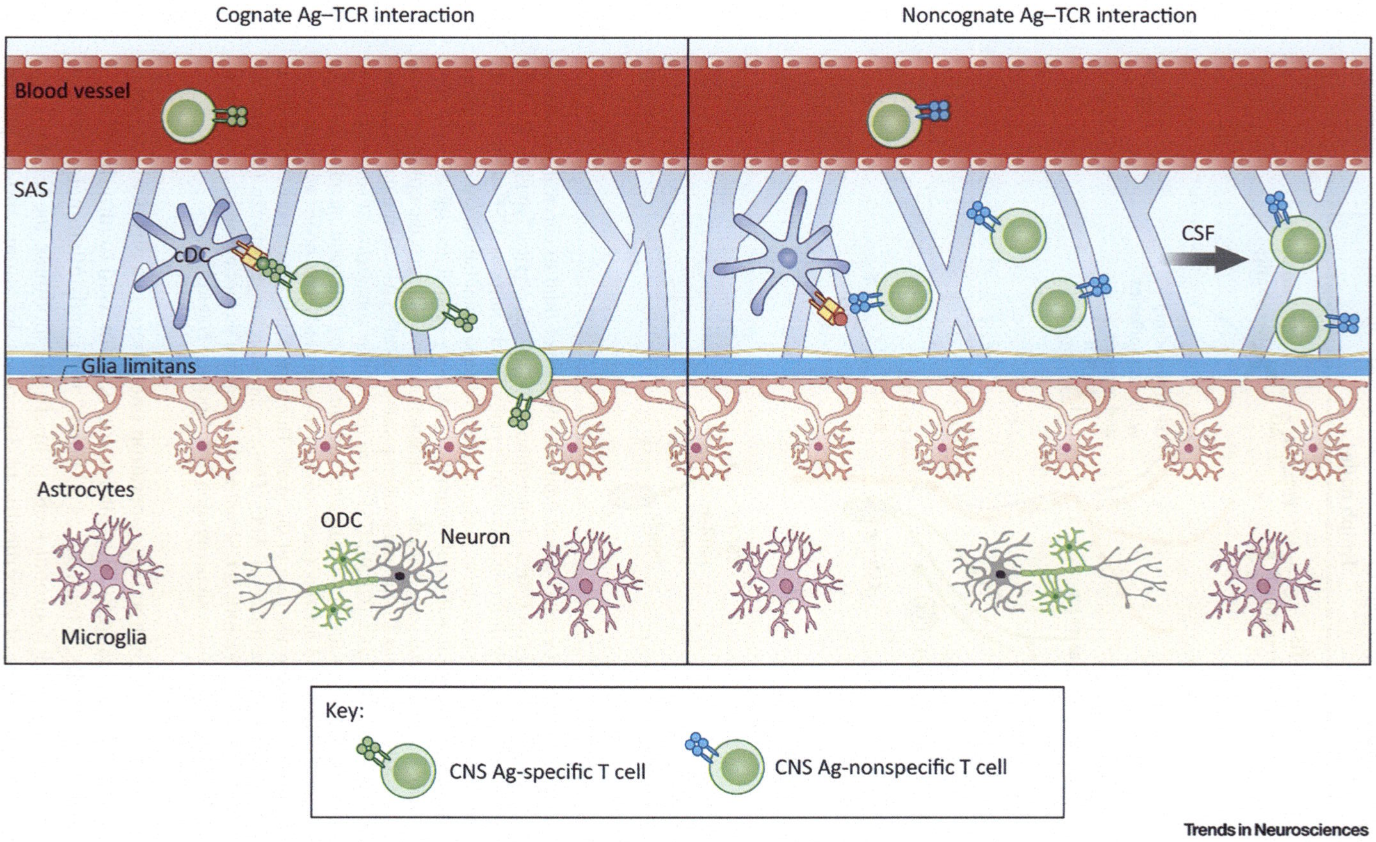

Fig. 8.2 Antigen (Ag)-specific T cell trafficking within the healthy brain. Activated T cells can leave the leptomeningeal blood vessels to screen the subarachnoid space (SAS) for their cognate Ag. CNS Ag-ignorant T cells remain in this location, where they might either undergo cell death or be released again to the periphery via the cerebrospinal fluid (CSF) flow. However, in the case of confirmatory cognate Ag presentation by CNS-associated Ag-presenting cells [APCs; particularly conventional dendritic cells (cDC2s)], CNS-specific T cells become reactivated and are able to breach the glia limitans of the blood–brain barrier (BBB) and infiltrate the CNS parenchyma. (Reprinted from Ref. (14). Reprinted with permission from Elsevier Ltd.)

massive tissue bystander damage. This is in contrast to microglia interactions with CD8+ T cells (see below). It is important to keep in mind that active neurons reduce microglia activation and this in turn will help regulatory T cells to keep the destructive action of immune responses at a minimum [9]. A key role seems to be played by microglial DAP12 (immunoreceptor tyrosine-based activating motif-bearing adaptor molecule). Other molecules involved in microglial-induced restraining of CD4+ T cells are the receptors DC-HIL (dendritic cell-associated heparan sulfate proteoglycan-dependent integrin ligand) and PD-1 (programmed cell death protein 1).

The CD8+ lymphocytes will also migrate into the brain parenchyma using similar mechanisms. While CD4+ lymphocytes are more important for the initiation of the adaptive immune response in the parenchyma and are mainly kept under close restrictive control by microglia, CD8+ lymphocytes are more involved in sustaining such a response. Reactive microglia also express the MHC I–antigen complex as do some neurons and astrocytes during inflammation. These CD8+ lymphocytes will interact with these cells [10]. Upon interaction, these CD8+ cells turn into activated cytotoxic lymphocytes. This involves the release of channel-forming perforin to increase calcium in targeted cells, enhanced glutamate release, and granzyme delivery. These actions result in the death of targeted cells, including neurons expressing the MHC I–antigen complex. Cytotoxic T cells induce microglia to secrete reactive oxygen and nitrogen species. In addition, CD8+ T cells secrete interferon, which upregulates MHC I molecules in surrounding cells (mainly microglia, but also neurons and maybe astrocytes) to sensitize these cells for the immune/neuroinflammatory actions [11]. These actions by cytotoxic T cells in synchronization with microglia are capable of simultaneously and sequentially killing targeted cells. In addition, a spillover effect can cause bystander damage to myelin, oligodendrocytes, and synapses [12]. A subpopulation of CD8+ T cells continues to exist in cuffs of the perivascular space as brain resident memory cells [13], which presumably can be reactivated if there is a need.

References

1. Spadoni I, Fornasa G, Rescigno M. Organ-specific protection mediated by cooperation between vascular and epithelial barriers. Nat Rev Immunol. 2017;17(12):761–73.
2. Forrester JV, McMenamin PG, Dando SJ. CNS infection and immune privilege. Nat Rev Neurosci. 2018;19(11):655–71.
3. Goldmann T, Wieghofer P, Jordão MJ, Prutek F, Hagemeyer N, Frenzel K, et al. Origin, fate and dynamics of macrophages at central nervous system interfaces. Nat Immunol. 2016;17(7):797–805.
4. Yang T, Guo R, Zhang F. Brain perivascular macrophages: recent advances and implications in health and diseases. CNS Neurosci Ther. 2019;25(12):1318–28.
5. Neumann H. Control of glial immune function by neurons. Glia. 2001;36(2):191–9.
6. Engelhardt B, Carare RO, Bechmann I, Flügel A, Laman JD, Weller RO. Vascular, glial, and lymphatic immune gateways of the central nervous system. Acta Neuropathol. 2016;132(3):317–38.

7. Carare RO, Hawkes CA, Weller RO. Afferent and efferent immunological pathways of the brain. Anatomy, function and failure. Brain Behav Immun. 2014;36:9–14.
8. Schetters STT, Gomez-Nicola D, Garcia-Vallejo JJ, Van Kooyk Y. Neuroinflammation: microglia and T cells get ready to tango. Front Immunol. 2017;8:1905.
9. Perry VH, Nicoll JA, Holmes C. Microglia in neurodegenerative disease. Nat Rev Neurol. 2010;6(4):193–201.
10. Stojić-Vukanić Z, Hadžibegović S, Nicole O, Nacka-Aleksić M, Leštarević S, Leposavić G. CD8+ T cell-mediated mechanisms contribute to the progression of neurocognitive impairment in both multiple sclerosis and Alzheimer's disease? Front Immunol. 2020;11:566225.
11. Vass K, Lassmann H. Intrathecal application of interferon gamma. Progressive appearance of MHC antigens within the rat nervous system. Am J Pathol. 1990;137(4):789–800.
12. Melzer N, Meuth SG, Wiendl H. CD8+ T cells and neuronal damage: direct and collateral mechanisms of cytotoxicity and impaired electrical excitability. FASEB J. 2009;23(11):3659–73.
13. Smolders J, Fransen NL, Hsiao CC, Hamann J, Huitinga I. Perivascular tissue resident memory T cells as therapeutic target in multiple sclerosis. Expert Rev Neurother. 2020;20(8):835–48.
14. Mundt S, Greter M, Flügel A, Becher B. The CNS immune landscape from the viewpoint of a T cell. Trends Neurosci. 2019;42(10):667–79.

Chapter 9
Viral and Bacterial Infections

Abstract There are several entrance ways for pathogens to enter the brain despite the blood–brain barrier. They can breach the tight junctions, cross through endothelial cells, or use the Trojan horse route by hiding in leukocytes. Less used are the nasal pathway and the meninges and subarachnoid space. All cells of the brain have intrinsic mechanisms to battle infections by viruses like other body cells. The difference is that infected neurons cannot easily be removed by immune cells as most of them have irreplaceable roles in circuits. Microglial cells have a key function in the defense. They are the first line of defense and one of their main priorities is to protect neurons that are infected. They even instruct invading T cells to avoid attacks on infected neurons. The downside is that latent virus infections of neurons is a common problem. Despite these protective mechanisms for neurons, direct damage by viruses is almost as common as neuronal damage by overreacting immune cells, for example, by a cytokine storm. In extreme cases, this can lead to encephalitis, an out-of-control inflammatory reaction due to infections.

Keywords Bacterial infections · Cytokine storm · Infection pathways · Latent infections · Lymphocytes · Microglia · Trojan horse route · Viral encephalitis · Viral infections

9.1 Major Infection Pathways

The entry pathways have been extensively reviewed by Forrester et al. [1]. The most direct entry pathway is from the blood through the endothelium, which constitutes the blood–brain barrier. There are various ways to cross the endothelial barrier lined by tight junctions. The most direct is around the cells by breaking the tight junction barrier. Both viruses and bacteria can use this pathway and the mechanisms are variable; however, they seem to involve matrix metalloproteinases in at least one step [2]. All mechanisms involve a displacement or cleavage of the tight junction proteins, leading to a gap between the endothelial cells. Both viruses and bacteria can also cross the endothelial barrier without leaving structural damage. They can act on

W. Walz, *The Gliocentric Brain*, https://doi.org/10.1007/978-3-031-48105-5_9

several endothelial receptors on the luminal surface and initiate pinocytosis/endocytosis. This will lead to the release of the invaders on the parenchymal site of the endothelial barrier. The third entry pathway directly from the blood is the so-called Trojan horse route. Pathogens can survive inside leukocytes of the bone marrow long after a systemic infection is resolved. They replicate intracellularly and inhibit leukocyte autophagy and phagocytosis. These infected leukocytes can cross the endothelial cell barrier and bring with them the pathogens into the parenchyma. Many viruses use this Trojan horse entry and develop individual mechanisms. They can hide inside the leukocytes in vesicles. These vesicles can then enter the endothelial cells through membrane fusion [3].

There are two additional pathways facilitating entry from contiguous tissue but not from blood. The pathogens can originate from the nasal mucosa or the meninges and subarachnoid space. The subarachnoid/meninges entry results often in meningitis. The nasal pathway is mainly used by viruses, but also by some bacteria and by prions. The pathogens enter the peripheral axons of the trigeminal nerve or olfactory nerve and migrate within the axons to the nerve cell body, where they can hide till activated.

9.2 Latent Infections

A latent infection is an infection where the infection (most likely from a virus) is in a host organ as a "non-replicating, but replication-competent, viable organism" [1]. This can occur in organs other than the nervous system, for example, the rabies virus in muscle tissue [4]. Due to its somewhat sheltered immune status, the brain offers an even more favorable environment for latent pathogens [5]. All brain cell types without exception are candidates for such latent infections. First, there are the various leukocytes that act as Trojan horses to get the virus across the blood–brain barrier into the brain parenchyma. Microglia and perivascular macrophages, considered part of the innate immune system of the brain, are targeted by pathogens. Some of them like the human immunodeficiency viruses (HIV) and simian immunodeficiency viruses (SIV) can survive long term in these cells and suppress apoptosis, which would normally occur [6]. Other cell types, which can host all kinds of pathogens, are endothelial cells and the epithelia of the choroid plexus. Neurons, astrocytes, and oligodendrocytes can serve as hosts to viruses such as HIV, herpesviruses, measles viruses, and John Cunningham viruses. Neurons are especially prone to latent infections as T cells are restricted in their cytolysis of neurons (see below). Neurons are part of a circuit and most of them cannot be replaced. This is the reason cytolysis of infected neurons is not as common as that of other brain cells.

9.3 Viral Infections

Viral infections of the brain are usually the result of a zoonotic infection. This means that the virus was transmitted from an animal host, or an intermediate animal host and the human infection is a "dead end." This is not necessarily in the interest of the widespread dissemination of the virus. The damage to the human nervous system might be so severe that the body cannot survive. An exception is the rabies virus (RABV) infection, which targets brain circuits to induce specific behavioral changes in the body that will facilitate the transmission of body fluids to other members of the same or other species. In general, the impairment of nervous system function can be due to damage induced by the virus, the overreaction of the innate ("cytokine storm") and adaptive immune systems, or all of these. Part of the systemic defense of virus-infected cells is the elimination of the infected cells. This is not an option in the central nervous system (CNS) if neurons are involved. Most neurons have a defined role in a circuit and cannot be replaced without serious functional deficits. This situation sets the brain apart from the rest of the body, as it implies a restrained immune response if neurons are affected. Neuronal cytolysis by T cells or other cells is therefore not a first option. It can also mean that the infection might be contained within neurons, but not eliminated. This will favor latent neuronal infections by viruses, and this could be problematic.

9.3.1 Viral Infections of Neurons

Some viruses are infecting only neurons in the CNS. Such an infection causes a strategic problem. It is not advisable to use lysis to destroy the virus-infected neurons as they are mostly non-renewable unlike epithelial cells. Therefore strategies, which preserve the function of the infected neuron, are used. Another consequence is that keeping the neurons alive, means many will have latent infections as the virus is only kept in check, but not eliminated. Many neuron-specific viruses enter through the nasal pathway. RABV, herpes simplex virus 1 (HSV-1), vesicular stomatitis virus (VSV), Borna disease virus (BDV), several influenza viruses, Hendra virus, and Nipah viruses are examples [7]. The viruses enter axons by endocytosis or membrane fusion and travel mostly in a retrograde direction from postsynaptic to presynaptic neuron and onto the cell body with the help of microtubule motors [7]. Thus, the virus spreads within neurons along neuronal and axonal circuits and not by random diffusion. However, virus nucleic acid replication, messenger ribonucleic acid (mRNA), and protein expression occur mainly in the cell body of the neurons. Deoxyribonucleic acid (DNA) viruses like alpha herpesvirus replicate and assemble in the nucleus, before leaving it and obtaining their envelope from the Golgi apparatus [8]. Alpha herpesviruses are an exception to the common rule of retrograde transport in a way that they can move in both directions, retrograde and anterograde, and exit at axon terminals. Enveloped viruses (like alpha herpesviruses) can leave neurons by crossing the cell membrane without leaving much or even any damage behind. The situation for nonenveloped viruses (poliovirus and adenoviruses) is not quite as simple, but not much is known about the mechanisms and possible neuronal damage [7].

The latent infection of neurons is best explained with the example of varicella-zoster virus (VZV or herpesvirus 3), a member of the alpha-herpesvirus family with double-stranded DNA. It only infects humans as animal hosts are unknown. The primary infection of the body causes varicella (chickenpox). Thereafter, it survives latent in the ganglia of the peripheral nervous system (PNS) [9]. In later stages of life, VZV can reactivate, depending on circumstances such as stress and immunocompromises, and cause zoster (shingles). During latency, less than 2% of ganglionic neurons are infected in patients with a varicella history [10]. In these latently infected neurons, the VZV DNA is circular, non-replicating, and located in the nucleus. There are viral proteins in the cytoplasm. It is not clear if this is indicative of a complete inhibition of viral gene expression or of repeated attempts of reactivation, which are subsequently aborted [9]. After reactivation and subsequent viral gene expression, viral proteins translocate to the nucleus. This reactivation causes upregulation of neuronal major histocompatibility complex (MHC) I and II, infiltration of CD4+ and CD8+ T cells (T helper cells and cytotoxic T lymphocytes), and ganglionitis (neuroinflammation of the ganglion). This results in necrosis and hemorrhagic damage.

Neurons exhibit Toll-like receptors (TLRs) and retinoic acid-inducible gene-I-like receptors (RLRs) for viral pathogen-associated molecular patterns (PAMPs). These viral PAMPs are mostly double-stranded (ds) RNA (ribonucleic acid) and DNA (desoxyribonucleic acid) motives specific for viruses. RNA lariat debranching enzyme (DBR1) converts lariat intron RNA into linear RNA. This process interferes with the reverse transcription of the viral RNA into DNA of HIV, HSV, influenza viruses, noroviruses, and others. It has been shown in humans that mutations of this enzyme cause viral infection in certain brain regions which could be ameliorated by the wildtype enzyme [11]. The human small nuclear RNA (snoRA31) was found to interfere in neurons with HSV1 replication and mutations of the snoRA31 gene can cause brain infections, as the interference is less effective [12]. The mechanism is not yet known. In addition, VZV and probably other viruses are eliminated in neurons by autophagy [13]. Activation of neuronal TLR3 causes the release of cytokines like tumor necrosis factor-alpha (TNFα) and interleukin-6 (IL-6), complement factors, chemokines, and antiviral molecules such as interferons as well as 2′–5′ oligoadenylate synthetase. The latter enzyme is involved in inhibiting viral replication in neurons depending on some other factors [14].

Most viruses that infect the CNS reach it from organs or body fluids outside of the CNS. Therefore, an adaptive immune response in regional peripheral lymph nodes is initiated early on. In addition, drainage of viral antigens out of the brain starts early on, and these antigens are detected by dendritic cells and an adaptive immune response is initiated at the cervical lymph nodes (see previous chapter). This leads to the invasion of reactivated T cells into the parenchyma. Far more activated CD8+ than CD4+ T cells invade during a viral brain infection. In addition, microglia and astrocytes in the vicinity of infected neurons are alerted by the release of neuronal interferons and chemokines. As well viral PAMPs activate pattern recognition receptors (PRRs) in microglia, astrocytes, and neurons. A large role is played by adenosine triphosphate (ATP) release into the extracellular space from infected neurons. This extracellular ATP gradient acts as a chemotactic signal for microglia. The microglial receptor responsible for this recruitment is the purinergic receptor P2Y12 [15]. In addition, microglia, although themselves not infected by the virus, acquire the antigens from the infected neurons and display them with MHC I on their cell membrane. This MHC I antigen complex is cross-presented to cytotoxic T cells, which display the same antigen [16]. The T cells have been reactivated by dendritic cells before entering the brain. They are attracted by the chemokine and cytokine gradients within the parenchyma. Healthy neurons do not express MHC I, except during development [17, 18]. However, after a virus infection, MHC I is expressed by the neurons together with viral antigen. The cross-presentation of viral antigen via MHC I by uninfected microglia to invading cytotoxic T cells seems to transform the T cells from a cell type capable of cytolysis to one that eliminates or restrains the virus without damaging the neuronal host. Granzymes are usually released by these T cells to activate pro-apoptotic proteins and therefore cause lysis of the infected cell. In the brain, however, the granzyme action does not lead to apoptosis but to the cleaving of viral proteins, which are needed for viral translation and transcription [19]. Due to the activation of their PRRs, these cell types (neurons, microglia, astrocytes, and T cells) release interferon I (IFNα and IFNβ), IFNγ, and tumor necrosis factor (TNF) α to act on the infected neurons in concert without causing cytolysis. The interferon I family modulates the immune responses of all these cells and, in addition, has direct inhibitory effects on the viral replication process. IFNγ activates 2′5′ oligoadenylate synthetase-induced RNAse L, which degrades viral RNA. It also activates dsRNA-activated protein kinase to inhibit viral protein synthesis. These antiviral actions by the IFN family are induced irrespective if the cells were infected or not. TNFα induces nitric oxide release, which inhibits viral protease activity and therefore interrupts viral replication.

The microglia–CD8+ T cell interaction via antigen-MHC I complexes has a restraining effect on the T cells, as pointed out above. Thus, the microglia are controlling the response to promote neuronal survival. Microglia secrete molecules like programmed cell death ligand 1 (PDL1) to control the T cells and protect neuronal integrity [20]. They are also responsible for transforming the T cells into regulatory, anti-inflammatory T cells once the threat diminishes and finally to induce apoptosis in T cells. Some T cells will survive in the tissue at the sites of the virus infection as regional memory T cells for these viral antigens. They keep latent infections in check and reactivate if the same infections are recurring. There is also the possibility that some non-memory T cells are not eliminated and are surviving after the infection is over. They can release IFNγ, which then activates microglia to eliminate presynaptic termini and cause neuronal apoptosis. These surviving T cells will prevent repair processes, and this situation can lead to cognitive deficits because of an unresolved immune response to viral infection of neurons [21].

9.3.2 Viral Infections of Non-neuronal Cells

Astrocytes form close contacts with interfaces and the glia limitans with their endfeet. They also form processes that extend close to synapses. They are therefore very susceptible to virus invasion. This can be best demonstrated with the example of the flaviviruses. Flaviviruses are a family of single positive-stranded RNA, which replicates in the cytoplasm of infected cells. They are also arboviruses, meaning that they are transmitted from insects to human hosts. Humans are normally a dead-end host for these viruses. Examples are Zika viruses (ZIKV), tick-borne encephalitis virus (TBEV), West Nile virus (WNV), and Japanese encephalitis virus (JEV). Once these viruses enter the brain, they infect primarily astrocytes. From astrocytes, they spread to other cell types, including neurons [22]. The viruses infect astrocytes through endocytosis and then move along the cytoskeleton [23]. Astrocytes have a high virus load when infected by flaviviruses, but they show a high resilience to the infection [22]. Thus, astrocytes act as a reservoir for these viruses. Despite the resilience, the astrocytic cytoskeleton and some intracellular organelles show some damage [24]. This is probably an indication that the astrocytic homeostatic and signal function is negatively affected. After infection, astrocytes also release vascular endothelial growth factor, IL-6, and matrix metalloproteinases, which will increase the permeability of the blood–brain barrier [25]. However, astrocytes have also a strong defense system. They release cytokines and chemokines as well they are a major source of type I interferons. All neural cell types are capable of releasing interferons, but astrocytes seem to be a major source [26]. Astrocytes can detect neuronal viral infections, probably due to TLR (Toll-like receptors) 2 and 3 on their cell membrane, which interact with viral glycoproteins. They also detect cytoplasmic viral RNA via RLRs. Additionally, dying cells surrounding astrocytes release DAMPs and viral nucleic acids which will be detected by PRRs. Together, these stimuli result in major interferon production, which seems to be a hallmark of astrocytes – at least after encountering viruses. The release leads to the expression of interferon-stimulated genes. The gene products and interferons directly or indirectly inhibit virus replication and lead to virus elimination. The interferon-stimulated gene products target every step of the virus life cycle from entry to release [27]. This is probably the reason astrocytes survive virus attacks reasonably well. The downside of this is, however, that at the same time, they can act as a virus reservoir in the parenchyma, with detrimental effects on other cell types and the blood–brain barrier. Astrocytic cytokines can also interfere with the function of adjacent neurons. However, there have been viruses evolved which contain mutations that help them to evade some of the interferon actions. An example is the ability of TEBV strains to hide their dsRNA in vesicles, thus reducing the time window during which the dsRNA can be sensed and therefore reducing the antiviral response [28]. Other examples are the suppression of interferon responses by Chikungunya virus and the production of IL-10, an anti-inflammatory cytokine, by various viruses [1].

Microglia are the first line of defense as they are the resident macrophages of the brain. Their PRRs react to viral components, and they can exhibit the antigen–MHC

I complex at their membrane and interact with reactivated and migrating T cells (see above). They have an immunosuppressive effect on T cells, turning them into regulatory (anti-inflammatory) cells and protecting infected neurons from lysis by T cells. Next to astrocytes, they produce large amounts of interferon, which are used to protect other cells. However, microglia can become themselves a target of viral infection and some viruses might use microglia as a reservoir during latent infections. Examples for viruses using microglia as a reservoir are HIV-1, SARS-CoV-2 (severe acute respiratory syndrome-associated coronavirus 2), ZIKV, JEV, and Measles virus [29]. HIV-1 infection of microglia is probably the most prominent as microglia act as a main reservoir for this virus for decades before an acute infection will surface [30]. HIV enters the parenchyma via invading macrophages and CD4+ T cells. HIV then spreads to microglia. HIV-1 is a retrovirus, and it integrates into the host's DNA (integrated HIV-1 provirus). The infected microglia are nonlytic and resistant to apoptosis and other cytopathic effects. The reason for the latency is not well understood but probably involves transcriptional repressors, epigenetic regulation, and formation of heterochromatin in the vicinity of the viral promoter [30]. In addition, neurons excrete factors that silence HIV expression and prevent spontaneous reactivation of HIV in microglia [31]. Prime candidates for factors are glucocorticoids and other substances that normally act as OFF signals to microglia (see previous chapters). In contrast, low-level reactivation of HIV in microglia would result in proinflammatory cytokine release by microglia, and this would decrease neuronal activity and therefore their release of silencing factors. Thus, there is a fine balance between neurons and microglia in preventing HIV provirus reactivation. Downregulation of the neuronal factors and release of neuronal DAMPS due to neuron injury can shift this balance and lead to HIV reactivation. Now, sustained, low-level production and release of the virus as well as resulting proinflammatory cytokine production by microglia can damage synaptic function and dendritic integration. This would lead to a further shift of the balance between neurons and microglia, due to the release of more DAMPS and less microglia silencing factors. In the end, a positive feedback loop would lead to a resurgence of the virus. This resurgence will contribute to HIV-associated cognitive disorders, as observed in many patients [32].

9.3.3 Viral Encephalitis

Encephalitis is a widespread inflammation of the brain parenchyma. It affects neuronal circuits and therefore causes neurological and often psychic symptoms. Virus infections are the main reason for encephalitis, accounting for about 20–50% of the cases [33]. The likelihood of viral encephalitis increases when the infection is caused by zoonotic RNA viruses with humans as a "dead end" host. An infection with the neurotropic virus herpes simplex type 1 (HSV-1) is the main reason for viral encephalitis in Western countries [34]. Most infections are mild and transient, leading to fever, headaches, and neck stiffness. More severe infections cause neurological and cognitive symptoms only in the acute phase. However, some of the more serious infections leading to encephalitis result in permanent damage, even after the infection is resolved. The damage manifests itself in motor and cognitive problems as well as recurrent seizures. In addition, there can be persistent activation of the inflammatory response [35]. In severe cases, death can result. The outcome of encephalitis depends very much on the virus, which is causing it. Infections with WNV, HIV-1, rabies, ZIKV, and Dengue virus (DenV) have a high mortality rate [36]. In many of these cases, autopsies show virus-infected neurons, heavy infiltration of T cells and perivascular macrophages as well as neurons being destroyed by phagocytosis and apoptosis. In addition, blood–brain barrier breakdown and micro-infarcts are causing edema and hemorrhage [37, 38]. Necrotic and apoptotic neurons are replaced by diffuse reactive gliosis (microgliosis and astrogliosis) and – if there are cavities – by reactive barrier astrocytes around the border of the cavities.

Experiments with mouse models suggest that the failure of microglia to coordinate and orchestrate an immune response against a virus infection is at the core of the causes of viral encephalitis [39]. In the end, encephalitis will cause transient, chronic, or terminal neuronal dysfunction. There are two main factors that can cause this: first, direct damage to neurons caused by the actions of the virus and second, an overreaction of the brain immune response, which will damage neurons. In the various encephalitis scenarios, both factors have different impacts.

Neuronal apoptosis is caused by viral proteins, whereas excessive viral budding leads to neuronal necrosis [40]. The release of glutamate leads to excitotoxicity of the neurons. Coxsackievirus B3 is interesting as it infects neural stem cells in the subventricular zone, then gets disseminated as these cells divide and move along the rostral migratory stream. Along the way, the virus spreads to other neurons in the parenchyma [41]. An extreme case of direct neuronal damage by a virus is the rabies virus. This virus causes active immunosuppression of the microglia, astrocytic, and T cell responses. This suppression has direct detrimental effects on neurons. One main result is neuronal apoptosis – a major cause of neuronal death in these situations [42].

The other major mechanism for neuronal damage in encephalitis is an excessive immune response. This path seems to be especially prevalent during infections with flaviviruses. The inhibition of CD8+ cells by microglia to prevent neuronal lysis seems to fail in these cases. Attack by T cells as well as excessive release of proinflammatory cytokines and oxygen radicals by macrophages is then a major cause of neuronal damage [43]. An extreme case in this group is HIV-1 infection. These viruses infect microglia and macrophages but not neurons [40]. Yet, the infections have a high likelihood of encephalitis and HIV-associated dementia due to neuronal death. The main reason is the release of proinflammatory cytokines, free radicals, and excitatory amino acids by macrophages and microglia. This toxic environment created by these immune cells is far more detrimental to the neurons than to the viruses [44].

9.4 Bacterial Infections

In general, bacterial infection can cause either one or both of the following pathologies: meningitis and brain abscesses. In addition, some bacteria, which infect first other body parts, can have secondary and serious effects on the brain after accessing it. The two most common and serious ones are *Treponema pallidum* (neurosyphilis) and *Borrelia burgdorferi* (Lyme disease or Lyme neuroborreliosis, LNB).

Meningitis is the most serious result of bacterial infections with a mortality rate of 26% [45]. *Streptococcus agalactiae* and *Escherichia coli K1* are the most frequent bacteria causing meningitis. Bacteria invade from meningeal blood vessels to subarachnoid space by transcytosis through endothelial cells, disruption of tight junctions around endothelial cells, or by use of the "Trojan horse" way, hiding in perivascular macrophages [46]. Once arrived in the subarachnoid space, the bacteria cause cerebral edema, a process that is dependent on aquaporin-4, indicating participation of astrocytic endfeet [47]. Bacterial toxins interact and damage endothelial cells, which causes inflammation at the walls of the blood vessels. Inflammation together with macrophage and neutrophil recruitment contributes to vascular occlusions. This in turn interferes with the blood supply to the brain causing ischemia. This will damage neurons depending on which vessels are affected. In addition, microglia release oxygen radicals, TNF-α, matrix metalloproteinases, and excitotoxins. This toxic environment, while detrimental to bacterial proliferation and survival, will damage neuronal function.

Bacterial abscesses are caused by bacteria that enter the parenchyma either because of ongoing bacterial meningitis or de novo from the systemic circulation. Microglia TLRs are the main responders, causing proinflammatory cytokine release, displaying antigens on the cell surface, and attacking the bacteria via phagocytosis. Neutrophils are recruited for antibacterial actions (phagocytosis, degranulation, and neutrophil extracellular traps). This results in edema and neuronal necrosis. Lymphocytes enter and are activated by microglial antigen presentation. Due to the antibacterial actions and neuronal necrosis, a vascularized capsule forms to insulate the infected region from healthy tissue. The capsule is encased by the border-forming reactive astrocyte type forming a true abscess.

Neurosyphilis and LNB are both based on spirochetal infections. Neurosyphilis develops after untreated or resistant systemic infection with *Treponema pallidum*. It is the result of bacteria attacking the vascular endothelium of the meninges. The resulting obstruction of the blood supply causes ischemia. This in turn damages mostly oligodendrocytes and therefore interferes with myelination [48]. LNB is caused by *Borrelia burgdorferi* infection of the meninges. It infects all cell types of the nervous tissue. This induces a strong inflammatory response, which in turn causes neurological pathologies (memory loss, dementia, mood disorders). *Borrelia burgdorferi* has well-developed mechanisms to evade the immune response of the host, like antigen variation and downregulation of components that react with pattern recognition receptors.

9.5 Conclusions

It turns out that the brain is not an immune-privileged site. However, this does not mean that there are no special arrangements not seen in other organs to protect neuronal circuits. The blood–brain barrier and the segregated fluid compartments warrant a more complicated process for lymphocyte activation. The lymphocytes are still recruited in the lymph nodes nearest to the brain. This is possible as the brain has effective drainage systems for foreign antigens, and the APCs (dendritic cells) are sampling the efflux from the brain at the periphery. Activated lymphocytes must be validated and reactivated before receiving permission to enter the brain. This reactivation process ensures that the brain is protected from reactions by the systemic immune system and adds to blood–brain barrier and leptomeninges in establishing a special brain environment. Microglia have a major role in coordinating all the immune responses with participating cells including T lymphocytes. A major function of the microglia, not seen with macrophages in other organs, is its protective function of neuronal circuits. Microglia prevent T lymphocytes from excessively lysing infected neurons. This role does not extend to other brain cells. Microglia therefore orchestrate anti-pathogen mechanisms that ensure the survival of neuronal circuits as much as possible.

References

1. Forrester JV, McMenamin PG, Dando SJ. CNS infection and immune privilege. Nat Rev Neurosci. 2018;19(11):655–71.
2. Schubert-Unkmeir A, Konrad C, Slanina H, Czapek F, Hebling S, Frosch M. Neisseria meningitidis induces brain microvascular endothelial cell detachment from the matrix and cleavage of occludin: a role for MMP-8. PLoS Pathog. 2010;6(4):e1000874.
3. Altan-Bonnet N. Extracellular vesicles are the Trojan horses of viral infection. Curr Opin Microbiol. 2016;32:77–81.
4. Yamaoka S, Okada K, Ito N, Okadera K, Mitake H, Nakagawa K, et al. Defect of rabies virus phosphoprotein in its interferon-antagonist activity negatively affects viral replication in muscle cells. J Vet Med Sci. 2017;79(8):1394–7.
5. McGavern DB, Kang SS. Illuminating viral infections in the nervous system. Nat Rev Immunol. 2011;11(5):318–29.
6. Castellano P, Prevedel L, Eugenin EA. HIV-infected macrophages and microglia that survive acute infection become viral reservoirs by a mechanism involving Bim. Sci Rep. 2017;7(1):12866.
7. Koyuncu OO, Hogue IB, Enquist LW. Virus infections in the nervous system. Cell Host Microbe. 2013;13(4):379–93.
8. Smith G. Herpesvirus transport to the nervous system and back again. Annu Rev Microbiol. 2012;66:153–76.
9. Gershon AA, Breuer J, Cohen JI, Cohrs RJ, Gershon MD, Gilden D, et al. Varicella zoster virus infection. Nat Rev Dis Primers. 2015;1:15016.
10. Levin MJ, Cai GY, Manchak MD, Pizer LI. Varicella-zoster virus DNA in cells isolated from human trigeminal ganglia. J Virol. 2003;77(12):6979–87.

11. Zhang SY, Clark NE, Freije CA, Pauwels E, Taggart AJ, Okada S, et al. Inborn errors of RNA lariat metabolism in humans with brainstem viral infection. Cell. 2018;172(5):952–65.e18.
12. Lafaille FG, Harschnitz O, Lee YS, Zhang P, Hasek ML, Kerner G, et al. Human SNORA31 variations impair cortical neuron-intrinsic immunity to HSV-1 and underlie herpes simplex encephalitis. Nat Med. 2019;25(12):1873–84.
13. Heinz J, Kennedy PGE, Mogensen TH. The role of autophagy in varicella zoster virus infection. Viruses. 2021;13(6)
14. Madden JC Jr, Cui D, Brinton MA. RNase L antiviral activity is not a critical component of the Oas1b-mediated Flavivirus resistance phenotype. J Virol. 2019;93(22)
15. Fekete R, Cserép C, Lénárt N, Tóth K, Orsolits B, Martinecz B, et al. Microglia control the spread of neurotropic virus infection via P2Y12 signalling and recruit monocytes through P2Y12-independent mechanisms. Acta Neuropathol. 2018;136(3):461–82.
16. Moseman EA, Blanchard AC, Nayak D, McGavern DB. T cell engagement of cross-presenting microglia protects the brain from a nasal virus infection. Sci Immunol. 2020;5(48) https://doi.org/10.1126/SCIIMMUNOL.ABB1817.
17. Huh GS, Boulanger LM, Du H, Riquelme PA, Brotz TM, Shatz CJ. Functional requirement for class I MHC in CNS development and plasticity. Science (New York, NY). 2000;290(5499):2155–9.
18. Goddard CA, Butts DA, Shatz CJ. Regulation of CNS synapses by neuronal MHC class I. Proc Natl Acad Sci U S A. 2007;104(16):6828–33.
19. Miller KD, Schnell MJ, Rall GF. Keeping it in check: chronic viral infection and antiviral immunity in the brain. Nat Rev Neurosci. 2016;17(12):766–76.
20. Jeon S, St Leger AJ, Cherpes TL, Sheridan BS, Hendricks RL. PD-L1/B7-H1 regulates the survival but not the function of CD8+ T cells in herpes simplex virus type 1 latently infected trigeminal ganglia. J Immunol (Baltimore, MD: 1950). 2013;190(12):6277–86.
21. Garber C, Soung A, Vollmer LL, Kanmogne M, Last A, Brown J, et al. T cells promote microglia-mediated synaptic elimination and cognitive dysfunction during recovery from neuropathogenic flaviviruses. Nat Neurosci. 2019;22(8):1276–88.
22. Potokar M, Korva M, Jorgačevski J, Avšič-Županc T, Zorec R. Tick-borne encephalitis virus infects rat astrocytes but does not affect their viability. PLoS One. 2014;9(1):e86219.
23. Potokar M, Jorgačevski J, Zorec R. Astrocytes in Flavivirus infections. Int J Mol Sci. 2019;20(3)
24. Palus M, Bílý T, Elsterová J, Langhansová H, Salát J, Vancová M, et al. Infection and injury of human astrocytes by tick-borne encephalitis virus. J Gen Virol. 2014;95(Pt 11):2411–26.
25. Chang CY, Li JR, Chen WY, Ou YC, Lai CY, Hu YH, et al. Disruption of in vitro endothelial barrier integrity by Japanese encephalitis virus-infected astrocytes. Glia. 2015;63(11):1915–32.
26. Pfefferkorn C, Kallfass C, Lienenklaus S, Spanier J, Kalinke U, Rieder M, et al. Abortively infected astrocytes appear to represent the main source of interferon beta in the virus-infected brain. J Virol. 2016;90(4):2031–8.
27. Schoggins JW, Rice CM. Interferon-stimulated genes and their antiviral effector functions. Curr Opin Virol. 2011;1(6):519–25.
28. Lindqvist R, Mundt F, Gilthorpe JD, Wölfel S, Gekara NO, Kröger A, et al. Fast type I interferon response protects astrocytes from flavivirus infection and virus-induced cytopathic effects. J Neuroinflammation. 2016;13(1):277.
29. Filgueira L, Larionov A, Lannes N. The influence of virus infection on microglia and accelerated brain aging. Cell. 2021;10(7)
30. Wallet C, De Rovere M, Van Assche J, Daouad F, De Wit S, Gautier V, et al. Microglial cells: the main HIV-1 reservoir in the brain. Front Cell Infect Microbiol. 2019;9:362.
31. Alvarez-Carbonell D, Ye F, Ramanath N, Garcia-Mesa Y, Knapp PE, Hauser KF, et al. Cross-talk between microglia and neurons regulates HIV latency. PLoS Pathog. 2019;15(12):e1008249.
32. Portilla I, Reus S, León R, Van Der Hofstadt C, Sánchez J, López N, et al. Neurocognitive impairment in well-controlled HIV-infected patients: a cross-sectional study. AIDS Res Hum Retrovir. 2019;35(7):634–41.

33. Chen Z, Zhong D, Li G. The role of microglia in viral encephalitis: a review. J Neuroinflammation. 2019;16(1):76.
34. Reinert LS, Rashidi AS, Tran DN, Katzilieris-Petras G, Hvidt AK, Gohr M, et al. Brain immune cells undergo cGAS/STING-dependent apoptosis during herpes simplex virus type 1 infection to limit type I IFN production. J Clin Invest. 2021;131(1)
35. Soung A, Klein RS. Viral encephalitis and neurologic diseases: focus on astrocytes. Trends Mol Med. 2018;24(11):950–62.
36. Chhatbar C, Prinz M. The roles of microglia in viral encephalitis: from sensome to therapeutic targeting. Cell Mol Immunol. 2021;18(2):250–8.
37. Jois D, Moorchung N, Gupta S, Mutreja D, Patil S. Autopsy in dengue encephalitis: an analysis of three cases. Neurol India. 2018;66(6):1721–5.
38. Wong KT, Munisamy B, Ong KC, Kojima H, Noriyo N, Chua KB, et al. The distribution of inflammation and virus in human enterovirus 71 encephalomyelitis suggests possible viral spread by neural pathways. J Neuropathol Exp Neurol. 2008;67(2):162–9.
39. Hatton CF, Duncan CJA. Microglia are essential to protective antiviral immunity: lessons from mouse models of viral encephalitis. Front Immunol. 2019;10:2656.
40. Swanson PA 2nd, McGavern DB. Viral diseases of the central nervous system. Curr Opin Virol. 2015;11:44–54.
41. Feuer R, Pagarigan RR, Harkins S, Liu F, Hunziker IP, Whitton JL. Coxsackievirus targets proliferating neuronal progenitor cells in the neonatal CNS. J Neurosci. 2005;25(9):2434–44.
42. Chelbi-Alix MK, Vidy A, El Bougrini J, Blondel D. Rabies viral mechanisms to escape the IFN system: the viral protein P interferes with IRF-3, Stat1, and PML nuclear bodies. J Interferon Cytokine Res. 2006;26(5):271–80.
43. Nazmi A, Dutta K, Das S, Basu A. Japanese encephalitis virus-infected macrophages induce neuronal death. J Neuroimmune Pharmacol. 2011;6(3):420–33.
44. Kaul M, Garden GA, Lipton SA. Pathways to neuronal injury and apoptosis in HIV-associated dementia. Nature. 2001;410(6831):988–94.
45. Rzaska M, Niewiadomski S, Karwacki Z. Molecular mechanisms of bacterial infections of the central nervous system. Anaesthesiol Intensive Ther. 2017;49(5):387–92.
46. Barichello T, Generoso JS, Milioli G, Elias SG, Teixeira AL. Pathophysiology of bacterial infection of the central nervous system and its putative role in the pathogenesis of behavioral changes. Revista brasileira de psiquiatria (Sao Paulo, Brazil: 1999). 2013;35(1):81–7.
47. Papadopoulos MC, Verkman AS. Aquaporin-4 gene disruption in mice reduces brain swelling and mortality in pneumococcal meningitis. J Biol Chem. 2005;280(14):13906–12.
48. Dersch R, Singh AE. Neurosyphilis and Lyme neuroborreliosis. Curr Opin Neurol. 2021;34(3):403–9.

Chapter 10
Autoimmune Diseases

Abstract For the phenomenon of autoimmune diseases, partial loss of immunological tolerance is key. The main disease is multiple sclerosis. It is still not clear what causes multiple sclerosis other than a fault in the tolerance mechanism. While the cause is still unknown, the major mechanisms in the progress of the disease are well-known. This is mainly due to animal models such as autoimmune encephalomyelitis and the cuprizone model of demyelination. The related phenomenon of autoimmune encephalitis is based on an immune reaction to the *N*-methyl-D-aspartate receptor. This receptor is often expressed in systemic cancer cells and the immune reaction against the tumor can spill over into the brain parenchyma. Invading T lymphocytes against myelin components seems to be at the root of multiple sclerosis-like phenomena. However, these cells seem not so much directly attack the myelin as to recruit microglia, which then do most of the damage. Reactive astrocytosis is involved in later stages as the myelin damage escalates into inflammatory processes. This is underlined by conclusions from the cuprizone model. Here, cuprizone has the role of activating microglia against myelin components and not by damaging myelin directly. In this group of diseases, neurons seem to take no active role but are victims of axonal conduction failure due to myelin damage.

Keywords Autoimmune encephalitis · Autoimmune encephalomyelitis · Cuprizone · Demyelination · Immunological tolerance · Multiple sclerosis

10.1 Introduction

Autoimmune diseases are the result of the immune system not tolerating some self-antigens and initiating an immune response against some of these antigens. With respect to the brain, the most common is multiple sclerosis (MS), which has a prevalence of roughly 1 in 1000. There are many more autoimmune brain diseases, but these are rare diseases with a prevalence of approximately 1 in 100,000 [1, 2]. Examples are autoimmune encephalitis (AE), Rasmussen encephalitis, cerebellitis, and neuromyelitis optica spectrum disorders. There are also several systemic inflammatory disorders with central nervous system (CNS) manifestations as well as

W. Walz, *The Gliocentric Brain*, https://doi.org/10.1007/978-3-031-48105-5_10

paraneoplastic neurological syndrome (PNS), where non-CNS cancer cells in the body are the target of an immune reaction with immunopathological effects on brain cells. It is thought that in PNS systemic tumor cells outside the brain ectopically express neuronal proteins on their surface and that in the aftermath T cells against these antigens enter the brain, causing neurological symptoms [3]. The underlying pathology of these autoimmune diseases is similar: brain cell autoreactive B cells, CD8+ cytotoxic, and CD4+ helper T cells invade the brain and in conjunction with an innate immune response result in damage to neuronal circuits. This chapter will start with reviewing autoimmune encephalitis followed by dealing with multiple sclerosis. But first, one needs to understand the basics of immune tolerance.

10.2 Immunological Tolerance

B and T lymphocytes are constantly exposed to antigens in conjunction with major histocompatibility complex (MHC). The most important key mechanism for a functioning immune system is the distinction between self- and non-self-antigens. A mistaken attack on the body's own components will lead to autoimmune reactions and chronic inflammation. Some of these attacks can result in serious and life-threatening diseases. What are the basic arrangements to prevent autoimmune reactions and to ensure immunological tolerance to self-antigens (autoantigens)?

Immune tolerance mechanisms are organized on two levels: central and peripheral tolerance [4]. Prevention of self-reactivity is a learned mechanism during development. Immature T cells migrate from bone marrow to the thymus. For these T cells, the central tolerance takes place in the thymus during development when the T cells are still immature. T cell receptors for antigen–MHC complexes are randomly generated during the development. Inevitably receptors are created which will react with the body's own antigens. Thymic epithelial cells and dendritic cells acting as antigen-presenting cells (APCs) express autoimmune regulator (AIRE), a transcription factor. Its action will lead to the expression in these APCs of antigens from other body organs. In addition, antigens from extra thymic sites are imported through the circulation. T cells whose receptors react with these self-antigens are eliminated by apoptosis or are kept in a non-reactive state. Some weakly reactive T cells are transformed into regulatory T cells, which during infections react as anti-inflammatory cells. Similar processes exist in the bone marrow for B cells, although autoreactive B cells alone are rare as they would need reconfirmation by interaction with autoreactive T cells [5]. Thus, this represents a "learning mechanism" for the elimination of T cells, whose receptors were randomly created to match self-antigens. Left are those T cells with receptors that were not matched for self-antigens [6]. Inevitably, some autoreactive T and B cells will escape this central tolerance mechanism and will be released out of the thymus and bone marrow. These cells are dealt with by peripheral tolerance processes, not all are well understood so far. The main mechanism for peripheral tolerance is the lack of a costimulatory signal. T cells in the periphery have constitutively expressed receptor CD28 (cluster of differentiation 28) or a similar molecule. Peripheral dendritic cells constantly express self-antigens bound to MHC. However, such dendritic cells with self-antigen-MHC complexes lack the corresponding receptor for CD28, usually CD80 or CD86. In this case, a corresponding antigen-MHC molecule on an APC but with no costimulatory engagement, the corresponding T cell will undergo apoptosis. This mechanism therefore represents another key to preventing self-reactivity. Only if the antigen is expressed by dendritic cells/APCs in conjunction with an innate immune response do these APCs express CD80 or CD86. If these T cells in the periphery react to an MHC–antigen complex and receive at the same time a costimulatory signal, they activate and this costimulation drives T cell clonal expansion [7]. In addition, the regulatory T cells will suppress a reaction to self-antigens. Release of IL-10 (interleukin 10) and some other cytokines is important in peripheral tolerance [8]. IL-10 deficient mice are prone to autoimmune diseases. In addition, certain organs, such as testes, eye chambers, and brain, are protected from exposure to the circulation by various barriers and special entry mechanisms (see previous chapter). These "privileged" compartments have therefore additional safeguards against autoreactivity.

10.3 Autoimmune Encephalitis

Autoimmune encephalitis (AE) syndrome is a group of diseases with inflammation of the brain parenchyma, with symptoms like viral encephalitis. AE is at least as common as the viral counterpart [9]. AE is distinct in that autoantibodies against neural cell surface proteins are detected in cerebrospinal fluid (CSF) and/or blood serum. AE is responsive to immunotherapy.

The most common isolated antibodies are against the *N*-methyl-D-aspartate receptor (NMDAR) and leucine-rich glioma inactivated 1 (LGI1) glycoprotein, which acts as a tumor suppressor and is also involved in glutamatergic synaptic transmission. LGI1-based AE involves drug-resistant seizures and cognitive impairment. Other, less common AEs are associated with antibodies against gamma-amino-butyric acid (GABA) receptors A and B, alpha-amino-3-hydroxy-5-methyl-4-isoxazolepropionic acid receptor (AMPAR), and the glycine receptor. There are many more (see Table 2 in ref. [9]). A reaction against NMDAR is the most common AE, and it is therefore dealt with here as an example of this group of diseases. In approximately half of the cases, NMDARs are expressed outside the brain on tumor cells, along with other non-tissue-specific antigens [10]. Common is NMDAR expression in ovarian teratomas [11]. The rate of female patients aged 18–45 years, with anti-NMDAR AE that suffer from ovarian teratoma, is 58% [12]. It is assumed that circulating antibodies against the ovarian NMDAR are finding their way through a leaky blood–brain barrier [13]. A previous viral infection is another large identifiable cause, especially after infection with herpes simplex [14]. Approximately one-quarter of patients with previous herpes simplex infection develop anti-NMDAR AE. In these cases, anti-NMDAR antibodies are present in the CSF, even so the CSF is clear of virus antigens [15]. The causal mechanism is unknown. Apart from these two groups, ovarian teratoma and previous herpes simplex infection, the triggers for the remainder of the patients are unknown. The isolated antibodies from the patients are against the NMDAR subunit GluN1 (glutamate subunit receptor zeta-1). The antibody cross-links NMDARs, which leads to their internalization. This reduces the glutamate-evoked NMDA-dependent synaptic currents. This results in turn in decreased long-term potentiation. In mouse models, this affects several glutamatergic synapses in the hippocampus and is probably responsible for spatial memory defects in these animals [16]. Seizures develop in most patients and are a major cause of morbidity and mortality. Mood-psychotic and movement disorders are other clinical symptoms. Not surprisingly, treatment with ketamine (an NMDAR antagonist) mimics the symptoms in humans. Oligodendrocytes possess NMDARs too and these receptors are responsible for calcium signaling and the insertion of GLUT1 (glucose transporter 1) into the myelin sheet during axonal transmission. Anti-NMDAR antibodies from AE patients' CSF interfere with the incorporation of the NMDA-dependent GLUT1 glucose transporter in a functional assay. This might explain some of the white matter damage, which is observed in anti-NMDAR AE patients [17]. This leaves the question of where the antibodies against NMDAR come from. There is no doubt that some of these diffuse through damaged tight junctions between the endothelial cells and in the choroid plexus. Cytokines released due to systemic or parenchymal inflammation would allow diffusion of the large antibody molecules. However, in anti-NMDAR AE this pathway seems to be a minor one. Recent studies and autopsies suggest that most of the antibodies are produced by B lymphocytes, which were cross-activated at cervical lymph nodes and entered the subarachnoid and CSF spaces and are probably also in the parenchyma [18].

10.4 Multiple Sclerosis and Its Animal Model – Experimental Autoimmune Encephalomyelitis

Multiple sclerosis (MS) is a chronic demyelinating disease, based on autoimmune inflammation. It is considered a T lymphocyte-mediated disease as these cells are found in autopsies and biopsies of plaques in MS patients [19]. The demyelination leads – together with toxic microglia activity – eventually to axonal damage. The disease has a prevalence of roughly 1 in 1000. The most common is the relapsing-remitting type. This involves symptoms developing over weeks and months with increasing intensity, only to improve or go into remission. This period of improvement can last for years. However, it is then interrupted by a new relapse period. The majority will experience secondary progressive MS. This involves a worsening of the symptoms during relapse periods and less frequent and shorter remission periods. This steady progression appears 10–20 years after the first diagnosis. There is a large individual difference in time course and intensity. A minority of patients do not fall into this pattern but exhibit primary progressive MS. In this case, there are no periods of relief, but a steady and progressive worsening of symptoms. Symptoms are the result of progressive axonal damage, which is due to progressively dysfunctional myelin and direct damaging activity of microglia. Numbness and weakness of limbs, unsteady gait, tremors, and visual disturbances are signs of MS. Slurred speech and problems with autonomous control are further progressive symptoms. Underlying these symptoms may be immune reactions against major myelin compounds by T and B lymphocytes. These cells are reactivated by APCs upon entry into the brain. However, no specific antigens for these lymphocytes have been identified yet [20]. In experimental animals, injections of myelin antigens as in EAE cause pathological events like MS. However, T lymphocytes against these antigens have also been found in normal healthy humans. Thus, although MS is categorized as an autoimmune disease for the above reasons, caution is advised as the evidence is not overwhelming. The triggering event or events for MS are still unknown. One possibility is that myelin-reactive T cells that have a low affinity for MHC escape the central tolerance mechanism by the APCs of the thymus epithelium. T and B lymphocyte invasion of the parenchyma and accumulation around demyelination lesions are other indications of MS having an autoimmune origin. Myelin-based antibodies are found in the CSF of most, but not all patients. A weak genetic linkage has been found to human leukocyte antigen genes, which encode for MHC II. Risk factors for developing MS are previous infection by Epstein–Barr virus, smoking, and a low serum concentration of vitamin D. Thus, the initial trigger that causes the disease is largely unknown. However, the downstream mechanisms are better researched, mostly thanks to EAE.

In pathological diagnostics, the MS hallmarks are the so-called plaques or multifocal lesions. In the acute phase of the MS disease time course, the plaques consist of an accumulation of reactive microglia/macrophages, which make up most of the cells, but the plaques contain also CD4+ T, CD8+ T, and B lymphocytes. They are mostly located near white matter tracts but can also appear in the gray matter around myelinated axons (see Fig. 10.1). The lymphocytes are more numerous in MS patients with a history of a faster and more progressive time course. Myelin and oligodendrocytes are damaged and myelin fragments can be seen taken up by phagocytosis by microglia/macrophages. As the disease progresses, gliotic tissue consisting of reactive astrocytes, develops around the foci. Later in the course of the disease, the myelin disappears, and this leaves unmyelinated axons surrounded by gliotic tissue. Surviving oligodendrocytes will attempt to remyelinate, but the gliotic tissue interferes with these attempts. Thus, remyelination attempts occur, but they are in most cases not successful. At this stage, there is axon loss and cavitation [19]. As stated above, the origin of MS is unknown and there are no clear genetic or environmental causes, just risk factors. EAE animal models give insight into disease development, but not into the causes of the disease. A very large problem is that no dominant T lymphocyte autoantigen has been found. However, clearly, T lymphocytes attack and phagocytose myelin components in the brain. There is in addition heavy involvement of B lymphocytes and macrophages, microglia, and astrocytes. Only CD4+ T lymphocytes from MS patients are able to create experimental EAE if injected into rodents. No other MS-derived cell types

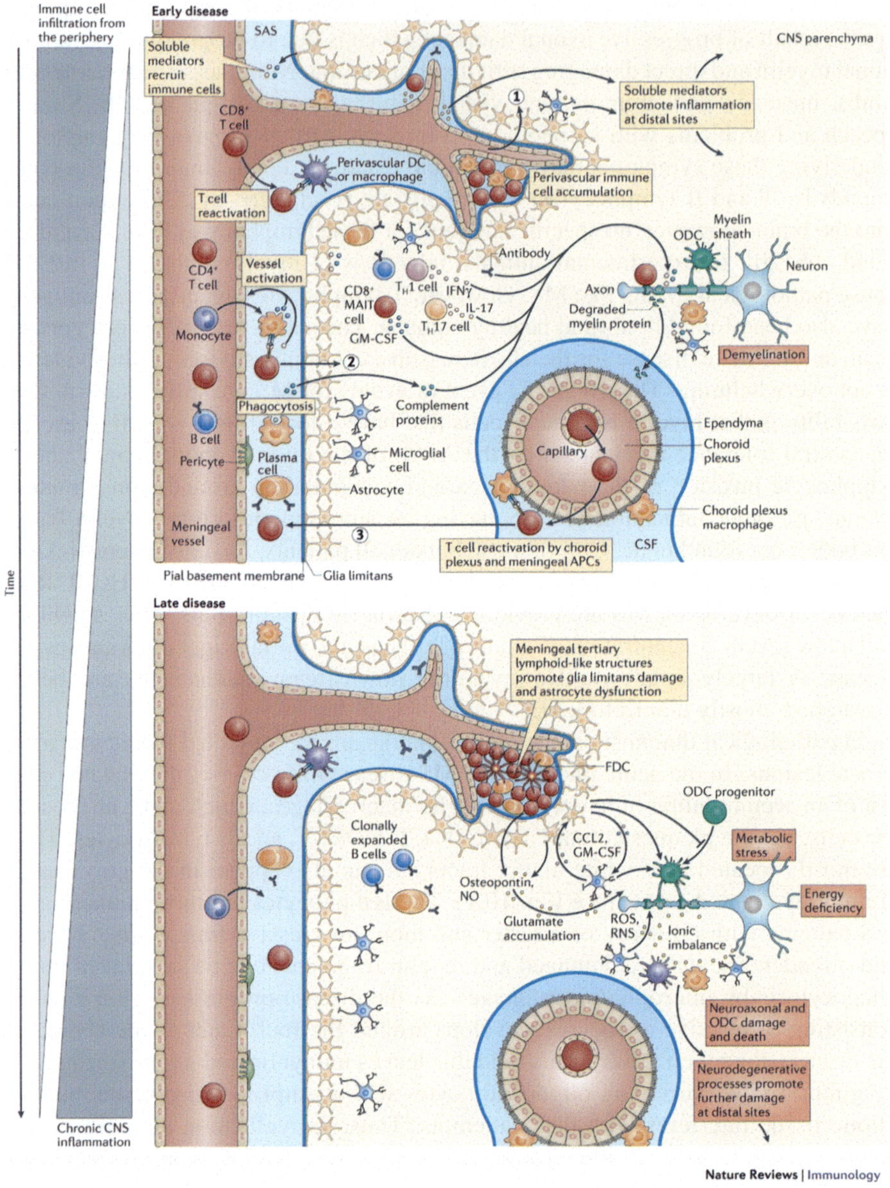

Fig. 10.1 Immune cell infiltration from the periphery is a prominent feature of early-stage multiple sclerosis (top panel) and can occur from the meningeal blood vessels by direct crossing of the blood–brain barrier (denoted "**1**" in the figure) or the subarachnoid space (SAS; denoted "**2**"), or from the choroid plexus across the blood–cerebrospinal fluid (CSF) barrier (denoted "**3**"). Peripheral innate and adaptive immune cells can accumulate in perivascular spaces and enter the central nervous system (CNS) parenchyma. These cells, along with activated CNS-resident microglia and astrocytes, promote demyelination and oligodendrocyte (ODC) and neuroaxonal injury through direct cell contact-dependent mechanisms and through the action of soluble inflammatory and

are capable of this feat. T and B cells interact in the peripheral circulation and lymph nodes to maintain activation and proliferation in MS. Almost all patients have B cell-derived antibodies in the CSF. The B lymphocytes release GM-CSF (granulocyte-macrophage colony-stimulating factor, an inflammatory cytokine) to induce a pro-inflammatory response by border-associated macrophages. However, atacicept, which targets B lymphocyte activation, increases the severity of the disease in MS patients. The mystery of the MS origin precipitated a creative controversy about the origins of MS, especially since the EAE model is not able to answer the question of the cause or causes of MS. The "outside in" hypothesis advances the view that CD4+ T cells are the driver of the disease: the cells react to myelin antigens and, after confirmation by APCs the reactivated lymphocytes invade the CNS to participate with microglia and astrocytes in the inflammatory processes. The "inside in" hypothesis advances an intrinsic origin by the innate immune system of the brain with microglia the major, but not only driver [21]. Neither of the hypotheses questions the autoimmune nature of the disease. To induce EAE, antigens are emulated with Complete Freund's Adjuvant to boost immune activation. Common antigens are all components of the myelin sheath, like myelin oligodendrocyte glycoprotein (MOG), myelin basic protein (MBP), proteolipid protein (PLP), myelin-associated glycoprotein (MAG), and myelin-associated oligodendrocyte basic protein (MOBP). They all have been found in MS patients. Pertussis toxin is injected together with the antigen–complete Freund's adjuvant mixture to cause blood–brain barrier breakdown and therefore facilitate invasion of the immune cells [22]. If T cell activation or invasion is prevented, EAE is not being initiated. Most drugs developed with the EAE model in rodents were found beneficial in MS patients. One exception is tumor necrosis factor alpha (TNFα) whose inhibition slows down EAE progression in rats. However, infliximab, a TNFα blocker, worsens the symptoms when used with MS patients. There are, however, various drugs that improve EAE and attenuate the MS disease course in patients. Minocycline, an unspecific inhibitor of microglia activation, fingolimod, an inhibitor of pro-inflammatory properties, and glatiramer acetate and interferon-β, both inducers of anti-inflammatory properties of microglia/macrophages, are all attenuating the MS disease course [23].

Fig. 10.1 (continued) neurotoxic mediators. Later in the disease (bottom panel), immune cell infiltration wanes, perhaps due to adaptive immune cell exhaustion from chronic antigen exposure. However, chronic CNS-intrinsic inflammation and neurodegeneration continue. Meningeal tertiary lymphoid-like structures, which have specifically been documented in secondary progressive disease, may contribute to late-stage inflammation in patients with this form of multiple sclerosis. The action of the CNS-resident innate cells may contribute to chronic inflammation irrespective of the precise disease subtype. Stimulated by the microglia, astrocytes can produce CC-chemokine ligand 2 (CCL2) and granulocyte–macrophage colony-stimulating factor (GM-CSF), leading to even further microglial recruitment and activation, and the astrocytes can prevent remyelination at sites of neuroaxonal injury by inhibiting progenitor cells from developing into mature ODCs. *APC* antigen-presenting cell, *CD8⁺* MAIT cell, *CD8⁺* mucosa-associated invariant T cell, *FDC*, follicular dendritic cell, *IFNγ* interferon-γ, IL-17 interleukin-17, *NO* nitric oxide, *RNS* reactive nitrogen species, *ROS* reactive oxygen species, T_H1 *cell* T helper 1 cell. (From Ref. [19], reprinted with permission from Macmillan Publishers Limited)

In MS, CD4+ T cells dominate the lesions. In EAE, CD8+ T cells are in the majority [24]. T cells interact physically with microglia, probably via MHC II. The microglia were previously shifted to a pro-inflammatory type by activation of pattern recognition receptors (PRRs) such as Toll-like receptors (TLRs), triggering receptors expressed in myeloid cells 2 (TREM2), and complement receptor 3. This makes them release toxic factors such as reactive oxygen species (ROS) and glutamate and present MHCII with costimulatory molecules CD40, CD80, and CD86. This is an indication that antigens are presented by the microglia to CD4+ T lymphocytes. However, microglia-mediated antigen presentation to T cells seems not directed against neurons. This is a similar microglia-mediated mechanism that during virus infections prevents T cell attack against neurons (see previous chapter). It is assumed that in pericellular cuffs, activated and invasive macrophages present antigens with MHC II to T cells. This interaction with macrophages, not microglia, increases T cell invasion and T cell toxicity to neurons [25, 26]. However, T cells release soluble factors, mostly cytokines, that shift microglia to a more inflammatory and toxic phenotype. In EAE, invasive macrophages appear more destructive to the myelin sheath than microglia or lymphocytes [27]. CD8+ lymphocytes interact with antigens presented on MHC I by macrophages and microglia. This cross-presentation for reactivation of CD8+ lymphocytes could be another source of myelin damage. The use of the EAE model in transgenic mice, which cannot initiate an NF-ĸB mediated inflammatory response in microglia and border-associated macrophages, prevents EAE initiation. The T cells secrete IFNγ (interferon gamma) and IL-17 (interleukin 17), which potentiate microglia activation. Microglia in turn release IL-6 which acts on endothelial cells to cause a leaky BBB and T and B cell invasion. However, for EAE, onset of microglia activation alone is not sufficient, macrophage invasion is necessary [28]. Knockout mice for transforming growth factor beta (TGFβ)-activated kinase 1 have an attenuated microglia activation. This results in a suppressed EAE course with reduced immune cell infiltration and demyelination [29].

There is no doubt that microglia play a central part in MS. In patients with no sign of demyelination yet in the pre-lesion state, microglia in the white matter express MHC II with pro-inflammatory properties (see above). Later in active white matter lesions with demyelination, microglia have even more upregulated pro-inflammatory properties and lose their homeostatic function as indicated by the loss of the purinergic receptor P2RY12, which is replaced by the pro-inflammatory purinergic receptor P2X7. Chronic lesions have no microglia in the center anymore, microglia are only in the rim of the lesion [22]. In MS autopsies, microglia express heavy upregulation of members of the tumor necrosis factor (TNF) family and of pro-inflammatory interleukins (IL-1β, IL-6, IL-12, IL-23, and IL-33). In addition, chemokines for the recruitment of T lymphocytes and macrophages are expressed by microglia [30]. Phagocytosis of myelin by microglia has a damaging effect early on in MS. This appears to go together with a reduction of the homeostatic role of microglia. At the onset of EAE in the parenchyma, microglia seem to be the major driving force of pathological processes. They proliferate very early with upregulated activation markers MHCII, CD40, and CD86. After this phase, proliferation declines and invaded T lymphocytes and microglia engage in a positive feedback loop [31]. The microglia activation includes the release of pro-inflammatory cytokines, chemokines complement factors, nitric oxide, and glutamate. At this stage, the microglia have direct damaging effects on neurons. TNF, IL-1β, and ROS release interferes with the function of glutamatergic synapses. Physical contact of microglia processes with the axon hillock (the area of axons where the action potential is initiated) causes axonal dysfunction. Several beneficial drugs used with MS patients interfere in EAE models with these direct pathological effects of microglia on neuronal function [22, 32]. In the later stages of the disease, microglia phagocytose to clear myelin debris, which seems to be beneficial in the remedial stages of the disease [22].

Astrocytes are also heavily involved in the MS pathology. Early, in the pre-lesion stage with leaky blood–brain barrier and lymphocyte infiltration, astrocytes appear reactive throughout the parenchyma with hypertrophy and upregulated GFAP [33]. In acute lesions, astrocyte reactivity is increased, and the reactive astrocytes close to the myelin lesion participate in myelin phagocytosis. Astrocyte endfeet are swollen and disconnected, contributing to the blood–brain barrier breakdown. In chronic lesions, astrocytes form reactive gliotic tissue right in the center of the lesion. This gliosis seems to have some beneficial effects on the survival of demyelinated axons, mainly mediated by brain-derived neurotrophic factor (BDNF) release [34]. In all stages, there is a clear gradient of astrocytic reactivity from the active margins of the demyelinating lesions into normal-appearing tissue. Reactive astrocytes near the center of the lesions express MHCII and CD80 as well as CD86 [35]. Reactive astrocytes interact with B lymphocytes, supporting their survival, maturation, and proliferation by secreting members of the TNF family [36]. A genetic MS risk variant, rs75665080, which increases NF-ĸB signaling in lymphocytes, increases lymphocyte load and lesion severity. It also shifts reactive astrocytes into a more pro-inflammatory and less homeostatic type [37]. It appears that early in the lesion, astrocytes support the maintenance of the blood–brain barrier and prevent lymphocytes from entering, in contrast to microglial activities. In the chronic later stages, this role changes significantly, as astrocytes contribute to inflammation and reduce their homeostatic and protective role substantially, especially close to the lesion core [22]. Administration of ganciclovir to GFAP-HSV-TK mice depletes proliferating reactive astrocytes. EAE in these mice, which were treated with ganciclovir before EAE induction, leads to increased immune cell infiltration and causes larger behavioral deficits. This indicates that astrocytes function as a barrier for immune cells entering the parenchyma. However, in the same transgenic mice, if ganciclovir was administered after chronic EAE treatment for more than 30 days, the symptoms improve in the transgenic mice compared with the wildtype. In the latter case, transgenic mice clearly have less lymphocyte infiltration than the wildtype. This indicates that astrocytes serve as a barrier function for immune cells. Once activated, they participate in the inflammatory response and potentiate it [38].

10.5 Cuprizone Model of Demyelination

The cuprizone (CPZ) model of demyelination has some interesting features. For once, T lymphocytes are not involved and during the CPZ treatment, mature oligodendrocytes, but not oligodendrocyte progenitor cells (OPCs) are being damaged and killed. CPZ is a copper chelator, which is fed to mice for 4–6 weeks. During this time, oligodendrocytes die, particularly in the corpus callosum. This results in demyelination. There is evidence that reactive microgliosis is the main cause of oligodendrocyte demise [22]. After 6 weeks, CPZ is usually withdrawn, and this is the peak of demyelination. Reactive microglia are now receding and OPCs are starting the remyelination process, which can be studied. In these phases, microglia have a similar gene expression pattern as in MS: an inflammatory expression in demyelination and an anti-inflammatory expression pattern in remyelination. It appears that the microglia and not CPZ are the cause of the death of the oligodendrocytes. If animal models are used that cannot create toxic astrocyte phenotypes, the microglia-driven demyelination process continues unhindered. This indicates that astrocytes are not a driver of the demyelination process [39]. Minocycline (an unspecific inhibitor of microglia activation) or experimental depletion of microglia is reducing the degree of demyelination [40]. The microglia are recruited by chemokines CCL 2 and 3 (chemokine ligands 2 and 3) as well as CXL1O (C-X-C motif chemokine ligand 10) [41]. If CPZ administration is not ceased after 6 weeks, but extended into 12 weeks, chronic demyelination occurs, and reactive microglia are sustained in high numbers. The conclusion from this more simplistic demyelination model of MS is that microglia are at least as important as T lymphocytes. Whereas T and B lymphocytes are important players during the initiation of MS, the sustenance during the often-long duration may be reactive microglia.

10.6 Conclusions

The origin of MS is highly controversial. It seems clear, however, that a major trigger is T lymphocytes, probably of the CD4+ type, which are autoreactive against myelin components, which are crucial. There, interaction at the blood–brain interface with macrophages and maybe yet unknown APCs is a key process. Lymphocytes do not seem to be the major aggressor against myelin. It seems a positive feedback loop between lymphocytes and microglia is another key factor. From then on, microglia are a key player, damaging myelin and directly attacking glutamatergic synapses. Myelin failure and direct attack by microglia are the main, if not only cause of axonal and neuronal damage and therefore of the major MS symptoms. To sustain the disease, microglia and astrocytic interaction is important. In later stages, microglia and astrocytes develop anti-inflammatory properties, reduce their damaging activities, and engage in repair processes. However, it appears that, for most MS patients, these beneficial transformations are too late and too few. In this disease group, neurons and their components appear as passive victims of their cellular environment including invading immune cells.

References

1. Sechi E, Flanagan EP. Antibody-mediated autoimmune diseases of the CNS: challenges and approaches to diagnosis and management. Front Neurol. 2021;12:673339.
2. Prüss H. Autoantibodies in neurological disease. Nat Rev Immunol. 2021;21:1–16.
3. Rosenfeld MR, Dalmau J. Paraneoplastic neurologic syndromes. Neurol Clin. 2018;36(3):675–85.
4. Rosenblum MD, Remedios KA, Abbas AK. Mechanisms of human autoimmunity. J Clin Invest. 2015;125(6):2228–33.
5. Vignali DA, Collison LW, Workman CJ. How regulatory T cells work. Nat Rev Immunol. 2008;8(7):523–32.
6. Alberti P, Handel AE. The contribution of thymic tolerance to central nervous system autoimmunity. Semin Immunopathol. 2021;43(1):135–57.
7. Azuma M. Co-signal molecules in T-cell activation: historical overview and perspective. Adv Exp Med Biol. 2019;1189:3–23.
8. Fujio K, Okamura T, Yamamoto K. The family of IL-10-secreting CD4+ T cells. Adv Immunol. 2010;105:99–130.
9. Ramanathan S, Al-Diwani A, Waters P, Irani SR. The autoantibody-mediated encephalitides: from clinical observations to molecular pathogenesis. J Neurol. 2021;268(5):1689–707.
10. Dalmau J, Lancaster E, Martinez-Hernandez E, Rosenfeld MR, Balice-Gordon R. Clinical experience and laboratory investigations in patients with anti-NMDAR encephalitis. Lancet Neurol. 2011;10(1):63–74.
11. Graus F, Dalmau J. Paraneoplastic neurological syndromes: diagnosis and treatment. Curr Opin Neurol. 2007;20(6):732–7.
12. Dalmau J, Armangué T, Planagumà J, Radosevic M, Mannara F, Leypoldt F, et al. An update on anti-NMDA receptor encephalitis for neurologists and psychiatrists: mechanisms and models. Lancet Neurol. 2019;18(11):1045–57.
13. Platt MP, Agalliu D, Cutforth T. Hello from the other side: how autoantibodies circumvent the blood-brain barrier in autoimmune encephalitis. Front Immunol. 2017;8:442.
14. Swayne A, Warren N, Prain K, Gillis D, Wong R, Blum S. Analysing triggers for anti-NMDA-receptor encephalitis including herpes simplex virus encephalitis and ovarian Teratoma: results from the Queensland autoimmune encephalitis cohort. Intern Med J. 2021;52:1943.
15. Linnoila J, Pulli B, Armangué T, Planagumà J, Narsimhan R, Schob S, et al. Mouse model of anti-NMDA receptor post-herpes simplex encephalitis. Neurology(R) Neuroimmunol Neuroinflamm. 2019;6(2):e529.
16. Huang YQ, Xiong H. Anti-NMDA receptor encephalitis: a review of mechanistic studies. Int J Physiol Pathophysiol Pharmacol. 2021;13(1):1–11.
17. Matute C, Palma A, Serrano-Regal MP, Maudes E, Barman S, Sánchez-Gómez MV, et al. N-methyl-D-aspartate receptor antibodies in autoimmune encephalopathy Alter oligodendrocyte function. Ann Neurol. 2020;87(5):670–6.
18. Dalmau J. NMDA receptor encephalitis and other antibody-mediated disorders of the synapse: the 2016 Cotzias lecture. Neurology. 2016;87(23):2471–82.
19. Dendrou CA, Fugger L, Friese MA. Immunopathology of multiple sclerosis. Nat Rev Immunol. 2015;15(9):545–58.
20. Hemmer B, Kerschensteiner M, Korn T. Role of the innate and adaptive immune responses in the course of multiple sclerosis. Lancet Neurol. 2015;14(4):406–19.
21. Stys PK, Tsutsui S. Recent advances in understanding multiple sclerosis. F1000Res. 2019;8:F1000.
22. Plastini MJ, Desu HL, Brambilla R. Dynamic responses of microglia in animal models of multiple sclerosis. Front Cell Neurosci. 2020;14:269.
23. Wesselingh R, Butzkueven H, Buzzard K, Tarlinton D, O'Brien TJ, Monif M. Innate immunity in the central nervous system: a missing piece of the autoimmune encephalitis puzzle? Front Immunol. 2019;10:2066.

24. Dong Y, Yong VW. When encephalitogenic T cells collaborate with microglia in multiple sclerosis. Nat Rev Neurol. 2019;15(12):704–17.
25. Sosa RA, Murphey C, Ji N, Cardona AE, Forsthuber TG. The kinetics of myelin antigen uptake by myeloid cells in the central nervous system during experimental autoimmune encephalomyelitis. J Immunol. (Baltimore, Md: 1950). 2013;191(12):5848–57.
26. Wolf Y, Shemer A, Levy-Efrati L, Gross M, Kim JS, Engel A, et al. Microglial MHC class II is dispensable for experimental autoimmune encephalomyelitis and cuprizone-induced demyelination. Eur J Immunol. 2018;48(8):1308–18.
27. Yamasaki R, Lu H, Butovsky O, Ohno N, Rietsch AM, Cialic R, et al. Differential roles of microglia and monocytes in the inflamed central nervous system. J Exp Med. 2014;211(8):1533–49.
28. Ajami B, Bennett JL, Krieger C, McNagny KM, Rossi FM. Infiltrating monocytes trigger EAE progression, but do not contribute to the resident microglia pool. Nat Neurosci. 2011;14(9):1142–9.
29. Goldmann T, Wieghofer P, Müller PF, Wolf Y, Varol D, Yona S, et al. A new type of microglia gene targeting shows TAK1 to be pivotal in CNS autoimmune inflammation. Nat Neurosci. 2013;16(11):1618–26.
30. Voet S, Prinz M, van Loo G. Microglia in central nervous system inflammation and multiple sclerosis pathology. Trends Mol Med. 2019;25(2):112–23.
31. Ponomarev ED, Shriver LP, Dittel BN. CD40 expression by microglial cells is required for their completion of a two-step activation process during central nervous system autoimmune inflammation. J Immunol. (Baltimore, Md: 1950). 2006;176(3):1402–10.
32. Parodi B, Rossi S, Morando S, Cordano C, Bragoni A, Motta C, et al. Fumarates modulate microglia activation through a novel HCAR2 signaling pathway and rescue synaptic dysregulation in inflamed CNS. Acta Neuropathol. 2015;130(2):279–95.
33. Ludwin SK, Rao V, Moore CS, Antel JP. Astrocytes in multiple sclerosis. Multiple Sclerosis (Houndmills, Basingstoke, England). 2016;22(9):1114–24.
34. Linker RA, Lee DH, Demir S, Wiese S, Kruse N, Siglienti I, et al. Functional role of brain-derived neurotrophic factor in neuroprotective autoimmunity: therapeutic implications in a model of multiple sclerosis. Brain J Neurol. 2010;133(Pt 8):2248–63.
35. Ponath G, Park C, Pitt D. The role of astrocytes in multiple sclerosis. Front Immunol. 2018;9:217.
36. Michel L, Touil H, Pikor NB, Gommerman JL, Prat A, Bar-Or A. B cells in the multiple sclerosis central nervous system: trafficking and contribution to CNS-compartmentalized inflammation. Front Immunol. 2015;6:636.
37. Ponath G, Lincoln MR, Levine-Ritterman M, Park C, Dahlawi S, Mubarak M, et al. Enhanced astrocyte responses are driven by a genetic risk allele associated with multiple sclerosis. Nat Commun. 2018;9(1):5337.
38. Mayo L, Trauger SA, Blain M, Nadeau M, Patel B, Alvarez JI, et al. Regulation of astrocyte activation by glycolipids drives chronic CNS inflammation. Nat Med. 2014;20(10):1147–56.
39. Marzan DE, Brügger-Verdon V, West BL, Liddelow S, Samanta J, Salzer JL. Activated microglia drive demyelination via CSF1R signaling. Glia. 2021;69(6):1583–604.
40. Taylor LC, Puranam K, Gilmore W, Ting JP, Matsushima GK. 17beta-estradiol protects male mice from cuprizone-induced demyelination and oligodendrocyte loss. Neurobiol Dis. 2010;39(2):127–37.
41. Masuda T, Sankowski R, Staszewski O, Böttcher C, Amann L, Sagar, et al. Spatial and temporal heterogeneity of mouse and human microglia at single-cell resolution. Nature. 2019;566(7744):388–92.

Chapter 11
Adult Glial Cell Proliferation and Neurogenesis

Abstract The nomenclature of stem and progenitor cells is reviewed. In the adult body, only glial cells and not neurons can proliferate. The glial cell with the highest proliferation rate is the oligodendrocyte progenitor followed by microglia. Astrocytes have a very low rate and brain macrophages do not divide inside the parenchyma. All these proliferation rates increase after injury, but they seem to be under neuronal control. This is most obvious in the case of oligodendrocyte progenitors which are on the receiving side of synapses with neurons. In three brain areas, glial cell-like stem cells can – through various proliferation steps and migration patterns – create new neurons. A minority of these neurons are incorporated into functional circuits. Injury can – depending on the species, location, and type of injury – increase adult neurogenesis. It is not clear if these mechanisms lead to functional replacement of damaged neurons. However, efforts are made to use these attempts for therapeutic interventions.

Keywords Adult neurogenesis · Brain cell proliferation · Hypothalamic ventricular zone · Injury-induced proliferation · Oligodendrocyte progenitor cells · Subependymal zone · Subgranular zone of the hippocampus

11.1 Introduction

Before going into detail, it is important to define the terminology of brain cell lineage. There are no generally agreed-upon rules, just widely used terms. The term *precursor cell* is used for both *stem cells* and *progenitor cells*. Stem cells have several unique properties. There is first the ability to self-renew. For this they have two optional cell division modes. One is a symmetrical cell division, leading to two new stem cells. The resulting new stem cells are used for maintaining the stem cell pool. The other division mode option is asymmetrical, leading to a new stem cell and a daughter cell. The daughter cell is called progenitor cell and is the first step to differentiation. Stem cells can be multipotent or pluripotent. The daughter cells of pluripotent stem cells can differentiate into all cell types of body organs and tissues. Multipotent stem cells are far more restricted, but they can create daughter cells that

W. Walz, *The Gliocentric Brain*, https://doi.org/10.1007/978-3-031-48105-5_11

can differentiate into at least two different cell types depending on local conditions. The daughter or progenitor cells can differentiate over several cell divisions into different cell type lineages. They also normally form a large pool by self-renewal. All these options are subject to regulatory signals and are therefore dependent on environmental factors. In the end a progenitor cell undergoes a symmetrical division into two differentiated cells and therefore ceases to exist as a functional unit. New progenitor cells can of course be provided by the stem cells if conditions are right.

The main cellular features of neural development were reviewed in a previous chapter. Radial glial cells are the main neural stem cells in the central nervous system. They generate all neurons, astrocytes, oligodendrocytes/NG2 (neuron-glial antigen 2) cells, tanycytes, and ependymal cells, but not microglia. In the adult brain, glial cells (astrocytes, oligodendrocyte lineage, and microglia) can divide to various degrees, depending on circumstances, exhibiting large differences among the different glial cell types (discussed later). In contrast neurons are not capable of dividing in the adult brain; they are terminally differentiated and post-mitotic. However, there are some exceptional zones in the adult brain, which stand out from this general pattern, all involving glial cells as main drivers. The radial glial cells are present in the whole embryonic brain and serve as neural stem cells. However, in the adult brain, there are three distinct niches where radial glia-like (RGL) cells survive and stay functional, albeit in quite a different context: the subependymal zone (SEZ) in the lateral wall of the lateral ventricle, the subgranular zone (SGZ) of the hippocampus and the hypothalamic ventricular zone (subependymal zone of the third ventricle). There are species differences: the above three zones pertain to rodents and primates. Humans seem to lose the SEZ in adolescence and only the SGZ is functional as a neurogenic zone throughout life [1].

11.2 Proliferation of Adult Glia

In the normal adult brain, the cell type with the highest proliferation rate is the NG2 cell type or OPC (oligodendrocyte progenitor cell) or polydendrocyte. This oligodendrocyte progenitor role is important as mature oligodendrocytes are post-mitotic and cannot divide. NG2 glia are homogeneously distributed throughout the whole adult brain and tend to keep a certain minimal distance from each other. Their proliferation rate is slow; it takes several weeks to finish a single cell cycle [2]. However, their cellular properties suggest different cell subsets [3]. Their function in the white matter of the brain is clearly the generation of myelinating oligodendrocytes. They have an additional function other than serving as OPCs in white matter. It has been shown that NG2 glia form synapses with neurons in some brain regions [4]. The neuronal partner forms the presynaptic part while the glial partner forms the postsynaptic part, leading to a signal transfer from neuron to glia. Glutamatergic and GABAergic synapses are found on the same OPC. Miniature excitatory synaptic potentials have also been recorded in NG2 glia as a response to neuronal stimulation. Thus, the OPCs are functioning on the receiving end of a neural circuit just like a neuron [5]. OPCs have the choice of self-renewal or differentiation into oligodendrocytes depending on the neuronal signals. Evidence is now accumulating that glutamatergic neuronal input promotes OPC proliferation, differentiation into oligodendrocytes and myelination for most, but maybe not all, OPCs. Some neurons need to activate OPCs with glutamatergic synaptic input to enhance myelination. Release of neuronal BDNF (brain-derived neurotrophic factor) as an additional stimulant seems to potentiate this mechanism [6]. Decreased neuronal activity reduces OPC differentiation into oligodendrocytes and myelination. After demyelinating injury in the adult animal, the increased neuronal activity on the neuron-OPC synapse, together with BDNF release, are crucial in enhancing remyelination [7]. Thus, the glial cell most widely distributed in the brain with the highest proliferation rate, is under strict neuronal control. In traumatic brain injury the OPCs divide readily near the lesion, but only once and they migrate short distances [2]. As will be seen in the subsequent chapter, a cell type with such a proliferation potential can escape regulation and cause malignancy. There are clues that OPC proliferation is heavily contributing to some forms of malignant glioma and that glutamatergic neuronal input is involved [8]. This fact only underlines that the cell type with the highest proliferation rate in the normal brain must be under tight control and integration of the overall brain function.

There is not much evidence for astrocyte proliferation in the adult, healthy brain. Each astrocyte occupies a domain with close contact to other astrocytic domains, but without overlap. However, thalamic and hypothalamic astrocytes seem to be an exception [9]. Here, the astrocytic proliferation rates in the absence of injury are significantly higher than of other parenchymal astrocytes. It seems that like adult neurogenesis (discussed later), not all newborn astrocytes reach functional maturity. There is no indication that adult astrocytes in the healthy brain can serve as neural stem cells either in vivo or in vitro. Hypothalamic astrocytes have a functional role reaching above local homeostasis. They show functional plasticity related to metabolic and homeostatic control of the whole body [10]. In the injured brain astrocyte proliferation is more widespread, but only in cases where astrogliosis is intense. During severe diffuse astrogliosis, there are indications of minor astrocyte proliferation. It is different in traumatic brain injury like a stab wound (but not puncture wound) with compact scar formation and the appearance of barrier astrocytes. Most of these reactive astrocytes accumulate at the lesion border due to proliferation and not due to migration [11]. The reactive astrocytes create two daughter cells which remain close together. Such stab wound injuries are accompanied by macrophage, but not T or B lymphocyte invasion. The macrophage invasion has an inhibitory effect on the proliferation of these reactive (barrier) astrocytes [12]. This might be the reason why after a stab wound the astrocyte proliferation leads only to an increase of 20% of the number of astrocytes in the lesion rim. This is lower than the corresponding OPC (NG2 cell) and microglia proliferation in this situation. Likewise, an experimentally induced increase in astrocyte proliferation rate reduces the macrophage invasion. The resulting increase in reactive (barrier) astrocytes around a stab wound lesion had a beneficial effect on the wound healing process [12]. In vitro some isolated astrocytes react differently: if exposed to special growth conditions they can turn into neural stem cells. In such an environment, but not in vivo, a small pool of reactive astrocytes can serve as neural stem cells for neurons, oligodendrocytes, and more astrocytes [13]. This means *some* reactive, but not normal astrocytes have the potential to create other cell types. This potential is not called up in the brain after injury. However, it might be used during manipulations for therapeutic purposes.

Microglia proliferate at a basic rate in the healthy and adult brain. This is due to self-renewal as there are no microglial progenitors in the adult brain [14]. Due to apoptosis the microglia density is kept constant throughout the brain's lifetime, although the mechanisms are not known. In the mouse brain approximately 0.7% of all microglia are in the cell cycle at any time. This translates into a complete renewal of the mouse brain microglia population every 95 days. In the human brain the rate is even higher with 2%. However, compared with other residential macrophage types in the body, this is still on the low side [15]. The proliferation rates are not uniformly distributed, the dentate gyrus (DG) has the highest rate. The death rate for newborn microglia (first 5 days after division) is higher than for longtime resident microglia (5.0 compared to 2.4%). In the healthy brain, microglial turnover must be finely balanced between proliferation and apoptosis. Interleukin-1 receptors and colony-stimulating factor 1 receptors seem to play an important role as their genetic ablation leads to unstable microglia populations [16].

Microglia proliferation is part of its activation process after injury and constitutes a major part of reactive microgliosis. This proliferation seems to be like the one in the healthy brain. It is more intense, involves migration and is oriented towards a lesion injury. In injuries where the integrity of the blood–brain barrier is compromised, macrophages invade the lesion, but they do not divide within the brain. They have a transient appearance and die within days of an injury due to apoptosis [17]. Microglia show local proliferation in addition to migration around a lesion. Within the first days of an ischemic lesion about 20–30% of the microglia around the lesion enter the cell cycle at any time [18]. Colony stimulating factor 1 and Toll-like receptor 2 activation are crucial for proliferation of microglia in seizure lesions [19].

11.3 Adult Neurogenesis

The concept that mature and terminally differentiated neurons could re-enter the cell cycle and divide is not accepted. As pointed out, neurons (and mature oligodendrocytes) are permanently post-mitotic. More recent is the concept that self-renewing and multipotent neural stem cells survive into adulthood [20]. They initiate a process that leads to new functional neurons. However, in its essence, adult neurogenesis (the ability to generate new neurons in the adult brain) is a glial process [21]. It creates highly restricted and specialized neurons, of which a minority can integrate into circuits. During embryonic development and in the adult brain it is the radial glial cell type that functions as neural stem cell. The marker profile and morphology of these adult neural stem cells is similar to embryonic radial glia cells. The cells are considered highly specialized astrocytes [21]. They are referred to as RGL stem cells, more specifically as type-1 cells in the SGZ, and as B cells in the SVZ. An RGL can generate an intermediate, transit-amplifying progenitor cell (TAP). The TAPs do not resemble RGCs or even glia cells anymore and they divide in order to increase their population to a large pool before further differentiation steps [22]. The accepted nomenclature uses letters for the SVZ and numbers for the SGZ. In the three adult neurogenic brain regions, the neural stem cells have, although being multipotent, a considerable fate restriction in vivo. They are mainly geared to the task of generating highly specialized neurons [13]. In vitro manipulation and in vivo treatment with growth factors results in a far more flexible reprogramming of the progeny with exciting avenues for therapeutic approaches [23].

11.3.1 Subependymal Zone

In the SVZ, the RGLs are called B cells and with their processes are in direct contact with the cerebrospinal fluid (CSF). They receive signals through the CSF from the choroid plexus (CP). The resulting TAPs are called C cells. The C cells are fast dividing, resulting in thousands of new cells per day in rodents. The daughter cells are almost all neuroblasts (called type A cells). A small amount of the daughter cells are OPCs which radially move out of the SVZ and migrate into adjacent areas to turn into functional oligodendrocytes. The rostral migratory stream (RMS) is a tubular structure, which connects the SVZ with the olfactory bulb (OB). The tube consists mainly of specialized astrocytes, which are elongated and connected to each other. The neuroblasts migrate within the RMS in a process called chain migration. The neuroblasts differentiate during the migration slowly under the influence of the tubular astrocytes. In the OB the neuroblasts terminally differentiate into GABAergic or glutamatergic neurons (with a few dopaminergic neurons) and integrate into the circuitry [24]. However, only a minority of the neuroblasts are integrated as functional neurons. There is an overproduction of neuroblasts with most being removed during the migration in the RMS and OB by apoptosis. There is also apoptosis of mature and older interneurons in the OB in order that they can be replaced by new neurons without increasing neuronal numbers. There is therefore a steady turnover. In the OB the neuroblasts use the blood vessels as guides for migration and to find the correct location. At most 10% of the olfactory neurons get replaced each month in rodents. Experimentally enriched odor exposure of rodents increases this percentage of neurons being replaced. Thus, the turnover of neurons is most likely connected to olfactory learning and memory. The migration and differentiation of the neuroblasts (A cells) in the RMS is controlled by attractant factors (netrin-1, prokineticin 2) and repulsive factors (Slit proteins). Other signals are ephrins, integrins, GABA, BDNF, GDNF (glial cell line-derived neurotrophic factor), and VEGF (vascular endothelial growth factor) [25].

11.3.2 Subgranular Zone of the Hippocampus

The hippocampal DG has three layers. The molecular layer consists of dentate granule cells and axons from the perforant path. The granule cell layer contains the granule cells and the hilus has various cell types including mossy cells. The SGZ is located between the granule cell layer and the hilus. The neuronal stem cells (NSCs) are RGLs (called type 1 cells in the SGZ) with processes in contact with blood vessels. Other cells in this interphase are the NPCs/TAPs (type 2 cells) and GABAergic basket cells. The basket cells release GABA (gamma-aminobutyric acid) to inhibit the activity of the type 1 cells and prevent them from dividing. The few that divide create – other than self-renewal – the type 2 cells. The type 2 cells are also under control by the basket cells, but here GABA release has an excitatory function and causes strong proliferation activity. The daughter cells are called type 3 cells and express neuronal markers, but the proliferation rate is low. Type 3 cells migrate into the granule cell layer and differentiate into post-mitotic granule cells. They continue differentiation and extend axons to other layers. Thus, through the basket cells the neurogenesis rate is controlled by hippocampal activity. New neurons are more likely to induce long-term potentiation (LTP) as they have lower thresholds than older resident neurons. Later, the GABAergic input from the basket cells switches from an excitatory to an inhibitory effect. Most of these new, low LTP thresholds exhibiting neurons are eliminated by apoptosis and only a minority can establish full mature contacts exhibiting normal LTP. Of the approximate 10,000 type 2 cells generated in rodents every day, only a small percentage gets permanently incorporated. Neurogenesis is important for novelty recognition, fear conditioning, spatial information processing, and memory formation. Inhibition of SGZ neurogenesis has detrimental effects on these processes [20].

11.3.3 Hypothalamic Ventricular Zone

There is another adult neurogenic area in the hypothalamic wall of the third ventricle and its vicinity. The NSCs are a small subpopulation of the tanycytes lining the ventricle and acting like RGLs. They have unlimited self-renewal rates and create progenitor cells which migrate into the hypothalamic parenchyma. They can reside there for a long time. They can also differentiate into other cell lineages and migrate to other hypothalamic destinations. They can differentiate into neuroblasts, astroblasts, or oligodendrocytic blasts. Existing neurons, which are replaced, are removed by apoptosis. New astrocytes are involved in regulation of the sleep-wake cycle and other whole-body homeostatic activities. New neurons differentiate into sleep-wake regulatory activities, circadian rhythms, energy, temperature, and body weight control. In short, they are incorporated into all the hypothalamic activities. Axonal endings in the hypothalamic parenchyma influence the differentiation, proliferation, and migration of the new cells. Therefore, they are efficiently integrated into existing functional building blocks. Proneuronal transcription factors are ASCL 1 (achaete-scute homolog 1) and neurogenin-2. Notch-1 shifts NSCs into proliferative states, while notch-2 inhibits this activity. Cytokines and growth factors serve as additional differentiation and proliferation signals. The molecular mechanisms responsible for guiding the new neurons and astrocytes to their location are still unknown. The neurogenic zone of the hypothalamus was only recently identified [26].

11.3.4 Adult Neurogenesis After Injury

There is evidence that after brain injury for both, the SVZ and SGZ dispatching differentiating neuroblasts toward the injured area. Evidence comes from lesions due to traumatic brain injury, seizures, Alzheimer's disease, and Huntington's disease. However, the best explored scenarios are those following a stroke in rodents and humans [27]. This paragraph therefore focuses on stroke-related adult neurogenesis.

From 2 to 28 days after a stroke, the SVZ increases the RGL division from asymmetric to symmetric to increase its population size. This division is not widespread. It was found that an RGL subpopulation, that is normally quiescent and dormant, is far more responsive in reacting with cell division to a remote ischemic lesion than the remainder of the population [28]. The proliferation of type-C cells is increased as is their differentiation into neuroblasts. This happens bilaterally even so the stroke is only in one hemisphere [29]. This points to the role of soluble factors released into the CSF. One known factor is high mobility group box 1 (HMGB1). Neuroblasts then leave the tubular structure of the RMS in migrate to the striatum if the lesion is in this brain region. If the lesion is in the cortex the neuroblasts move along the corpus callosum into the cortex. The migration is similar to the RMS chain migration and along blood vessels and astrocytes [30]. Factors involved in the attraction to the site of injury are stromal-derived factor 1α and monocyte chemoattractant factor 1. While the processes in the SVZ occur bilateral despite the unilateral injury, the migration out of the RMS appears only in the stroke-affected hemisphere. The destiny of these neuroblasts is highly varied. Most disappear during the migration probably due to apoptosis. A minority forms new astrocytes which become part of the glial scar tissue around the lesion. Only a very small proportion (less than 1%) incorporates as fully functional neurons into the damaged circuits. Yet, this small incorporation seems to be of functional significance. Inhibition of SVZ neurogenesis after stroke interferes negatively with recovery and is detrimental to neurological outcome after recovery. Transplantation of neural precursors on the other hand improves this outcome [30].

The situation in the SGZ is quite different. Different unilateral ischemia models all cause an up to tenfold bilateral increase of proliferation within the SGZ neurogenic zone. Again, many of the new cells are eliminated, but a minority differentiate into granular cells. These granular cells migrate into the granular cell layer. Interestingly, irrespective of the location of the ischemic lesion, the differentiating neurons migrate into the granular cell layer. This happens irrespectively if the hippocampus is damaged by the stroke and if the ischemic lesion is quite remote from the hippocampus. Unlike the neuroblasts from the SVZ, these SGZ-derived neuroblasts do not migrate to the boundary of the ischemic lesion. The neurons which incorporate into the granular cell layer are functional and participate in the function of circuits. However, many of these incorporated neurons show abnormal functional properties and are aberrantly integrated into the hippocampal circuits. The affected neurons have abnormal dendritic structures. They contribute to hyperexcitability of the circuits. It was found that the percentage of aberrant neurons and the stroke lesion size are positively correlated irrespective of the lesion site [31]. In contrast to SVZ, where migration toward the lesion occurs only ipsilaterally, the SGZ migration and incorporation into hippocampal circuits is bilateral [32]. In an experimental ischemia mouse model, Cuartero et al. [33] abolished SGZ neurogenesis without affecting other basal neurogenesis processes. When tested long-term after the ischemia, the mice with inhibited SGZ neurogenesis had a better spatial memory recall than controls. Thus, this SGZ-based injury-induced neurogenesis has a detrimental effect and might contribute to the cognitive deficits observed after a stroke. It must be considered as aberrant post-stroke neurogenesis. Interestingly, in a seizure model the SGZ RGLs display symmetric division and instead of maintaining the neuron progenitor pool, the daughter cells differentiate into astrocytes, contributing to reactive gliosis and not aberrant neural circuits [34]. This is an indication that the injury-induced neurogenesis is complicated and highly dependent on the context.

11.4 Conclusions

Differentiated neurons and oligodendrocytes are post-mitotic and do not divide. To increase or replenish their numbers, they need glial progenitors. These are normally either OPCs or RGLs. The creation of new astrocytes and OPCs is more plastic and therefore flexible. In contrast microglia are in a different league due to their immune lineage. To no surprise, the incorporation of newly differentiated neurons into established circuits is difficult and therefore the exception. Problems leading to aberrant circuits are common.

There are two important consequences from these conclusions. First, as restricted the lineages appear, the microenvironment has an influence on the lineage development. This opens the doors to therapeutic manipulation in vitro and in vivo. Second, these controlled stereotype-like proliferation and differentiation processes have the potential to get out of control and cause malignant growth. This will be a problem due to the compartmentation and restricted cavity of the brain.

References

1. Bonfanti L, Peretto P. Adult neurogenesis in mammals–a theme with many variations. Eur J Neurosci. 2011;34(6):930–50.
2. von Streitberg A, Jäkel S, Eugenin von Bernhardi J, Straube C, Buggenthin F, Marr C, et al. NG2-glia transiently overcome their homeostatic network and contribute to wound closure after brain injury. Front Cell Dev Biol. 2021;9:662056.
3. Vigano F, Mobius W, Gotz M, Dimou L. Transplantation reveals regional differences in oligodendrocyte differentiation in the adult brain. Nat Neurosci. 2013;16(10):1370–2.
4. Eugenin-von Bernhardi J, Dimou L. NG2-glia, more than progenitor cells. Adv Exp Med Biol. 2016;949:27–45.
5. Monje M, Káradóttir RT. The bright and the dark side of myelin plasticity: neuron-glial interactions in health and disease. Semin Cell Dev Biol. 2021;116:10–5.
6. Gibson EM, Purger D, Mount CW, Goldstein AK, Lin GL, Wood LS, et al. Neuronal activity promotes oligodendrogenesis and adaptive myelination in the mammalian brain. Science (New York, NY). 2014;344(6183):1252304.
7. Ortiz FC, Habermacher C, Graciarena M, Houry PY, Nishiyama A, Nait Oumesmar B, et al. Neuronal activity in vivo enhances functional myelin repair. JCI insight. 2019;5(9):e123434.
8. Galvao RP, Kasina A, McNeill RS, Harbin JE, Foreman O, Verhaak RG, et al. Transformation of quiescent adult oligodendrocyte precursor cells into malignant glioma through a multistep reactivation process. Proc Natl Acad Sci U S A. 2014;111(40):E4214–23.
9. Ohlig S, Clavreul S, Thorwirth M, Simon-Ebert T, Bocchi R, Ulbricht S, et al. Molecular diversity of diencephalic astrocytes reveals adult astrogenesis regulated by Smad4. EMBO J. 2021;40:e107532.
10. García-Cáceres C, Balland E, Prevot V, Luquet S, Woods SC, Koch M, et al. Role of astrocytes, microglia, and tanycytes in brain control of systemic metabolism. Nat Neurosci. 2019;22(1):7–14.
11. Bardehle S, Krüger M, Buggenthin F, Schwausch J, Ninkovic J, Clevers H, et al. Live imaging of astrocyte responses to acute injury reveals selective juxtavascular proliferation. Nat Neurosci. 2013;16(5):580–6.
12. Frik J, Merl-Pham J, Plesnila N, Mattugini N, Kjell J, Kraska J, et al. Cross-talk between monocyte invasion and astrocyte proliferation regulates scarring in brain injury. EMBO Rep. 2018;19(5):e45294.
13. Götz M, Sirko S, Beckers J, Irmler M. Reactive astrocytes as neural stem or progenitor cells: In vivo lineage, In vitro potential, and Genome-wide expression analysis. Glia. 2015;63(8):1452–68.
14. Huang Y, Xu Z, Xiong S, Sun F, Qin G, Hu G, et al. Author correction: repopulated microglia are solely derived from the proliferation of residual microglia after acute depletion. Nat Neurosci. 2021;24(2):288.
15. Askew K, Li K, Olmos-Alonso A, Garcia-Moreno F, Liang Y, Richardson P, et al. Coupled proliferation and apoptosis maintain the rapid turnover of microglia in the adult brain. Cell Rep. 2017;18(2):391–405.
16. Bruttger J, Karram K, Wörtge S, Regen T, Marini F, Hoppmann N, et al. Genetic cell ablation reveals clusters of local self-renewing microglia in the mammalian central nervous system. Immunity. 2015;43(1):92–106.
17. Li T, Pang S, Yu Y, Wu X, Guo J, Zhang S. Proliferation of parenchymal microglia is the main source of microgliosis after ischaemic stroke. Brain J Neurol. 2013;136(Pt 12):3578–88.
18. Khan A, Ju F, Xie W, Tariq Hafeez M, Cheng X, Yang Z, et al. Transcriptomic analysis reveals differential activation of microglial genes after ischemic stroke in mice. Neuroscience. 2017;348:212–27.
19. Hong J, Cho IH, Kwak KI, Suh EC, Seo J, Min HJ, et al. Microglial Toll-like receptor 2 contributes to kainic acid-induced glial activation and hippocampal neuronal cell death. J Biol Chem. 2010;285(50):39447–57.

20. Ribeiro FF, Xapelli S. An overview of adult neurogenesis. Adv Exp Med Biol. 2021;1331:77–94.
21. Schneider J, Karpf J, Beckervordersandforth R. Role of astrocytes in the neurogenic niches. Methods Mol Biol (Clifton, NJ). 2019;1938:19–33.
22. Götz M. Revising concepts about adult stem cells. Science (New York, NY). 2018;359(6376):639–40.
23. Matsubara S, Matsuda T, Nakashima K. Regulation of adult mammalian neural stem cells and neurogenesis by cell extrinsic and intrinsic factors. Cell. 2021;10(5):1145.
24. Gengatharan A, Bammann RR, Saghatelyan A. The role of astrocytes in the generation, migration, and integration of new neurons in the adult olfactory bulb. Front Neurosci. 2016;10:149.
25. Uzquiano A, Gladwyn-Ng I, Nguyen L, Reiner O, Götz M, Matsuzaki F, et al. Cortical progenitor biology: key features mediating proliferation versus differentiation. J Neurochem. 2018;146(5):500–25.
26. Kostin A, Alam MA, McGinty D, Alam MN. Adult hypothalamic neurogenesis and sleep-wake dysfunction in aging. Sleep. 2021;44(2):zsaa173.
27. Ceanga M, Dahab M, Witte OW, Keiner S. Adult neurogenesis and stroke: a tale of two neurogenic niches. Front Neurosci. 2021;15:700297.
28. Llorens-Bobadilla E, Zhao S, Baser A, Saiz-Castro G, Zwadlo K, Martin-Villalba A. Single-cell transcriptomics reveals a population of dormant neural stem cells that become activated upon brain injury. Cell Stem Cell. 2015;17(3):329–40.
29. Palma-Tortosa S, Hurtado O, Pradillo JM, Ferreras-Martín R, García-Yébenes I, García-Culebras A, et al. Toll-like receptor 4 regulates subventricular zone proliferation and neuroblast migration after experimental stroke. Brain Behav Immun. 2019;80:573–82.
30. Cuartero MI, García-Culebras A, Torres-López C, Medina V, Fraga E, Vázquez-Reyes S, et al. Post-stroke Neurogenesis: friend or foe? Front Cell Dev Biol. 2021;9:657846.
31. Niv F, Keiner S, Krishna WOW, Lie DC, Redecker C. Aberrant neurogenesis after stroke: a retroviral cell labeling study. Stroke. 2012;43(9):2468–75.
32. Toni N, Schinder AF. Maturation and functional integration of new granule cells into the adult hippocampus. Cold Spring Harb Perspect Biol. 2015;8(1):a018903.
33. Cuartero MI, de la Parra J, Pérez-Ruiz A, Bravo-Ferrer I, Durán-Laforet V, García-Culebras A, et al. Abolition of aberrant neurogenesis ameliorates cognitive impairment after stroke in mice. J Clin Invest. 2019;129(4):1536–50.
34. Sierra A, Martín-Suárez S, Valcárcel-Martín R, Pascual-Brazo J, Aelvoet SA, Abiega O, et al. Neuronal hyperactivity accelerates depletion of neural stem cells and impairs hippocampal neurogenesis. Cell Stem Cell. 2015;16(5):488–503.

Chapter 12
Glioma

Abstract Most adult neurons have exited the cell cycle. Therefore, almost all primary brain cancers in the adult body are related to glial cells. There exists a multitude of glioma characterizations; most of them are based on not only malignancy but also on genotypes and epigenetic changes. There is a genomic instability during the growth of the most aggressive cancers, mostly caused by a molecular subtype migration. Molecular classification is nowadays standard for selecting a treatment schedule. There are two competing, but not exclusive, theories for the genesis of brain cancer cells. The most established is the glioma stem cell theory that promotes that a very small population gives rise to differentiated and genetically unstable progenitor cells. These in turn give rise to many heterogeneous subtypes. Another theory focuses on the oligodendrocyte progenitor cell, the cell type in the brain with the largest proliferative potential. The tumor microenvironment is very supportive of tumor cell growth. The reason is that microglia, macrophages, astrocytes, neurons, blood vessels, and T cells are coopted by the cancer cells in providing a supportive environment. This environment is coordinated by "hijacked" microglia cells. Metastases from body cancer cells can invade the brain. Before they can enter the brain, these circulating cancer cells must undergo a transition phase. After this transition the invading cells change the properties of the normal blood–brain barrier into a functionally different blood tumor barrier. After entering, a small number of these cancer cells migrate within the parenchyma, create colonies, and manipulate the glial cells and their microenvironment to support the growing colonies.

Keywords Epigenetic changes · Glioma stem cells · Metastatic colonization · Molecular subtype migration · Oligodendrocyte progenitor cells · Primary brain tumors · Tumor microenvironment

12.1 Introduction

Cellular proliferation in the adult and healthy brain is – as shown in the previous chapter – not very prominent. For proper brain function it is only necessary in some exceptional cases. Thus, cell proliferation in the adult and healthy brain is well

W. Walz, *The Gliocentric Brain*, https://doi.org/10.1007/978-3-031-48105-5_12

controlled and restricted. Yet, oncogenic transformation of the cell cycle of adult brain cells exists. It affects mainly glial cells as most adult neurons exited the cell cycle. Brain cancers account for approximately 1% of all cancer cases, which might indicate that an aberrant brain cell cycle is not a big problem. However, brain cancers are responsible for approximately 3% of cancer deaths. Whereas the brain cancer rate cannot be rated as high, the mortality rate, once a cancerous growth is initiated, is very high. This results in a 5-year relative survival rate of about 33%. However, brain cancers are very heterogeneous and come in many forms. It is not justified to put them all into one category. There are various classifications, many by clinicians and surgeons to facilitate a treatment plan. For this chapter, the information of the previous chapter given on proliferation and neural stem cells (NSCs) will serve as a starting point. Some rare genetic syndromes are associated with brain tumors, but there is no family history for brain tumors other than genetic susceptibility to cancers in general. The only environmental factor found so far to be associated with brain cancer risk is ionizing radiation.

12.2 Primary Brain Tumors

These are tumors that originate from brain cells. There are central nervous system (CNS) embryonal tumors, many of them of a neuronal nature. As the name implies, these tumors are more frequent in children and rare in adults. These tumors are mainly connected to mutations in the cell lineage during brain development. By far the most common tumor in this group is medulloblastoma, a neuronal cancer which together with leukemia is the biggest threat to children's health. It arises from oncogenic transformation of neuronal precursors in the cerebellum. There are at least four subgroups with different precursors in defined areas of the cerebellum and with different mutations [1]. Other cancers in this group of embryonal tumors are multilayered rosettes (mainly cerebrum), medulloepitheliomas (brain and spinal cord), and atypical teratoid/rhabdoid tumors (cerebellum). Meningiomas are the most common brain cancer, but these meningiomas are largely nonmalignant with a very good survival rate. Gliomas are found in all age groups, but in the adult brain, glioma is the most prevalent malignant cancer type. They make up at least half of all malignant brain cancers and the survival rate for the deadliest form (glioblastoma) is 8–15 months after diagnosis depending on the treatment [2]. This chapter therefore deals mainly with glioma.

12.3 Gliomas

12.3.1 Characterization of Gliomas

Gliomas are a very heterogeneous group of cancers and their origin is highly controversial. The World Health Organization (WHO) developed a grading system to distinguish benignity from malignancy. This system was originally based on pathological and histological analysis, but now includes molecular markers [3]. Grades 1 and 2 are low grades. Grade 1 is benign, slow growing, noninfiltrative, and curable with surgery alone. Grade 2 has similar criteria, but this grade has the potential to develop into a higher grade. The other two grades are considered high grade. Grade 3 is malignant and infiltrative and has the tendency to develop over time into grade 4. Grade 4 is the most malignant and very aggressive. It is necrosis-prone and widely infiltrated with rapid recurrence. The survival rates are dismal. Grade 4 is called glioblastoma multiforme (GBM). GBM can develop de novo (most common and called primary glioblastoma) or from pre-existing lower-grade gliomas (less frequent and called secondary glioblastoma). These two forms are histologically indistinguishable but have different genomic alterations.

Two theories, which are not mutually exclusive with some overlapping, are being discussed: cancer stem cells and dedifferentiation of committed brain cells.

Almost all adult gliomas are diffusely infiltrating surrounding tissues, except grade 1, which is not a diffuse glioma cell growth [4]. The gliomas are categorized into astrocytomas, oligodendrogliomas, and mixed oligoastrocytomas (grades 2 and 3), as well as diffuse midline glioma (H3 K27M-mutant; most common in children). However, 75% of all gliomas are astrocytomas [5]; this includes the highest grade, GBM. More nondiffuse and circumscribed gliomas are occurring mainly in children and young adults. Ependymomas are included in this group of circumscribed (i.e., nondiffuse) gliomas and appear in all age cohorts.

Genome, epigenetics (deoxyribonucleic acid [DNA] methylation), and tumor microenvironment together cause the glioma growth to be heterogeneous. Almost all intra-tumor glioma growths consist of cancerous cells with genotypes that differ from each other. Therefore, many tumors are heterogeneous, meaning that there is not one cancerous cell type, but a mixture of different tumor cells, due to genomic instability. This genomic instability is a robust phenomenon in almost all gliomas [6]. It is not clear what causes it, but it is especially prominent in recurrent GBM and named *molecular subtype migration*. It is speculated that the widely used chemotherapeutic agent temozolomide and radiation therapy are contributing to this migration. Another hypothesis is that neurosurgery in low-grade early tumor development may switch the proliferative nature to a more migratory pattern [7].

The heterogeneity of GBM is based on the following factors affecting transcription factors and genome [8]: chromosome changes, somatic mutations, and DNA methylation. There are widespread chromosome changes including amplification and deletion within chromosomes. Examples are amplification of cyclin-dependent kinase 6 and deletion of phosphatase and tensin homolog (PTEN). Major somatic mutations in GBM affect tumor protein p53 (34% of mutations), epidermal growth factor receptor (EGFR; 33%), PTEN (32%), neurofibromin 1 (14%), and phosphatidylinositol 4,5-bisphosphate 3-kinase catalytic subunit alpha isoform (12%). There up to 500 further mutations are so far described in GBM [8]. DNA methylation creates a heterogeneous glioma growth pattern [9] and the most important are methylation of the gene for GATA binding protein 6 (occurs in 68% of analyzed patients) and of the promoter of the gene for O′-methylguanine-DNA methyltransferase (MGMT), which occurs in 49% of analyzed GBM patients.

Molecular classification is now standard for selecting a treatment for glioma. There are several overlapping and competing classifications. These are reviewed in reference [8]. Subtypes are classified according to those involving DNA methylation (epigenetics) and mutations to transcription factors as well as isocitrate dehydrogenase (IDH). Mutations other than IDH are based on the promoter for telomerase reverse transcriptase, EGFR amplification, and PTEN mutation. The classifications multiplied in the last 10 years.

Most of the low-grade gliomas are based on mutations of the genes for IDH. All three isozymes (cytosolic IDH1 and mitochondrial matrix IDH2 and IDH3) are involved in NADPH (nicotinamide adenine dinucleotide phosphate) production. NADPH is key for glutathione's role in neutralizing reactive oxygen species and DNA (deoxyribonucleic acid) damage. The IDH mutations are closely associated with two epigenetic consequences. They lead to glioma CpG island methylator phenotype and methylation for the promoter of the gene for MGMT. MGMT repairs some DNA damage. IDH mutations are associated with a favorable treatment prognosis and are absent in GBM, except for some secondary glioblastoma [4]. However, the classification based on IDH is not helpful for primary GBM as these glioma types express the IDH wildtype.

In this chapter I follow the subtype classification by Wang et al. [10]. The authors distinguish three subtypes by analyzing only tumor-intrinsic genes. Therefore, they excluded a neural subtype described by other authors. The reason is that this subtype is not expressed in tumor cells but represents a normal neural lineage contamination. Instead, their transcriptional glioma subtypes cluster into three subtypes called mesenchymal, proneural, and classical. The proneural subtype hallmark is a high expression of the gene for pathways associated with cellular processes like cell circle regulation. Platelet-derived growth factor receptor alpha is highly expressed. The subtype has a relatively favorable outcome due to the prevalence of IDH mutations and is found more in younger patients. The favorable outcome is, however, not based on a better response to chemo- or radiation therapy. The classical subtype has a high expression of the marker for stem cell and neural precursor cell marker nestin. It has prominent aberrations like chromosome 7 amplification, chromosome 10 loss, and inactivation of the retinoblastoma-associated protein. Sonic hedgehog pathways and notch signaling pathways are highly expressed. Untreated, the classical subtype has a poorer outcome than the proneural one, but it responds well to chemo- and radiation therapy. The mesenchymal subtype has a high expression of the genes for pathways associated with the immune response. Vascular endothelial growth factors (VEGF), angiopoietin 1 and 24, tumor necrosis factor superfamily, and nuclear factor kappa B are highly expressed. In addition, it has deletions of several tumor suppressor genes. It is characterized by strong necrosis and inflammation in the microenvironment. Although it has the worst prognosis of all subtypes, it is responsive to aggressive therapy [8].

There is a molecular subtype migration in recurrent GBM as pointed out earlier. In GBM the tumor reoccurs after treatment. Two-thirds of patients with primary GBM switched transcriptional subtype after recurrence [11] with epidermal growth factor expression showing the most reduction. However, the mesenchymal subtype is the most stable.

Temozolomide arrests the cell cycle and therefore causes apoptotic tumor cells. Surgery, radiation therapy, and temozolomide are the main tools in glioma therapy, either alone or in combination. If temozolomide is used in addition to radiation therapy, it lengthens the mean survival time by 4 months compared to radiation alone. However, patients with no MGMT methylation present develop temozolomide resistance and the drug treatment cannot be used.

12.3.2 Glioma Stem Cell Versus Cell-of-Origin

A well-established theory states that cancerous growth is driven by a cancer stem cell in the apex. This stem cell with a potential for self-renewal and proliferation is acting as a progenitor and generator of all cells found in a tumor [12]. The cancer stem cells are relatively resistant to radiation and chemotherapy [13]. They therefore survive aggressive treatment for GBM and are the seeds for renewed and recurrent cancer growth. This fact makes the concept of a glioma stem cell (GSC) more than an academic exercise. The GSC is therefore studied in detail in order to focus therapeutic efforts on it. GSCs are a very small population within a glioma growth, but they give rise to differentiated and genetically unstable progenitor cells. The progenitors of GSCs consist of multiple cellular lineages, which give rise to many of the heterogeneous subtypes as discussed earlier. This means that each GBM has a distinct subgroup pattern which can be traced back to a GSC. The GSCs have developmental programs like expression of transcription factors and chromatin regulators like normal stem cells. These developmental programs act in these GSCs as oncogenes [14]. In the IDH wildtype the GSCs are identified by a varied group of surface markers. The most important is the pentaspan transmembrane glycoprotein CD133 (prominin-1) [15]. The functional properties of the GSCs gave of course rise to the concept that these GSCs are derived from NSCs (neural stem cells). Indeed, it was shown that many gliomas had physical contact with the subventricular zone (SVZ). In radiographic examinations of cancer patients, that showed that GBM in the glioma growth had physical contact with the SVZ, these patients had a worse prognosis than those without contact of the two zones [16]. The theory of an NSC–SVZ–GSC axis is based on very strong evidence. These three cell phenotypes are similar in their gene expression [17] and ablation of tumor suppression genes. In addition, upregulation of growth factors in rodent SVZ leads to glioma growth [18]. GBMs are frequently spreading through the human SVZ [19]. The normal route taken by SVZ progenitor cells is to develop into neurons. This route is, however, never taken. It is speculated that this may be due to IDH1/2 mutations [20]. Inducing mutations in specific cells in mouse models by retroviral injections showed that NSCs are readily turned into glioma cells. Using this approach outside the SVZ proliferative niche did not result in glioma growth [21].

One might think that the existence of GSCs and the above arguments settle the question of the cell-of-origin for a glioma growth in favor of a neural stem or progenitor cells from the SVZ. However, this is not so. Another potential route to arrive at glioma growth is through dedifferentiation of adult glial cells. The most likely candidate is oligodendrocyte progenitor cells (OPCs). In the adult brain under normal conditions OPCs are the parenchyma cell type that shows the highest proliferative potential (see Chap. 11). Human gliomas as well as murine glioma models contain prominent expressions of OPC markers, like neuron-glia antigen 2, oligodendrocyte transcription factor 2, and oligodendrocyte marker 4. The gene expression profile of proneural subtype glioma is close to the OPC profile. Platelet-derived growth factor is a specific mitogen for OPCs in the brain. If it is induced in various progenitors by retroviral vectors, glioma tumors develop as a result [17]. These are strong arguments in favor of some glioma types evolving from OPCs.

Adult astrocytes can divide during strong gliotic responses. They are another candidate as cell-of-origin due to dedifferentiation. The case of astrocytes as cells-of-origin for some gliomas is not as strong as for NPCs and OPCs. In some of the mouse models the specific deletion of tumor suppressor genes in astrocytes causes glioma cell growth in the proliferative niches (SVZ and subgranular layer). Surprisingly, however, about one-fifth of the resulting glioma growth created astrocytoma cells. These astrocytoma cells were outside of the proliferative niches and spread into many brain regions [22].

There seems to be not much doubt that NPCs create a large part of primary glioma, however the case for OPCs is also very strong. This means that gliomas could well have different origins within the brain.

12.3.3 Tumor Microenvironment

Differentiated glioblastoma cells, the fast-dividing daughter cells of the GSCs [23], migrate along blood vessels and invade deep into the brain parenchyma irrespective of their origin. In order to succeed GSCs coopt the support cells of the brain to support the growth of their daughter cells and to shield them from the immune system. Nontransformed cells constitute most of the cells in glioma tissue. The GSCs and the differentiated glioma cells seem to represent only about 20% of the cell mass in a glioma tumor [24]. The largest part consists of microglia, macrophages, and reactive astrocytes. The tumor also induces irregular vascular growth for support. These abnormal blood vessels are the site of a peri-vascular niche for abnormal glioma growth [25]. Blood-borne macrophages, neutrophils, and stromal cells are recruited. This microenvironment serves to support and protect the glioma growth and migration. It is now a major focus of experimental therapeutic efforts. The cancer cells themselves are heterogeneous consisting almost always of all three subtypes (proneural, classical, and mesenchymal). Interestingly, the higher the grade and aggressive the glioma growth, the more abundant are recruited macrophages, neutrophils, and inflammatory status [26].

Normally, cancer cells in the body display mutated molecules on their surface with MHC I (major histocompatibility complex I) receptors. They are recognized by dendritic cells which migrate to secondary lymph nodes and stimulate cytotoxic CD8+ T lymphocytes. These T cells move into the tumor and link up with the antigens displayed on the MHC I molecules. Upon specific contact the cytotoxic CD8+ lymphocytes destroy the cancer cells. Some cancer cells do not display antigens which are recognized as nonself or – due to adaptation of surviving cancer cells – have a low expression of MHC I molecules. They are therefore not targeted by cytotoxic T cells. However, the body has an alternative strategy for these cases. Natural killer cells (NK cells) are T lymphocytes which mature in the bone marrow and secondary lymph nodes. They are independent of non-self-antigens and the MHC I system. They recognize lipid molecules on tumor or stressed cells, but not normal cells, without exposure to antigens. They are activated if their target cells do not display MHC I molecules and they destroy them. This is the basic concept of the immune system dealing with cancerous cells in the body. The question is whether the same principles apply to the central nervous system. As we will see, this concept applies to metastatic brain tumors in the brain as the CD8+ T cells are primed beforehand by the primary tumor outside the brain. These cytotoxic T cells enter in these tumors the brain parenchyma and are numerous in the tumor microenvironment. The immune response to glioblastoma is largely determined by the IDH mutation status. Gliomas with enriched IDH mutant cells are devoid of CD8+ T cells. This may be due to the abundance of microglia cells, which are highly concentrated in the tumor microenvironment and act as suppressors of T cell responses [27]. In mostly IDH wildtype gliomas, macrophages from the periphery are highly abundant, there are less microglia, but still, cytotoxic T cells are lacking. There are indications that NK cells are CNS resident cells, but a majority is probably recruited from the periphery [28]. Glioma cells have altered lipid metabolism. They produce large amounts of ganglioside GD2 (disialoganglioside 2), which causes increased motility and proliferation, as well as GD3 which leads to T cell apoptosis. NK cells are activated by these lipids and become cytotoxic. They are assisted in this task by invading neutrophils [28].

The tumor cells produce VEGF, which leads to increased angiogenesis to supply the glioma growth [29]. Despite this, there is a hypoxic area within the tumor causing necrosis of cells. Angiogenesis, together with physical distortion and inflammation, results in a leaky blood–brain barrier. This permissive blood–brain barrier favors invasion of immune cells. This invasion occurs despite VEGF release resulting in a decrease of vascular adhesion molecules [29].

There is an unexpectedly novel nontumorigenic cell type within the tumor microenvironment. It resembles fibroblasts and is located inside the tumor and in the perivascular area. It lacks specific surface antigens but expresses some fibroblast and mesenchymal markers. It has been given the name glioma-associated stromal cell (GASC) [30]. Its origin is not clear. However, it is speculated that either one of the reactive astrocytes, pericytes/vascular smooth muscle cells, or mesenchymal stem cells might be the cell type of origin. The role of GASCs seems to be to enhance self-renewal and proliferation of GSCs as well as promoting angiogenesis [30].

Glioma-associated microglia (GAMs) and infiltrated macrophages constitute up to half of the cells in the tumor and its microenvironment. The macrophages can enter from the circulation through the blood–brain barrier. They can also use recently discovered [31] vascular channels, which cross the inner skull cortex and connect skull bone marrow and brain tissue. This connection provides an alternative migration path for leukocytes during neuroinflammation. In animal glioma models with experimental microglia depletion glioma growth is decreased [32]. Within tumors GAMs are ameboid shaped, whereas outside they have a ramified appearance [33]. GAMs release factors to stimulate glioma growth (transforming growth factor β, stress-inducible protein-1, and various interleukins), stimulate the migration of tumor cells (induction of platelet-derived growth factor receptor expression on the tumor cells), and suppress apoptosis of tumor cells [32]. This is accomplished by the release of various chemokines by the tumor cells which by acting as chemoattractants recruit GAMs into the tumor microenvironment. Other signals released by the tumor cells lead to a reprogramming of GAMs. This functional reprogramming causes cytokine production, phagocytosis, and secretion of matrix metalloproteinases (MMPs) by GAMs. The upregulated microglial phagocytosis is not directed against tumor cells. This may be mainly due to the expression of CD47 (cluster of differentiation 47) on the tumor cell surface, which acts as a "do not eat me" signal. The MMPs are used to degrade the extracellular matrix to facilitate tumor cell invasion and migration. Different molecular subtypes of glioma tumor cells recruit different mixtures and reprogrammed GAMs. A major vehicle is microglial secretion of interleukin 6, which acts as a mitogen for tumor cells. Thus, different glioma mutations cause different plasticity changes in the function, amount, and type of recruited GAMs [34]. The reprogrammed GAMs are manipulated to support glioma growth and maintenance. This is somewhat surprising, as naïve microglia, that is microglia not previously exposed to tumor cells, have anti-tumorigenic activities [35]. Strategies, which deplete or inhibit microglia within tumors are promising in animal models, but so far, all clinical trials using this approach, failed [32]. This reprogramming of GAMs involves transformation to an anti-inflammatory and immunosuppressive type. The major signals accomplishing this are osteopontin, lactadherin, and fibrinogen-like protein 2 [33]. Interleukin 10 (IL 10) and transforming growth factor β are released by GAMs and downregulate MHC II expression. IL10 also acts on T lymphocytes and causes their apoptosis. Glioma cells release extracellular vesicles that contain messenger ribonucleic acid (mRNA), microRNAs (miRNAs), and proteins. The extracellular space in the tumor microenvironment is quite acidic. This leads the extracellular vesicles to lyse and release their content. One miRNA, miR-21, is an especially powerful signal for reprogramming the microglia into a tumor-supportive type [36]. Some of the proteins released by the extracellular vesicles have proangiogenic effects which promote the sprouting of new blood vessels into the tumor mass.

Not unlike microglia, normal astrocytes are recruited by the glioma tumor cells to assist in the maintenance, growth, and migration of the tumor as well as in the immunosuppression of the innate and T cell-assisted immune reactions [37]. Under this influence, they turn into reactive subtypes. The reactive astrocytes surround the tumor and intermingle with CD8+ and CD4+ lymphocytes [38]. GBM patients with a gene expression pattern, that points to a strong reactive gliosis, have a worse survival rate than those with a corresponding less reactive gene expression. Consequently, the most malignant mesenchymal glioma subtype is associated with highly reactive astrocytes [39]. Furthermore, it has been shown that the different molecular glioma subtypes induce reactive astrocytes with different, but matching properties. The general functional repercussions are not yet clear, except for some correlations. The mesenchymal glioma cell type seems to be associated with a reactive astrocytic pattern known for supporting synaptogenesis during development. It also seems to be correlated with tumor-induced seizures [40]. Another reactive astrocyte subtype, which expresses osteopontin, is located mainly around the perivascular region of a tumor. Osteopontin is a ligand for CD44 on GSCs. Both high osteopontin and CD44 expression correlate with a more negative survival prognosis [41]. There are also physical cell-to-cell contacts between GSCs and reactive astrocytes. The signals exchanged by these contacts induce astrocytes to release cytokines which in turn promote tumor growth [42]. Tumor cells and microglia seem to induce a phosphorylated STAT3+ (signal transducer and activator of transcription 3) immunosuppressive reactive astrocyte type, which among other actions releases immunosuppressive cytokines (IL10, TGFβ; transforming growth factor beta) [43]. Microglia and reactive astrocytes in the tumor microenvironment have positive feedback loops: both release cytokines that promote each other's activation and as a result of this activation, promote the expansion of the tumor growth [44].

What about the health and activity of neuronal circuits? The impact of the tumor growth on these activities will cause the neurological symptoms associated with brain tumors. One might assume that in contrast to all the other involved cell types, neurons are passive bystanders at the receiving end of the impact of tumor growth. However surprisingly, the case is not quite as simple. It has been demonstrated with ex vivo human glioma tissue and in vivo animal models that neuronal activity promotes glioma growth. The main factor released by active neurons is synaptic protein neuroligin-3. Its release is activity-dependent, and it has a mitogenic effect on glioma tumors. Glioma xenografts have arrested growth in mouse brains with deleted neuroligin-3 [45]. Another activity-related secretion product by neurons with a similar impact on glioma is brain-derived neurotropic factor (BDNF). Glioblastoma patients are at high risk for developing seizures [46]. It turns out that the development of seizures in glioma patients and animal models is correlated with the appearance of a reactive astrocyte subtype, which promotes synaptogenesis, as discussed earlier [40]. The appearance of this reactive astrocyte type correlates with specific molecular patterns in GSCs and with synaptogenesis. It appears likely that in certain gliomas GSCs cause the reprogramming of some astrocytes to a reactive subtype that promotes synaptogenesis between neurons. This synaptogenesis causes hyperexcitability in neurons, which in turn will eventually lead to seizures [40]. There is another mechanism that contributes to neuronal hyperexcitability: glutamate release by GSCs [47]. Thus, this is another positive feedback loop based on glutamate. Growing glioma cells release glutamate and promote synaptogenesis-promoting astrocytes. This glutamate and increased synaptogenesis in turn cause increased

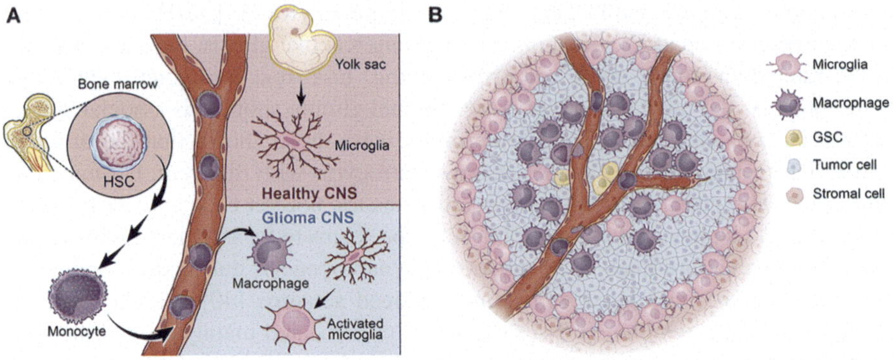

Fig. 12.1 Schematic outline of the architecture of a proneural glioblastoma growth. In this type of GBM, invasive macrophages rather than microglia dominate. The majority of macrophages is in the perivascular space together with GSCs, whereas microglia are mainly in the peritumoral region. (**a**) Pathway of tumor-associated macrophage recruitment. (**b**) Architecture of GBM microenvironment. HSC = hematopoietic stem cells, GSC = glioma stem cells. (From Ref. [48]. Reprinted with permission from Frontiers Media SA)

neuronal activity. This activity, in turn, promotes further glioma growth and invasion. This expansion of the tumor will increase the glioma contribution to neuronal activity till hyperexcitability and seizures occur [45].

As pointed out earlier, the architectural arrangement of the GAMs and its relative proportion of macrophages and microglia depend on the molecular subtype(s) of the tumor (see Fig. 12.1). In general, macrophages are positioned in the perivascular area next to GSCs and promote gliomagenesis, whereas microglia are mostly in the periphery and are mostly concerned with promoting tumor invasion [48]. Mesenchymal GBM subtypes involve a higher number of macrophages than the proneural subtype.

12.4 Metastatic Colonization

At first glance one would be surprised that the well-shielded brain parenchyma permits the colonization by cancer cells from growths located outside the brain. Once, however, a metastatic cancer enters, one would expect similar mechanisms as in primary brain cancers to promote maintenance, growth, and invasion of secondary cancer. Yet, in autopsies of a cross section of cancer patients, one-fifths of all patients had brain metastases [49]. The brain metastases are highest for lung cancer, breast cancer, and melanoma. Brain metastases have an even worse outcome prognosis than stage 4 GBM. The average survival time is somewhat more than 3 months after diagnosis [50]. The colonization leads to brain mass compression. Headaches are a major clinical symptom. There are focal neurological deficits depending on what parts of the brain the metastases spread. Numbness, seizures, and hemorrhage are other common symptoms. If there is a significant intracranial pressure increase, edema or cerebrospinal fluid obstruction and impaired cognition can occur [51].The mechanisms of colonization of the brain parenchyma by cancer cells are not well understood. This is mainly because there is no accessible in vitro system to study the molecular details. It seems that the different cancer types use all similar mechanisms to cross the blood–brain barrier and invade the brain parenchyma. The metastases have their origin from intravasation of tumor cells from their organ of origin into the circulation or – occasionally – into the lymphatic system. The circulating tumor cells have undergone a transition from an epithelial cell type to a more mesenchymal one. They acquire more stem cell-like qualities and are therefore more invasive and migratory [52]. They are well prepared to invade different organs, including crossing the blood–brain barrier into the brain parenchyma. Metastatic cancer cells are larger than red blood cells, often up to three times. In the parts of the brain circulation where the flow rate slows down, for example, capillary branching points, the cancer cells can adhere to the endothelial cell layer [53]. Extravasation in the brain occurs mainly through destruction of the tight junctions between the endothelial cells. This is the result of the release of VEGF and MMPs, which together also promote new vascular growth and extracellular matrix destruction. These actions not only facilitate cancer cell migration but also increase the permeability of the blood–brain barrier. Once the cancer cells are inside the brain parenchyma, the blood–brain barrier is functionally different. It is now called the blood–tumor barrier and consists of swollen endothelial cells, loose connections between endothelial cells, an altered basement membrane, and severe impairment of astrocyte endfeet function [54]. The resultant increased permeability of the barrier is currently used in clinical trials to target the metastases inside the parenchyma [55]. As it turns out, the metastatic tumor cells in the brain have substantially different molecular properties than the parent tumor cells in their tissue of origin [56]. This is due to epigenetic changes and transcriptional reprogramming. The cancer cells have now more neuronal characteristics resembling those expressed during neurogenesis [57]. It is assumed that the tumor microenvironment is the cause for this reprogramming. An alternative would be a selection process for colonizing cancer cells due to their high mutation rate. Most metastatic cancer cells entering the brain do not survive, many are dormant, and only a small part gives rise to new colonies [51]. Certain cancer cells, like melanoma cells, can migrate long distances in the brain. Multiple lesions are often not the result of multiple entries into the brain, but of a high motility. This motility is the result of active migration along blood vessels and compartment interphases as well as of passive distribution by the cerebrospinal fluid [58].There are interactions with the other brain cells. Astrocytes and their endfeet will at first defend the integrity of the blood–brain barrier. The invading tumor cells, however, establish gap junctions with astrocytes, enabling cyclic guanosine monophosphate-adenosine monophosphate to cross into the astrocytes. This turns the astrocytes into a reactive subtype with dysfunctional endfeet and proinflammatory properties [42]. The astrocytes release two major factors, type 1 interferons alpha and tumor necrosis factor, which promote tumor growth. The tumor cells secrete neurotrophin 3, which blocks the transformation of the microglia to a cytotoxic subtype. GAMs consist mainly of macrophages with few microglia and they are reprogrammed into immunosuppressive and tumor-promoting cells in the periphery of the metastatic growth, similar to GAMs in glioma cell growth [59]. T lymphocytes are specifically recruited to attack the cancer cells of the original tumor. Therefore, there are circulating lymphocytes that will be recruited into the brain despite the molecular changes of the metastatic tumor cells. The T lymphocytes migrate toward the metastatic growth area in the brain. Here, they are suppressed by various secretions from astrocytes and GAMs. A major factor is the release of programmed cell death ligand 1 by the cancer cells which inactivates cytotoxic T lymphocytes and leads to immune escape by the metastatic growth [60]. Thus, the spread and survival of the metastases and interactions with their microenvironment are fundamentally like those of glioma cells. The brain support cells are reprogrammed and coopted by the invaders.

12.5 Conclusion

The major sources of glioma cancer cells are the cells in the adult brain, which are capable of proliferation: NSCs and OPCs. They must undergo mutations or DNA methylation to key cell cycle or tumor suppression genes to escape the tight control mechanisms. Yet, these glioma cells would not be successful if they were not able to manipulate all other brain cells, including neurons and invading macrophages and lymphocytes into supporting their growth. This recruitment is key in understanding the success of glioma cell growth. Metastatic growth involves a selection in the properties of the original cancer cells, which enables the metastatic cells to escape the primary tumor, circulate, and then breach the blood–brain barrier. A minority of these metastatic cells will succeed in colonizing the brain. Yet, in this case, the key for this colonization is the successful reprogramming of all other cell types, including neurons, similar to primary cancers. This cooption of mainly glial cells but also neurons is very similar to glioma cells.

References

1. Azzarelli R, Simons BD, Philpott A. The developmental origin of brain tumours: a cellular and molecular framework. Development. 2018;145(10):dev162693.
2. Francis SS, Ostrom QT, Cote DJ, Smith TR, Claus E, Barnholtz-Sloan JS. The epidemiology of central nervous system tumors. Hematol Oncol Clin North Am. 2022;36(1):23–42.
3. Komori T. Grading of adult diffuse gliomas according to the 2021 WHO Classification of Tumors of the Central Nervous System. Lab Invest. 2022;102(2):126–33.
4. Waker CA, Lober RM. Brain tumors of glial origin. Adv Exp Med Biol. 2019;1190:281–97.
5. Ferris SP, Hofmann JW, Solomon DA, Perry A. Characterization of gliomas: from morphology to molecules. Virchows Arch. 2017;471(2):257–69.
6. Duffau H. Dynamic interplay between lower-grade glioma instability and brain metaplasticity: proposal of an original model to guide the therapeutic strategy. Cancers (Basel). 2021;13(19):4759.

7. Touat M, Li YY, Boynton AN, Spurr LF, Iorgulescu JB, Bohrson CL, et al. Mechanisms and therapeutic implications of hypermutation in gliomas. Nature. 2020;580(7804):517–23.
8. Zhang P, Xia Q, Liu L, Li S, Dong L. Current opinion on molecular characterization for GBM classification in guiding clinical diagnosis, prognosis, and therapy. Front Mol Biosci. 2020;7:562798.
9. Brennan CW, Verhaak RG, McKenna A, Campos B, Noushmehr H, Salama SR, et al. The somatic genomic landscape of glioblastoma. Cell. 2013;155(2):462–77.
10. Wang Q, Hu B, Hu X, Kim H, Squatrito M, Scarpace L, et al. Tumor evolution of glioma-intrinsic gene expression subtypes associates with immunological changes in the microenvironment. Cancer Cell. 2017;32(1):42–56.e6.
11. Wang J, Cazzato E, Ladewig E, Frattini V, Rosenbloom DI, Zairis S, et al. Clonal evolution of glioblastoma under therapy. Nat Genet. 2016;48(7):768–76.
12. Sinnaeve J, Mobley BC, Ihrie RA. Space invaders: brain tumor exploitation of the stem cell niche. Am J Pathol. 2018;188(1):29–38.
13. Muftuoglu Y, Pajonk F. Targeting glioma stem cells. Neurosurg Clin N Am. 2021;32(2):283–9.
14. Suvà ML, Tirosh I. The glioma stem cell model in the era of single-cell genomics. Cancer Cell. 2020;37(5):630–6.
15. Ludwig K, Kornblum HI. Molecular markers in glioma. J Neuro-Oncol. 2017;134(3):505–12.
16. Mistry AM. Clinical correlates of subventricular zone-contacting glioblastomas: a meta-analysis. J Neurosurg Sci. 2019;63(5):581–7.
17. Zong H, Parada LF, Baker SJ. Cell of origin for malignant gliomas and its implication in therapeutic development. Cold Spring Harb Perspect Biol. 2015;7(5):a020610.
18. Chen J, McKay RM, Parada LF. Malignant glioma: lessons from genomics, mouse models, and stem cells. Cell. 2012;149(1):36–47.
19. Lim DA, Cha S, Mayo MC, Chen MH, Keles E, VandenBerg S, et al. Relationship of glioblastoma multiforme to neural stem cell regions predicts invasive and multifocal tumor phenotype. Neuro-Oncology. 2007;9(4):424–9.
20. Turcan S, Makarov V, Taranda J, Wang Y, Fabius AWM, Wu W, et al. Mutant-IDH1-dependent chromatin state reprogramming, reversibility, and persistence. Nat Genet. 2018;50(1):62–72.
21. Jacques TS, Swales A, Brzozowski MJ, Henriquez NV, Linehan JM, Mirzadeh Z, et al. Combinations of genetic mutations in the adult neural stem cell compartment determine brain tumour phenotypes. EMBO J. 2010;29(1):222–35.
22. Chow LM, Endersby R, Zhu X, Rankin S, Qu C, Zhang J, et al. Cooperativity within and among Pten, p53, and Rb pathways induces high-grade astrocytoma in adult brain. Cancer Cell. 2011;19(3):305–16.
23. Ahmed AU, Auffinger B, Lesniak MS. Understanding glioma stem cells: rationale, clinical relevance and therapeutic strategies. Expert Rev Neurother. 2013;13(5):545–55.
24. Stiles CD, Rowitch DH. Glioma stem cells: a midterm exam. Neuron. 2008;58(6):832–46.
25. Hambardzumyan D, Bergers G. Glioblastoma: defining tumor niches. Trends Cancer. 2015;1(4):252–65.
26. Perus LJM, Walsh LA. Microenvironmental heterogeneity in brain malignancies. Front Immunol. 2019;10:2294.
27. Najem H, Khasraw M, Heimberger AB. Immune microenvironment landscape in CNS tumors and role in responses to immunotherapy. Cells. 2021;10(8):2032.
28. Brettschneider EES, Terabe M. The role of NKT cells in glioblastoma. Cells. 2021;10(7):1641.
29. Dapash M, Hou D, Castro B, Lee-Chang C, Lesniak MS. The interplay between glioblastoma and its microenvironment. Cells. 2021;10(9):2257.
30. Clavreul A, Menei P. Mesenchymal stromal-like cells in the glioma microenvironment: what are these cells? Cancers (Basel). 2020;12(9):2628.
31. Herisson F, Frodermann V, Courties G, Rohde D, Sun Y, Vandoorne K, et al. Direct vascular channels connect skull bone marrow and the brain surface enabling myeloid cell migration. Nat Neurosci. 2018;21(9):1209–17.

32. Gutmann DH, Kettenmann H. Microglia/brain macrophages as central drivers of brain tumor pathobiology. Neuron. 2019;104(3):442–9.
33. Catalano M, D'Alessandro G, Trettel F, Limatola C. Role of infiltrating microglia/macrophages in glioma. Adv Exp Med Biol. 2020;1202:281–98.
34. Guo X, Pan Y, Gutmann DH. Genetic and genomic alterations differentially dictate low-grade glioma growth through cancer stem cell-specific chemokine recruitment of T cells and microglia. Neuro-Oncology. 2019;21(10):1250–62.
35. Sarkar S, Döring A, Zemp FJ, Silva C, Lun X, Wang X, et al. Therapeutic activation of macrophages and microglia to suppress brain tumor-initiating cells. Nat Neurosci. 2014;17(1):46–55.
36. van der Vos KE, Abels ER, Zhang X, Lai C, Carrizosa E, Oakley D, et al. Directly visualized glioblastoma-derived extracellular vesicles transfer RNA to microglia/macrophages in the brain. Neuro-Oncology. 2016;18(1):58–69.
37. Parmigiani E, Scalera M, Mori E, Tantillo E, Vannini E. Old stars and new players in the brain tumor microenvironment. Front Cell Neurosci. 2021;15:709917.
38. Priego N, Zhu L, Monteiro C, Mulders M, Wasilewski D, Bindeman W, et al. STAT3 labels a subpopulation of reactive astrocytes required for brain metastasis. Nat Med. 2018;24(7):1024–35.
39. Mega A, Hartmark Nilsen M, Leiss LW, Tobin NP, Miletic H, Sleire L, et al. Astrocytes enhance glioblastoma growth. Glia. 2020;68(2):316–27.
40. John Lin CC, Yu K, Hatcher A, Huang TW, Lee HK, Carlson J, et al. Identification of diverse astrocyte populations and their malignant analogs. Nat Neurosci. 2017;20(3):396–405.
41. Pietras A, Katz AM, Ekström EJ, Wee B, Halliday JJ, Pitter KL, et al. Osteopontin-CD44 signaling in the glioma perivascular niche enhances cancer stem cell phenotypes and promotes aggressive tumor growth. Cell Stem Cell. 2014;14(3):357–69.
42. Chen Q, Boire A, Jin X, Valiente M, Er EE, Lopez-Soto A, et al. Carcinoma-astrocyte gap junctions promote brain metastasis by cGAMP transfer. Nature. 2016;533(7604):493–8.
43. Brandao M, Simon T, Critchley G, Giamas G. Astrocytes, the rising stars of the glioblastoma microenvironment. Glia. 2019;67(5):779–90.
44. Liddelow SA, Guttenplan KA, Clarke LE, Bennett FC, Bohlen CJ, Schirmer L, et al. Neurotoxic reactive astrocytes are induced by activated microglia. Nature. 2017;541(7638):481–7.
45. Johung T, Monje M. Neuronal activity in the glioma microenvironment. Curr Opin Neurobiol. 2017;47:156–61.
46. Samudra N, Zacharias T, Plitt A, Lega B, Pan E. Seizures in glioma patients: an overview of incidence, etiology, and therapies. J Neurol Sci. 2019;404:80–5.
47. Campbell SL, Buckingham SC, Sontheimer H. Human glioma cells induce hyperexcitability in cortical networks. Epilepsia. 2012;53(8):1360–70.
48. Chen Z, Hambardzumyan D. Immune microenvironment in glioblastoma subtypes. Front Immunol. 2018;9:1004.
49. Budczies J, von Winterfeld M, Klauschen F, Bockmayr M, Lennerz JK, Denkert C, et al. The landscape of metastatic progression patterns across major human cancers. Oncotarget. 2015;6(1):570–83.
50. Hatiboglu MA, Chang EL, Suki D, Sawaya R, Wildrick DM, Weinberg JS. Outcomes and prognostic factors for patients with brainstem metastases undergoing stereotactic radiosurgery. Neurosurgery. 2011;69(4):796–806; discussion 806.
51. Achrol AS, Rennert RC, Anders C, Soffietti R, Ahluwalia MS, Nayak L, et al. Brain metastases. Nat Rev Dis Primers. 2019;5(1):5.
52. Lah TT, Novak M, Breznik B. Brain malignancies: glioblastoma and brain metastases. Semin Cancer Biol. 2020;60:262–73.
53. Kienast Y, von Baumgarten L, Fuhrmann M, Klinkert WE, Goldbrunner R, Herms J, et al. Real-time imaging reveals the single steps of brain metastasis formation. Nat Med. 2010;16(1):116–22.
54. Connell JJ, Chatain G, Cornelissen B, Vallis KA, Hamilton A, Seymour L, et al. Selective permeabilization of the blood-brain barrier at sites of metastasis. J Natl Cancer Inst. 2013;105(21):1634–43.

55. Nduom EK, Yang C, Merrill MJ, Zhuang Z, Lonser RR. Characterization of the blood-brain barrier of metastatic and primary malignant neoplasms. J Neurosurg. 2013;119(2):427–33.
56. Brastianos PK, Carter SL, Santagata S, Cahill DP, Taylor-Weiner A, Jones RT, et al. Genomic characterization of brain metastases reveals branched evolution and potential therapeutic targets. Cancer Discov. 2015;5(11):1164–77.
57. Park ES, Kim SJ, Kim SW, Yoon SL, Leem SH, Kim SB, et al. Cross-species hybridization of microarrays for studying tumor transcriptome of brain metastasis. Proc Natl Acad Sci U S A. 2011;108(42):17456–61.
58. Simonsen TG, Gaustad JV, Rofstad EK. Intracranial tumor cell migration and the development of multiple brain metastases in malignant melanoma. Transl Oncol. 2016;9(3):211–8.
59. Luo L, Liu P, Zhao K, Zhao W, Zhang X. The immune microenvironment in brain metastases of non-small cell lung cancer. Front Oncol. 2021;11:698844.
60. Mansfield AS, Aubry MC, Moser JC, Harrington SM, Dronca RS, Park SS, et al. Temporal and spatial discordance of programmed cell death-ligand 1 expression and lymphocyte tumor infiltration between paired primary lesions and brain metastases in lung cancer. Ann Oncol. 2016;27(10):1953–8.

Chapter 13
Neurodegenerative Disorders

Abstract This chapter deals with the most common neurodegenerative diseases: Alzheimer's disease, Parkinson's disease, Huntington's disease (HD), amyotrophic lateral sclerosis (ALS), and prion disease. Each of these diseases has a different role for non-neuronal cells in the neurodegeneration processes and therefore the cells do not have a standardized response pattern in all neurodegenerative processes and diseases. Instead, each disease must be judged separately. The roles range from drivers of the disease process to modifiers and uninvolved bystanders. What can be generalized, however, is that both astrocytes and microglia have protective and damaging roles for neurons. It seems that early in disease development the cells are more protective and then, as the pathology develops, they turn into a more damaging phenotype. Microglia develop a subset, called disease-associated microglia (DAM), during Alzheimer's disease, ALS, and some demyelinating diseases. The DAMs are prone to excessive replication, which is the reason for exhibiting senescence, and then contribute to pathological processes. It may be one of the major contributors to aging as a risk factor in these diseases. In some Alzheimer's disease animal models, microglia seem to be the initiators of the disease. It is not clear if this fact has any connection to DAM subtype and how much the findings can be generalized. Another interesting fact is the appearance of immune-competent (major histocompatibility complex [MHC] II expressing) astrocytes in Parkinson's disease, which involves interaction with CD4+ T lymphocytes. In this disease astrocytes appear to be a major driver if not a cause. Despite the strong inflammation accompanying ALS, microglia are more of a bystander and modifier of the disease. CD4+ T lymphocytes appear, except in HD and prion disease. They seem to be kept in check by healthy microglia to conduct a beneficial function. Once microglia fail or turn toxic, the T cells transform into a more damaging function. In Parkinson's disease there is a unique interaction of the T cells with an astrocyte subset.

Keywords Activated microglia · Alzheimer's disease · Amyotrophic lateral sclerosis · Disease-associated microglia · Huntington's disease · Immune-competent astrocytes · Parkinson's disease · Prion disease · Reactive astrocytes · T lymphocytes

W. Walz, *The Gliocentric Brain*, https://doi.org/10.1007/978-3-031-48105-5_13

13.1 Introduction

Neurodegeneration is a disease process with the symptoms caused by dying neurons. There is usually no apparent external cause. It normally leads to a progressive loss of function and structure of a specific subset of neurons in a defined area. There are hundreds of defined diseases within this group, many of them exceedingly rare. The major ones and the ones dealt with in this chapter are Alzheimer's disease, Parkinson's disease, Huntington's disease (HD), and amyotrophic lateral sclerosis (ALS). Neurodegenerative diseases are grouped into various categories, but the most reliable are those focusing on affected areas and therefore symptoms [1]. Dementia is based mainly on the cerebral cortex and Alzheimer's disease is the most prominent representative. However, there are other diseases which cause dementia, some not based on degeneration per se as are seen after stroke, neural trauma, and infections. Degenerative processes affecting the basal ganglia are manifest as movement disorders (Parkinson's and Huntingdon's disease as well as others). Other degenerative diseases affect the cerebellum and lead to various atrophies and ataxias. Spinal cord degeneration leads to ALS and muscular atrophies. There are many more. The causes are mostly unknown, but some cases point clearly to genetic mutations as drivers (Huntingdon's disease, familial Alzheimer's disease). Some toxic substances are also known to cause neurodegeneration; examples are a toxin from *Cycas circinalis* and 1-methyl-4-phenyl-1,2,3,6-tetrahydropyridine. An outstanding fact is that advancing age is a very strong risk factor for almost all neurodegenerative diseases. However, for most patients with degenerative diseases, there are no clearly delineated genetic causes and environmental factors are elusive.

Neurodegeneration involves neuronal death, which is accompanied by strong reactive gliosis. The many forms of neuronal death in neurodegeneration were recently reviewed [2]. Neurodegeneration does not kill neurons by necrosis or unregulated cell death. The exception is secondary death in the periphery by necrosis, but in general degenerating neurons are dying of a programmed cell death. There are many kinds of regulated neuronal deaths and often the death is a mixture of several established cell death forms and pathways [3, 4]. Intrinsic (or mitochondrial) apoptosis plays a role in ALS mouse models and contributes to Parkinson's disease, but there is limited evidence in Alzheimer's disease. Necroptosis, a lytic form involving inflammation, is well presented in Alzheimer's and Parkinson's disease. Autophagy, involving lysosomes, also has a strong presence in all neurodegenerative diseases presented here. There are currently clinical trials in Parkinson's disease patients targeting ferroptosis, which is iron-dependent necrotic programmed cell death. And finally, there is evidence for pyroptosis, which involves the inflammasome, occurring in most neurodegenerative diseases.

In all neurodegenerative processes, activated microglia and reactive astrocytes are involved [5]. Whereas heavy reactive gliotic involvement by both glia types is not in question, the issues are: is the nonneuronal involvement solely a bystander effect and a reaction to neuronal distress? Or are glial/immune cells more involved and are a major driver in some forms of neurodegeneration? Maybe they are even the cause of some of the neurodegenerative diseases? [6].

13.2 Alzheimer's Disease

Alzheimer's disease is the most common and lethal form of dementia worldwide. Incidence rates double every 6 years of age starting from age 60, when it is about 4 per 1000 person-years [7]. The amyloid hypothesis has been very influential in the last 40 years of research. Amyloid-β precursor protein (APP) is an integral membrane glycoprotein. It is not specific to the nervous system and is expressed in the brain mainly in synaptic membranes. It has a multifunctional role in synapse formation and repair as well as anterograde neuronal and iron transport [8]. In humans it is processed in two pathways. The non-amyloidogenic pathway involves subsequent proteolytic steps by α-secretase and then a γ-secretase to get two extracellular and one intracellular domain. The γ-secretase is a complicated complex. It needs presenilin 1 and 2 to function properly. The two presenilin forms are transmembrane proteins and are involved in synaptic calcium homeostasis. In association with γ-secretase, they form a complex and are the catalytic subunits to cleave APP. In the amyloidogenic pathway, β-secretase cleaves the ectodomain of APP and sAPPβ (soluble peptide APPβ), from APP. In the next step, the γ-secretase complex first releases the cytosolic APP intracellular domain (AICD). Then the remainder of the membrane-bound APP glycoprotein is cleaved in heterogeneous proteolytic steps into several extracellular amyloid-β (Aβ) residues [9]. These residues have about 39–49 amino acids. The major variant is, however, $A\beta_{40}$. The $A\beta_{40}$ monomer aggregates into soluble oligomers. Another frequent product is $A\beta_{42}$, which is more prone to aggregate. However, normally the relation of $A\beta_{40}$ and $A\beta_{42}$ is 10:1. This is important as $A\beta_{42}$ tends to aggregate into higher-order oligomers, protofibrils, and fibrils. These extracellular aggregates form plaques and initiate a pathophysiological process. The amyloid hypothesis contends that mutations in presenilin 1 or 2 as well as APP will cause the formation of Aβ aggregates as the equilibrium of the Aβ isoforms will shift to $A\beta_{42}$ with all the consequences for aggregation and plaque formation. The six isoforms of tau protein (tubulin binding protein) are mainly in axons and are necessary for the stability of the microtubules. Amyloid aggregation reacts with various receptors on the neuronal cell membrane [8]. This induces hyperphosphorylation of the intracellular tau proteins. The hyperphosphorylated tau proteins form neurofibrillary tangles (NFTs). The accumulation of intracellular NFTs leads to synaptic and neuronal dysfunction, resulting in neuronal degeneration. There is strong evidence for such a pathophysiological mechanism in Down syndrome and familial (early-onset) Alzheimer's disease [7]. Familial Alzheimer's disease, however, accounts for only about 2% of Alzheimer cases. In animal models of Alzheimer's disease and in clinical trials with late-onset Alzheimer patients, the sequence of events predicted by the amyloid hypothesis could not be clearly verified. However, this does not mean that amyloid plaques or NFTs do not play a role, either in combination or by themselves. Rather, the sequence of the amyloid hypothesis must be modified in these cases. Evidence points to the involvement of plaques and/or NFTs depending on the animal model used [7]. In many patients the amyloid–tau cascade is not linear. Aβ and/or tau deposits are found in nondemented people [10]. There are also alternative pathophysiological mechanisms hypothesized. The most attractive alternate hypothesis points to the APOE (apolipoprotein E) genotype. Carriers of the APOE ε4 isoform have a clear increased risk of late-onset (or sporadic) Alzheimer's disease [11]. APOE plays a major role in clearance of Aβ across the blood–brain barrier and APOE ε4 is the least efficient isoform in this clearance process. In addition, for both the appearance of tau protein pathology and α-synuclein pathology (discussed later), being an APOE ε4 carrier is a risk factor. Cortical hypoperfusion as well as compromised blood–brain barrier and vessel integrity are part of the Alzheimer pathology. This has led to the vascular hypothesis, which claims that cerebrovascular diseases are part of the comorbidities in the disease and may be linked to impaired Aβ clearance [12]. In genome-wide association studies, about 40 gene loci have been associated with increased risk for Alzheimer's disease. Interestingly these risk alleles are especially enriched in microglia [13]. Apart from these genetic risk factors, there are lifestyle and vascular risk factors for sporadic Alzheimer's disease [14].

In the past few years evidence has surfaced that implicated nonneuronal cells and peripheral immune cells in crucial roles for the development of Alzheimer's disease [15]. A key role in the pathogenesis of the disease seems to involve microglia as is evident from the genome-wide association studies. It is now thought that microglia play a major role not only in the development of the pathology but even in the critical step that initiates the start of neuronal damage. Still, the precise pathological mechanisms are controversial: microglia are seen to have beneficial as well as detrimental roles in the disease development. Table 2 in the research work of McFarland and Chakrabarty (2022) [15] gives an overview of the impact of microglia manipulation in animal models of Alzheimer's disease, confirming that there is no obvious clear pattern. Rather, the microglia impact is context-dependent. Aβ plaques are not created by microglial activity but not all authors agree [16]. The presence of the plaques does not necessarily lead to Alzheimer symptoms in neither humans nor animal models. Some observations suggest that the symptoms are worsened or even created in response to Aβ–microglia interaction. Activated microglia and reactive astrocytes surround Aβ plaques. It is known from normal and Alzheimer human brains, as well as mouse models, that microglia lysosomes contain Aβ materials. This is shown even for microglia not associated with plaques. The microglia surrounding the plaques stain positively for inflammatory markers as well as MHC (major histocompatibility complex) II. During Aβ accumulation, at the point of the appearance of tau pathology, microglia start with avid proliferation around the deposits. If in animal models, this proliferation is blocked, amyloid and tau pathology are reduced [17]. If after the appearance of the first plaques in the 5xfAD Alzheimer mouse model, most microglia are chronically eliminated, the plaque load does not change. However, dendritic spine and neuronal loss are prevented, and inflammatory processes are dramatically reduced. Contextual memory tests in these microglia-reduced mice show marked improvement compared with mice with non-manipulated microglia [18]. In a landmark study [16], which used the same mouse model, the authors showed that eliminating microglia, before Aβ plaques appear, prevents the appearance of the plaques. Plaques appeared around the microglia in the very few brain areas with residual microglia. This is a clear indication that – at least in this mouse model – microglia not only sustain Aβ pathology but also play a crucial role in its origin. The microglia, which proliferate in the various animal models due to plaque deposition, change their phenotype over time due to unknown mechanisms (see Fig. 13.1) [19]. This new phenotype is called disease-associated microglia (DAM). It is assumed that this type of microglia contributes to or may be even initiating neuronal damage [20]. In postmortem brain samples of Alzheimer patients DAMs were found with phagocytosed Aβ particles included [21]. An important final transition step in the development of DAMs is played by activation of triggering receptor expressed on myeloid cells 2 (TREM2). APOE, produced by microglia, and binding to TREM2 seems to be involved in this important pathway. A new study [20] suggests strongly that the proliferation of microglia is critical. In a subpopulation of the DAMs the early and excessive proliferation around plaques leads to replicative senescence. This phenomenon is characterized by telomere shortening, a special transcriptional signature, and increases in β-galactosidase. This pattern of DAM is not only observed in animal models but also in human patients' postmortem tissue [22]. Inhibition of microglial proliferation prevents the Alzheimer-like pathology in animal models. Thus, the possible crucial involvement of microglial excessive proliferation and senescence would explain why aging is a risk factor. Microglia are the cell type which proliferates widely in the normal healthy brain. An additional strong proliferative stimulus in the aging brain would

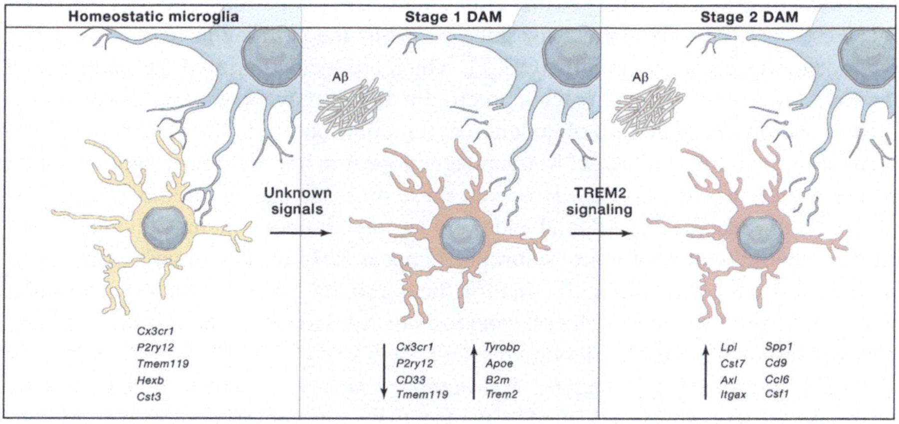

Fig. 13.1 A two-step model of induction of DAM. Unknown signals promote transition from homeostatic to stage 1 DAM, while TREM2 signaling is required for stage 2 induction. Each stage is characterized by a unique transcriptional signature. The abbreviations refer to the genes responsible for the following molecules: TREM2 (triggering receptor expressed on myeloid cells 2), Cx3cr1 (Cx3 chemokine or fractalkine receptor), P2ry12 (chemoreceptor for adenosine diphosphate), Tmem119 (transmembrane protein 119), Hexb (beta-hexosaminidase subunit beta), Cst3 (cystatin3), cd33 (cluster of differentiation 33), Tyrobp (protein tyrosine kinase-binding protein), Apoe (apolipoprotein E), B2m (beta-2 microtubulin), Lpl (lipoprotein lipase), Cst7 (cystatin7), Axl (receptor tyrosine kinase), Itgax (complement component 2 receptor 4 subunit), Spp1 (secreted phosphoprotein 1), Cd9 (cluster of differentiation 9), Ccl6 (chemokine C-C motif ligand 6), and Csf1 (colony-stimulating factor 1). (Reprinted from Deczkowska et al. [19] with permission. Copyright Elsevier Inc.)

explain this phenomenon in a DAM subpopulation. Simon et al. [23] (see Fig. 13.2) developed a model of how these dysfunctional DAMs could contribute directly to the Alzheimer's disease pathology. Aβ accumulation will lead to microglia proliferation and proinflammatory microglia (DAMs). CSFR1 (colony-stimulating factor 1 receptor) inhibition on microglia prevents this proliferation and inflammatory shift. It also prevents synaptic degeneration and deterioration of behavioral performance without changing the Aβ load. This is an indication that this microglia subpopulation reverts to an earlier stage in brain development when microglia are phagocytosing synapses in keeping with their role in synaptic pruning [24]. There are indications from animal models and postmortem Alzheimer patient tissue that microglia are additionally involved in tau spreading [25]. Tau proteins are phagocytosed by microglia at the synaptic cleft and then released within exosomes. Exosomes are released by microglia (and other cell types) to transport various substances including genetic materials over long distances. In this situation microglia release these tau-loaded exosomes to spread the tau load around (Fig. 13.2). Inhibition of exosome synthesis suppresses tau spreading [26].

The above considerations put microglia right into the focus of origin and pathogenesis of Alzheimer's disease. Other nonneuronal cells in the CNS are involved as well, although not in a such commanding position as the microglia. Reactive astrocytes accumulate around Aβ plaques and $A\beta_{42}$ interacts with the astrocytic glutamate transporter-1 and impairs its function [27]. This inhibition leads to improper glutamate clearance around synapses and contributes to excitotoxicity. APOE is released by these reactive astrocytes to interact with the microglia transition [28]. Oligodendrocyte precursor cells (OPCs), which are located around the plaques and the associated activated and reactive microglia and astrocytes, are less likely to proliferate and differentiate into oligodendrocytes, contributing to the observed demyelination in Alzheimer's disease. Macrophages appear in the Alzheimer patients' brain and start clearing Aβ deposits. However, these macrophages are not as effective as microglia in reducing the Aβ load and seem to play a minor role in Alzheimer's disease compared with microglia. Neutrophils appear early in Alzheimer's disease and seem to be recruited by microglia. In turn, the microglia–proinflammatory neutrophil interaction stimulates microglia phagocytosis of Aβ. However, the proinflammatory neutrophils damage the blood–brain barrier [29]. Cytotoxic T cells are also migrating into the brain toward Aβ plaques. The numbers, however, seem to be kept low by microglia activity. This is indicated by the fact that experimental depletion of microglia increases the number of cytotoxic T cells in Alzheimer's disease models [30].

13.3 Parkinson's Disease

After Alzheimer's disease, Parkinson's disease is the most common neurodegenerative disease, afflicting about 1–2% of the over 65-year-old population. The disease focuses on the degeneration of dopaminergic neurons in the substantia nigra of the midbrain and leads to the loss of their axons along the nigrostriatal pathway. Once

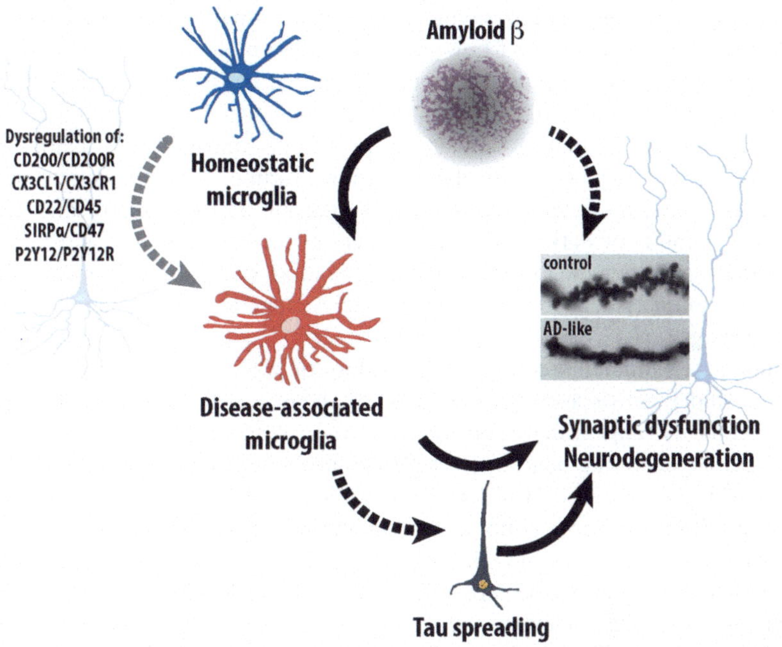

Fig. 13.2 Microglial cells as necessary effectors of pathology in Alzheimer's disease. Through direct interactions, neurons contribute to maintaining microglia in a homeostatic phenotype. A dysregulation of the key pathways depicted in the figure leads to a shift in the microglial phenotype to an activated inflammatory profile. In the context of Alzheimer's disease, microglia can have indirect effects on neurons through the interaction with the main pathological hallmarks (Aβ and tau). On one hand, Aβ can have indirect effects on neurons (right; black dotted line), causing abnormal synaptic function, initiating or pre-conditioning synaptic pathology. On the other hand, the accumulation of Aβ causes a progressive shift in microglia (left; gray dotted line), inducing a disease-associated phenotype that accelerates the progression of the pathology (direct effect, solid black line). Additionally, microglia are capable of promoting the spread of misfolded tau (indirect effect, dotted black line), propagating the pathology in Alzheimer's disease (direct effect). In this model, microglia would lead the executive phase of synaptic dysfunction and neurodegeneration, evidenced by recent data suggesting an uncoupling of Aβ from the beneficial effects observed after targeting microglia. Therefore, targeting the different steps of the sequence summarized in the left side of the figure (microglial route) provides a tantalizing therapeutic opportunity, applicable to advanced stages of AD. CD200R (receptor for cluster of differentiation 200), CX3CL1 (fractalkine), CX3CR1 (fractalkine receptor), CD 22, 45, 47 (cluster of differentiation 22, 45, 47), SIRPα (signal regulatory protein alpha), and P2Y12R (purinergic receptor for adenosine diphosphate Y12). (Reprinted from Simon et al. [23] with permission as by Creative Commons license Attribution 4.0 International. Copyright Holder is Elsevier Ltd.)

about 70% of the dopaminergic neurons of the substantia nigra are degenerated, motor symptoms become noticeable. These are tremors, bradykinesia, ataxia, rigidity, and postural instability. There are several mutations, which are involved in the pathogenesis. The most prominent mutation affects α-synuclein, which is in the presynaptic terminal and involved in synaptic vesicle recycling. Alpha-synuclein has a multi-functional role with actions not only inside the presynaptic terminal but also in the synaptic cleft and exosomes [31]. Indeed, a hallmark of the Parkinson's disease pathology is deposits of Lewy bodies, whose major constituent is the misfolded and aggregated α-synuclein protein. Aging is the major known risk factor. Environmental risk factors include certain pesticides, which damage mitochondria, and in humans and animal models are suspected of contributing to pathogenesis [32]. Widespread Lewy bodies are also known in a form of dementia called "dementia with Lewy bodies." In these cases, motor symptoms develop mostly after the first signs of dementia. In Parkinson's disease, cognitive problems can develop after motor symptoms when the Lewy bodies become more widespread. In many of these cases of dementia associated with Lewy bodies, Aβ and tau deposits are also eventually appearing.

In human α-synucleinopathies, the postmortem brain samples exhibit activated microglia and reactive astrocytes with internalized α-synuclein [33]. CD8+ and CD4+ T cells also appear in this postmortem tissue with CD4+ T cells closely associated with degenerating neurons [34]. Blood–brain barrier disruption is also common in Parkinson's disease. However, the evidence for a driving role of nonneuronal cells in the disease process is not as convincing as for Alzheimer's disease. There are about 20 genes implicated with Parkinson's disease in human patients; some of the monogenetic mutations are upregulated to a higher level in astrocytes. Several of these mutations are also expressed by microglia [35, 36]. Thus, these cells are not innocent bystanders. In the α-synuclein PFFS rat model (pre-formed α-synuclein fibrils model of Parkinson's disease in rats), intrastriatal injection of PFFS leads to MHC II expressing and α-synuclein containing activated microglia within 2 months. Only 3 months later nigral dopamine neurons degenerate. This observation suggests that microglia actively contribute to neuronal degeneration in this model [37]. Indeed, α-synuclein oligomers act as DAMPs (damage-associated molecular patterns) on microglia receptors, including various toll-like receptors (TLRs). The oligomers initiate an inflammatory response and reduce phagocytosis. The oligomers also activate the NLRP3 (nucleotide-binding leucine-rich repeat receptor family pyrin domain containing 3) gene, which encodes cryopyrin, a major component of the inflammasome. This activation in turn leads to massive release of IL-1β (interleukin 1β) [38]. In the MPTP (1-methyl-4phenyl-1,2,3,6-tetrahydropyridine) neurotoxin mouse model a TLR9 agonist caused dopamine neuronal loss, whereas in control mice with deleted TLR9 the neurons did well. TLR9 is a microglial, but not neuronal receptor [39]. DJ-1 (protein deglycase or Parkinson's disease protein 7) relocates into the mitochondria during periods of oxidative stress. DJ-1 mutations are a risk factor for Parkinson's disease. In microglia the mutations lead to inhibition of α-synuclein uptake and degradation as well as inhibited autophagy. Microglia with DJ-1 mutation have a potentiated proinflammatory response [40]. Another Parkinson-risk mutation linked to microglia is for LRRK2 (leucine-rich repeat serine/threonine-protein kinase 2). A mutation in LRRK2 leads to deformed microglial mitochondria and impaired α-synuclein uptake [36].

As pointed out, astrocytes react with reactive gliosis to α-synuclein accumulation. This reaction is caused by factors secreted from microglia, like IL-1α, TNFα (tumor necrosis factor α), and C1q (complement 1, q subcomponent) [41]. In the

PFFS model, inhibition of the astrocytic GLP1R (glucagon-like peptide-1 receptor) prevents the reactive transformation of astrocytes by activated microglia, which in turn prevents dopaminergic neuron degeneration and behavioral deficits [42]. Cultured human astrocytes take up α-synuclein and distribute it in the syncytium via tunneling nanotubes, thus potentially contributing to its spread [43].

The disruption of the blood–brain barrier in Parkinson's disease opens an opportunity for lymphocytes to invade the brain. In most cases the invaders are CD4+ T cells. They are observed near degenerating neurons in postmortem tissue and in animal models [34, 44, 45]. In the presence of these CD4+ T lymphocytes, microglia phagocytosis of aggregated α-synuclein is reduced. This seems to indicate a detrimental impact of CD4+ lymphocytes in the disease, at least as seen in a model of human wildtype α-synuclein transgenic mice, crossed with mice lacking mature lymphocytes [46]. Importantly, it was recently shown in postmortem tissue from Parkinson patients that astrocytes loaded with pathological α-synuclein expressed high levels of MHC II. These MHC-II-positive astrocytes were surrounded by CD4+ T lymphocytes [47]. Cultured human astrocytes exposed to and loaded with α-synuclein showed strong antigen presentation capacity, whereas cultured human microglia did not. This must be seen in conjunction with the discovery of tunneling nanotubes for α-synuclein distribution in the astrocytic syncytium. These new facts about immunocompetent astrocytes in Parkinson's disease suggest that this glial cell type is a major player in Parkinson's disease pathogenesis.

13.4 Huntington's Disease

Huntington's disease (HD) is a late-onset neurodegenerative disease. Huntingtin is a protein expressed in many cells of the body, but the highest density is in brain cells. Its function in the brain is not known, but it plays a role in development and in the adult nervous system. Huntingtin is encoded by the HTT gene (also called IT15 gene as in "interesting transcript 15"). The nonpathological gene has 6–35 glutamate residues. Mutations cause these residues to increase in number (now called mHTT from mutant HTT). Any repeat extension above 36 is a risk factor for HD. The higher the number, the earlier the age of disease onset [48]. The disease manifests itself with a diffuse neurodegeneration in the striatum to progress to cortical atrophy. Thus, the disease starts normally with motor symptoms, especially unwanted movements, and turns into cognitive symptoms and later ends with dementia.

Influx of peripheral immune cells is usually not occurring in patients with HD. Presymptomatic HD gene carriers were routinely screened with positron emission tomography (PET) and found that microglia activated before symptom onset [49]. Such an early microglia activation is also seen in animal models and involves release of proinflammatory cytokines, like IL-6, IL-8, IL-1β and TNF-α. Elevated cytokine levels were also found in the cerebrospinal fluid and blood of presymptomatic human HD gene carriers [50]. These observations suggest an early involvement by microglia in the pathogenesis. Of course, mHTT fragments are not only present in neurons but also in microglia and astrocytes. The BACHD (bacterial artificial chromosome-mediated transgenic) mouse model contains a full-length human mutant Huntingtin gene. A study [51] used this model to selectively deplete mHTT in microglia and in a converse experiment to deplete mHTT in most brain cells, but not microglia. Microglia-specific knockout did not lead to any changes from the normal disease process. Neural-specific knockout, however, showed significant rescue from the normal disease process. This confirms that mHTT expression in microglia is not directly connected to the HD origin.

In contrast to microglia, no astrocyte changes are detected in presymptomatic HD patients and mouse models. Reactive astrocytosis is observed after onset of neurodegeneration and its intensity is correlated with the severity of the neurodegeneration in the striatum. However, in the HD cortex, despite neuronal loss and microglial activation, no reactive astrogliosis can be detected [48]. In HD patients the astrocytic glutamate transporter GLT1 (glutamate transporter 1) is reduced, and the deficiency increases with increasing HD symptoms. However, the GLT1 loss does not precede the symptoms, but in animal models it precedes reactive astrocytosis [52]. Selective induction of mHTT in astrocytes, but not neurons, reduces the astrocytic potassium homeostatic properties and in turn changes the neuronal excitability due to compromised potassium, glutamate, and GABA homeostasis of astrocytes. It also reduces the release of ascorbic acid, causing neuronal oxidative stress. The selective expression of mHTT reduces also the astrocytic release of BDNF (brain-derived neurotrophic factor) which promotes neuroprotection [48].

In presymptomatic HD patients, diffusion tensor imaging revealed myelin damage. Interestingly, selective ablation of mHTT in OPCs but not other brain cells in the BACHD model restored the normal morphology of myelin sheaths [53].

13.5 Amyotrophic Lateral Sclerosis

Amyotrophic lateral sclerosis (ALS) is the third most common neurodegenerative disease. It is based on the degeneration of motor neurons in the cortex, brainstem, and spinal cord. It leads usually to the death of the patient within 2–5 years after the onset of symptoms. The symptoms start with muscle weakness and progress to more serious paralysis. In the end, in most patients, death occurs through respiratory failure. The underlying cause for the sporadic disease is not known, neither are risk factors, except aging. Around 10% of ALS cases are based on familial forms (fALS). In these cases, about 25 genes have been associated with the disease [54]. The most common are listed here. SOD1 (superoxide dismutase 1) is involved in the dismutation of the superoxide radical into molecular oxygen and hydrogen peroxide. The superoxide radical is an aggressive compound causing cellular damage and SOD1 acts as an antioxidant. Another genetic mutation associated with fALS concerns the gene encoding for TDP-43 (transactive response DNA binding protein 43 kDa). It is a multifunctional protein involved in transcriptional control. It is involved in dendritic mRNA regulation and in DNA repair. A third common mutation affects the gene for C9orf72 (chromosome 9 open reading frame 72). Its mutation is not only involved in fALS but also in FTD (frontotemporal dementia). These two disorders share not only a common genetic origin but also clinical and neuropathological features [55]. The C9orf72 protein is involved in intracellular membrane traffic. Some of the mutations of C9orf72 can lead to DNA damage in motor neurons. These neurons activate a DNA damage response, which if not successful, leads to apoptosis. Research into ALS has focused on these genetic alterations in animal models as a window into ALS pathogenesis.

The pathology of ALS is accompanied by strong inflammatory processes. This has led to a closer investigation of the role of nonneuronal cells, especially microglia in the disease progress. In fALS patients (SOD1 mutation), PET imaging found widespread microglial activation not only in symptomatic but also in asymptomatic carriers. The activated microglia extended into non-motor areas [56]. Reactive astrocytosis was evident in postmortem tissue of ALS patients and in PET studies of patients in vivo [57]. In a landmark study, Song et al. [58] isolated astrocytes from postmortem ALS patients. If these astrocytes were cultured with human neurons, they reduced the MHC I expression on these cultured neurons. This indeed was toxic for the neurons, as experimental upregulation of MHC I rescued neuronal survival. CD4+ and CD8+ T lymphocytes and peripheral (but not brain associated) macrophages are found in postmortem ALS tissue [59].

In SOD1 animal models, microglia activation occurs before the onset of symptoms. Creating various forms of mSOD1 in neurons or microglia alone results in various outcomes as the nature of the mutation seems to make a difference. However, the findings from all SOD1 models taken together indicate that microglia are not drivers, but modifiers of the disease once it occurs. It appears that microglia are neuroprotective at the early stage of the disease. In the later course they exacerbate damage, at least in SOD1 models [59]. Overexpression of the microglial anti-inflammatory cytokine IL-10 delayed the disease onset in these animals [60]. The damaging effect of microglia late in the SOD1 model disease might be based on loss of trophic support and neurotoxic actions. It is not clear which of the two, trophic support loss or neurotoxicity, is more important. Introducing mSOD1 selectively exclusively into astrocytes, results in reactive astrocytes with damaging effects on neurons, due to loss of protective function and neurotoxic action [61]. In the SOD1 model oligodendrocytes show early dysfunction and contribute to neuronal damage through reduced metabolic support. In contrast, T cell invasion in the SOD1 model seems to have protective effects, as the interaction with microglia and astrocytes shifts these T cells into a protective phenotype [62].

Cytoplasmic TDP-43 aggregates are found in almost all ALS patients, whereas SOD1 animal models do not feature such aggregation. Mutated TDP-43 models therefore gained recently in acceptance. However, the results with these mouse models have so far not been very convincing. The selective induction of mTDP-43 in microglia alone seems to result in a neurotoxic phenotype. Astrocyte-restricted expression of mTDP-43 leads to astrocytosis, causing neuronal death and paralysis, mainly due to downregulation of glutamate clearance, lipocalin 2 release and activation of microglia [63]. Of all the brain cells, microglia have the highest expression of C9orf72. However, results from studies in C9orf72 animal models in vivo are scarce and have so far not shown conclusive results.

13.6 Prion Disease

The prion (proteinaceous infectious particle) protein is widespread in the body and is found in cell membranes. In the nervous system its function is not clear, other than that it is somehow involved in myelination. PrP^{C} (cellular prion protein) refers to the normal protein which can fold into several isoforms, whereas PrP^{Sc} (scrapie prion protein) denotes a special property of this protein. This is an isoform that is resistant to proteases and can aggregate into long fibers and plaques. Only prion proteins with this isoform, but not PrP^{C}, can add to this aggregate. The PrP^{Sc} accumulation causes neurodegenerative diseases. It is highly infectious in that once it appears, it will grow and cause pathological changes to the brain tissue. The disease can appear in a sporadic form (approximately 85%) with the most common form called Creutzfeldt–Jacob's disease (CJD). One of the less common familial prion diseases is Gerstmann-Sträussler-Scheinker syndrome. An example of an acquired form of the disease is kuru. The disease has a very long nonsymptomatic incubation period; but once the first symptoms appear, the progression is very fast. Dementia appears followed by other neurological symptoms [64]. Death usually occurs within months of symptom onset. There are animal forms like scrapie, BSE (bovine spongiform encephalopathy, also called "mad cow disease"), and chronic wasting disease. Cross-species transmission is possible but not common.

Neuropathological highlights are prion aggregates with accompanying spongiosis (extra-neuronal vacuoles replacing lost neurons), extensive reactive astrogliosis, and moderately activated microglia. In contrast, peripheral immune cell invasion is not common. Mouse models with selective astrocyte or neuron-restricted PrP^{C} expression for conversion to PrP^{Sc} resulted in two major findings [65]. First, for prion pathology to develop, a neuronal PrP^{C} expression is necessary. Second, PrP^{Sc} expression in astrocytes does not lead to pathological changes. Even astrogliosis does not result from restrictive PrP^{Sc} expression in astrocytes. This means that the reactive gliosis and microglia activation accompanying prion disease must be triggered by neuronal distress and is therefore a reaction to PrP^{Sc} accumulation in neurons. Activated microglia appear in animal models at about the same time as changes in the neuronal appearance. In these early stages the microglia seem to have a neuroprotective function [66]. However, in later stages, the phagocytic microglia are no longer capable of clearing the PrP^{Sc} deposits [64]. The question is: is the widespread and severe astrogliosis independent of microglia activation? It was found in animal models [67] that activated microglial release TNF-α, IL-1α, and C1qa (complement C1q subcomponent subunit A) and this combined release created a new subtype of astrocyte, the so-called $C3^{+}$-PrP^{Sc}-specific astrocyte. This subtype can be detected in postmortem human tissue and in animal models. However, suppression of this subtype in animal models by inhibition of the astrocytic receptors results in accelerated disease progress. Thus, both microglia and astrocytes modify the course of prion disease in a context-dependent way.

13.7 Conclusion

The role of nonneuronal cells in neurodegeneration processes cannot be generalized for all processes and diseases. Instead, each disease must be judged separately. What can be generalized, however, is that both astrocytes and microglia have protective and damaging roles for neurons. In most cases they modify the disease process without causing it. It seems that early in disease development the cells are more protective and then, as the pathology develops, they turn into a more damaging phenotype. The damage is caused by a variation of toxic impact and loss of homeostatic function. Microglia develop a subset, called DAMs, during Alzheimer's disease, ALS, and some demyelinating diseases. The DAMs are prone to excessive replication, which is the reason for exhibiting senescence and then contribute to pathological processes. It may be one of the major contributors to aging as a risk factor in these diseases [19, 20]. In some Alzheimer's disease animal models, microglia seem to be the initiators of the disease. It is not clear if this fact has any connection to DAMs and how much the findings can be generalized. Another very interesting fact is the appearance of immune-competent (MHC II expressing) astrocytes in Parkinson's disease, which involves interaction with CD4+ T lymphocytes. In this disease astrocytes appear to be a major driver if not a cause. Despite the strong inflammation accompanying ALS, microglia are more of a bystander and modifier of the disease. CD4+ T lymphocytes appear, except in HD and prion disease. They seem to be kept in check by healthy microglia to conduct a beneficial function. Once microglia fail or turn toxic, the T cells transform into a more damaging function. In Parkinson's disease there is a unique interaction of the T cells with an astrocyte subset.

References

1. Przedborski S, Vila M, Jackson-Lewis V. Neurodegeneration: what is it and where are we? J Clin Invest. 2003;111(1):3–10.
2. Moujalled D, Strasser A, Liddell JR. Molecular mechanisms of cell death in neurological diseases. Cell Death Differ. 2021;28(7):2029–44.
3. Memou A, Dimitrakopoulos L, Kedariti M, Kentros M, Lamprou A, Petropoulou-Vathi L, et al. Defining (and blocking) neuronal death in Parkinson's disease: does it matter what we call it? Brain Res. 2021;1771:147639.
4. Tolkovsky AM, Spillantini MG. Tau aggregation and its relation to selected forms of neuronal cell death. Essays Biochem. 2021;65(7):847–57.
5. Bennett ML, Viaene AN. What are activated and reactive glia and what is their role in neurodegeneration? Neurobiol Dis. 2021;148:105172.
6. Gleichman AJ, Carmichael ST. Glia in neurodegeneration: drivers of disease or along for the ride? Neurobiol Dis. 2020;142:104957.
7. Frisoni GB, Altomare D, Thal DR, Ribaldi F, van der Kant R, Ossenkoppele R, et al. The probabilistic model of Alzheimer disease: the amyloid hypothesis revised. Nat Rev Neurosci. 2022;23(1):53–66.

8. Chen GF, Xu TH, Yan Y, Zhou YR, Jiang Y, Melcher K, et al. Amyloid beta: structure, biology and structure-based therapeutic development. Acta Pharmacol Sin. 2017;38(9):1205–35.
9. Sanches MN, Knapp K, Oliveira AB Jr, Wolynes PG, Onuchic JN, Leite VBP. Examining the ensembles of amyloid-β monomer variants and their propensities to form fibers using an energy landscape visualization method. J Phys Chem B. 2022;126(1):93–9.
10. Jagust W. Imaging the evolution and pathophysiology of Alzheimer disease. Nat Rev Neurosci. 2018;19(11):687–700.
11. Yamazaki Y, Zhao N, Caulfield TR, Liu CC, Bu G. Apolipoprotein E and Alzheimer disease: pathobiology and targeting strategies. Nat Rev Neurol. 2019;15(9):501–18.
12. Zhang Y, Bander ED, Lee Y, Muoser C, Schaffer CB, Nishimura N. Microvessel occlusions alter amyloid-beta plaque morphology in a mouse model of Alzheimer's disease. J Cereb Blood Flow Metab. 2020;40(10):2115–31.
13. Novikova G, Kapoor M, Tcw J, Abud EM, Efthymiou AG, Chen SX, et al. Integration of Alzheimer's disease genetics and myeloid genomics identifies disease risk regulatory elements and genes. Nat Commun. 2021;12(1):1610.
14. Lourida I, Hannon E, Littlejohns TJ, Langa KM, Hyppönen E, Kuzma E, et al. Association of lifestyle and genetic risk with incidence of dementia. JAMA. 2019;322(5):430–7.
15. McFarland KN, Chakrabarty P. Microglia in Alzheimer's disease: a key player in the transition between homeostasis and pathogenesis. Neurotherapeutics. 2022;19:186–208.
16. Spangenberg E, Severson PL, Hohsfield LA, Crapser J, Zhang J, Burton EA, et al. Sustained microglial depletion with CSF1R inhibitor impairs parenchymal plaque development in an Alzheimer's disease model. Nat Commun. 2019;10(1):3758.
17. Mancuso R, Fryatt G, Cleal M, Obst J, Pipi E, Monzón-Sandoval J, et al. CSF1R inhibitor JNJ-40346527 attenuates microglial proliferation and neurodegeneration in P301S mice. Brain. 2019;142(10):3243–64.
18. Spangenberg EE, Lee RJ, Najafi AR, Rice RA, Elmore MR, Blurton-Jones M, et al. Eliminating microglia in Alzheimer's mice prevents neuronal loss without modulating amyloid-β pathology. Brain. 2016;139(Pt 4):1265–81.
19. Deczkowska A, Keren-Shaul H, Weiner A, Colonna M, Schwartz M, Amit I. Disease-associated microglia: a universal immune sensor of neurodegeneration. Cell. 2018;173(5):1073–81.
20. Hu Y, Fryatt GL, Ghorbani M, Obst J, Menassa DA, Martin-Estebane M, et al. Replicative senescence dictates the emergence of disease-associated microglia and contributes to Aβ pathology. Cell Rep. 2021;35(10):109228.
21. Keren-Shaul H, Spinrad A, Weiner A, Matcovitch-Natan O, Dvir-Szternfeld R, Ulland TK, et al. A unique microglia type associated with restricting development of Alzheimer's disease. Cell. 2017;169(7):1276–90.e17.
22. Flanary BE, Sammons NW, Nguyen C, Walker D, Streit WJ. Evidence that aging and amyloid promote microglial cell senescence. Rejuvenation Res. 2007;10(1):61–74.
23. Simon E, Obst J, Gomez-Nicola D. The evolving dialogue of microglia and neurons in Alzheimer's disease: microglia as necessary transducers of pathology. Neuroscience. 2019;405:24–34.
24. Shi Q, Chowdhury S, Ma R, Le KX, Hong S, Caldarone BJ, et al. Complement C3 deficiency protects against neurodegeneration in aged plaque-rich APP/PS1 mice. Sci Transl Med. 2017;9(392):eaaf6295.
25. Bolós M, Llorens-Martín M, Jurado-Arjona J, Hernández F, Rábano A, Avila J. Direct evidence of internalization of tau by microglia in vitro and in vivo. J Alzheimers Dis. 2016;50(1):77–87.
26. Asai H, Ikezu S, Tsunoda S, Medalla M, Luebke J, Haydar T, et al. Depletion of microglia and inhibition of exosome synthesis halt tau propagation. Nat Neurosci. 2015;18(11):1584–93.
27. Bandyopadhyay S. Role of neuron and glia in Alzheimer's disease and associated vascular dysfunction. Front Aging Neurosci. 2021;13:653334.
28. Wyatt-Johnson SK, Brutkiewicz RR. The complexity of microglial interactions with innate and adaptive immune cells in Alzheimer's disease. Front Aging Neurosci. 2020;12:592359.

29. Rossi B, Constantin G, Zenaro E. The emerging role of neutrophils in neurodegeneration. Immunobiology. 2020;225(1):151865.
30. Unger MS, Schernthaner P, Marschallinger J, Mrowetz H, Aigner L. Microglia prevent peripheral immune cell invasion and promote an anti-inflammatory environment in the brain of APP-PS1 transgenic mice. J Neuroinflammation. 2018;15(1):274.
31. Mavroeidi P, Xilouri M. Neurons and glia interplay in α-synucleinopathies. Int J Mol Sci. 2021;22(9):4994.
32. Grünewald A, Kumar KR, Sue CM. New insights into the complex role of mitochondria in Parkinson's disease. Prog Neurobiol. 2019;177:73–93.
33. Doorn KJ, Moors T, Drukarch B, van de Berg W, Lucassen PJ, van Dam AM. Microglial phenotypes and toll-like receptor 2 in the substantia nigra and hippocampus of incidental Lewy body disease cases and Parkinson's disease patients. Acta Neuropathol Commun. 2014;2:90.
34. Brochard V, Combadière B, Prigent A, Laouar Y, Perrin A, Beray-Berthat V, et al. Infiltration of CD4+ lymphocytes into the brain contributes to neurodegeneration in a mouse model of Parkinson disease. J Clin Invest. 2009;119(1):182–92.
35. Booth HDE, Hirst WD, Wade-Martins R. The role of astrocyte dysfunction in Parkinson's disease pathogenesis. Trends Neurosci. 2017;40(6):358–70.
36. MacMahon Copas AN, McComish SF, Fletcher JM, Caldwell MA. The pathogenesis of Parkinson's disease: a complex interplay between astrocytes, microglia, and T lymphocytes? Front Neurol. 2021;12:666737.
37. Duffy MF, Collier TJ, Patterson JR, Kemp CJ, Luk KC, Tansey MG, et al. Lewy body-like alpha-synuclein inclusions trigger reactive microgliosis prior to nigral degeneration. J Neuroinflammation. 2018;15(1):129.
38. Kam TI, Hinkle JT, Dawson TM, Dawson VL. Microglia and astrocyte dysfunction in Parkinson's disease. Neurobiol Dis. 2020;144:105028.
39. Maatouk L, Compagnion AC, Sauvage MC, Bemelmans AP, Leclere-Turbant S, Cirotteau V, et al. TLR9 activation via microglial glucocorticoid receptors contributes to degeneration of midbrain dopamine neurons. Nat Commun. 2018;9(1):2450.
40. Wang T, Zhao N, Peng L, Li Y, Huang X, Zhu J, et al. DJ-1 regulates microglial polarization through P62-mediated TRAF6/IRF5 signaling in cerebral ischemia-reperfusion. Front Cell Dev Biol. 2020;8:593890.
41. Liddelow SA, Guttenplan KA, Clarke LE, Bennett FC, Bohlen CJ, Schirmer L, et al. Neurotoxic reactive astrocytes are induced by activated microglia. Nature. 2017;541(7638):481–7.
42. Yun SP, Kam TI, Panicker N, Kim S, Oh Y, Park JS, et al. Block of A1 astrocyte conversion by microglia is neuroprotective in models of Parkinson's disease. Nat Med. 2018;24(7):931–8.
43. Rostami J, Holmqvist S, Lindström V, Sigvardson J, Westermark GT, Ingelsson M, et al. Human astrocytes transfer aggregated alpha-synuclein via tunneling nanotubes. J Neurosci. 2017;37(49):11835–53.
44. Al-Bachari S, Naish JH, Parker GJM, Emsley HCA, Parkes LM. Blood-brain barrier leakage is increased in Parkinson's disease. Front Physiol. 2020;11:593026.
45. Baba Y, Kuroiwa A, Uitti RJ, Wszolek ZK, Yamada T. Alterations of T-lymphocyte populations in Parkinson disease. Parkinsonism Relat Disord. 2005;11(8):493–8.
46. Sommer A, Fadler T, Dorfmeister E, Hoffmann AC, Xiang W, Winner B, et al. Infiltrating T lymphocytes reduce myeloid phagocytosis activity in synucleinopathy model. J Neuroinflammation. 2016;13(1):174.
47. Rostami J, Fotaki G, Sirois J, Mzezewa R, Bergström J, Essand M, et al. Astrocytes have the capacity to act as antigen-presenting cells in the Parkinson's disease brain. J Neuroinflammation. 2020;17(1):119.
48. Palpagama TH, Waldvogel HJ, Faull RLM, Kwakowsky A. The role of microglia and astrocytes in Huntington's disease. Front Mol Neurosci. 2019;12:258.
49. Politis M, Lahiri N, Niccolini F, Su P, Wu K, Giannetti P, et al. Increased central microglial activation associated with peripheral cytokine levels in premanifest Huntington's disease gene carriers. Neurobiol Dis. 2015;83:115–21.

50. Rodrigues FB, Byrne LM, McColgan P, Robertson N, Tabrizi SJ, Zetterberg H, et al. Cerebrospinal fluid inflammatory biomarkers reflect clinical severity in Huntington's disease. PLoS One. 2016;11(9):e0163479.
51. Petkau TL, Hill A, Connolly C, Lu G, Wagner P, Kosior N, et al. Mutant huntingtin expression in microglia is neither required nor sufficient to cause the Huntington's disease-like phenotype in BACHD mice. Hum Mol Genet. 2019;28(10):1661–70.
52. Khakh BS, Beaumont V, Cachope R, Munoz-Sanjuan I, Goldman SA, Grantyn R. Unravelling and exploiting astrocyte dysfunction in Huntington's disease. Trends Neurosci. 2017;40(7):422–37.
53. Ferrari Bardile C, Garcia-Miralles M, Caron NS, Rayan NA, Langley SR, Harmston N, et al. Intrinsic mutant HTT-mediated defects in oligodendroglia cause myelination deficits and behavioral abnormalities in Huntington disease. Proc Natl Acad Sci U S A. 2019;116(19):9622–7.
54. Rodrigues Lima-Junior J, Sulzer D, Lindestam Arlehamn CS, Sette A. The role of immune-mediated alterations and disorders in ALS disease. Hum Immunol. 2021;82(3):155–61.
55. Ferrari R, Kapogiannis D, Huey ED, Momeni P. FTD and ALS: a tale of two diseases. Curr Alzheimer Res. 2011;8(3):273–94.
56. Tondo G, Iaccarino L, Cerami C, Vanoli GE, Presotto L, Masiello V, et al. (11) C-PK11195 PET-based molecular study of microglia activation in SOD1 amyotrophic lateral sclerosis. Ann Clin Transl Neurol. 2020;7(9):1513–23.
57. Johansson A, Engler H, Blomquist G, Scott B, Wall A, Aquilonius SM, et al. Evidence for astrocytosis in ALS demonstrated by [11C](L)-deprenyl-D2 PET. J Neurol Sci. 2007;255(1–2):17–22.
58. Song S, Miranda CJ, Braun L, Meyer K, Frakes AE, Ferraiuolo L, et al. Major histocompatibility complex class I molecules protect motor neurons from astrocyte-induced toxicity in amyotrophic lateral sclerosis. Nat Med. 2016;22(4):397–403.
59. Vahsen BF, Gray E, Thompson AG, Ansorge O, Anthony DC, Cowley SA, et al. Non-neuronal cells in amyotrophic lateral sclerosis – from pathogenesis to biomarkers. Nat Rev Neurol. 2021;17(6):333–48.
60. Gravel M, Béland LC, Soucy G, Abdelhamid E, Rahimian R, Gravel C, et al. IL-10 controls early microglial phenotypes and disease onset in ALS caused by misfolded superoxide dismutase 1. J Neurosci. 2016;36(3):1031–48.
61. Yamanaka K, Chun SJ, Boillee S, Fujimori-Tonou N, Yamashita H, Gutmann DH, et al. Astrocytes as determinants of disease progression in inherited amyotrophic lateral sclerosis. Nat Neurosci. 2008;11(3):251–3.
62. Beers DR, Henkel JS, Zhao W, Wang J, Appel SH. CD4+ T cells support glial neuroprotection, slow disease progression, and modify glial morphology in an animal model of inherited ALS. Proc Natl Acad Sci U S A. 2008;105(40):15558–63.
63. Tong J, Huang C, Bi F, Wu Q, Huang B, Liu X, et al. Expression of ALS-linked TDP-43 mutant in astrocytes causes non-cell-autonomous motor neuron death in rats. EMBO J. 2013;32(13):1917–26.
64. Tahir W, Thapa S, Schatzl HM. Astrocyte in prion disease: a double-edged sword. Neural Regen Res. 2022;17(8):1659–65.
65. Lakkaraju AKK, Sorce S, Senatore A, Nuvolone M, Guo J, Schwarz P, et al. Glial activation in prion diseases is selectively triggered by neuronal PrP(Sc). Brain Pathol. 2022;32:e13056.
66. Guijarro IM, Garcés M, Andrés-Benito P, Marín B, Otero A, Barrio T, et al. Assessment of glial activation response in the progress of natural scrapie after chronic dexamethasone treatment. Int J Mol Sci. 2020;21(9):3231.
67. Hartmann K, Sepulveda-Falla D, Rose IVL, Madore C, Muth C, Matschke J, et al. Complement 3(+)-astrocytes are highly abundant in prion diseases, but their abolishment led to an accelerated disease course and early dysregulation of microglia. Acta Neuropathol Commun. 2019;7(1):83.

Chapter 14
Vascular Diseases

Abstract This group of diseases is caused by defects in the blood supply. Brain cells are only affected secondarily. In ischemic stroke, all cells die in the core area, but it is in the penumbra that key processes occur. Microglia are activated within minutes of a stroke and in the initial period their activation is beneficial. A subset of macrophages enters the brain, and its presence is essential for vascular integrity and angiogenesis. Cytokines transform the astrocytes into reactive subsets. One such subset is the border-forming reactive astrocyte, which is key in sealing and stabilizing the core–penumbra interface. Within the core, another reactive astrocyte subset serves a different function. They exhibit a graded response with decreased reactivity with increased distance from the fibrotic scar. Under the influence of microglial secretions, at first, they display an inflammatory phenotype, but with time they transform into protective subtypes. They have a large role in subsequent repair processes. The core turns into a fibrotic scar with the involvement of pericytes, macrophages, T lymphocytes, and fibroblasts. In white matter, oligodendrocytes are damaged and die, but progenitor cells move in and proliferate. For yet unknown reasons, the progenitors do not differentiate into mature oligodendrocytes. The features and mechanisms of ischemic preconditioning are described as are the differences between ischemic and hemorrhagic stroke for these cellular processes. A delayed complication of these strokes is vascular cognitive impairment, where microglia play a major driving role.

Keywords Border-forming reactive astrocyte · Cytokines · Fibrotic scar · Hemorrhagic stroke · Ischemic preconditioning · Ischemic stroke · Penumbra · Macrophages · Microglia · Neuroinflammation · Oligodendrocyte progenitor cells · Reactive astrocytes · Vascular cognitive impairment

14.1 Introduction

This group of diseases is based on the consequences of a reduction or disruption of the blood flow to the brain or parts of it. Brain vascular diseases are the leading neurological cause of death. These diseases are occurring because of the

W. Walz, *The Gliocentric Brain*, https://doi.org/10.1007/978-3-031-48105-5_14

dysfunction of blood vessels or circulation. As pointed out in an earlier chapter, neurons are very susceptible to a reduction in oxygen supply. This is because the brain has no significant energy stores and is reliant on continuously matching blood supply as its energy source. The matching is accomplished by two mechanisms: metabolic feedback as in other organs and a neurovascular unit, which acts as a feedforward control unit and is a more powerful homeostatic mechanism [1]. The neurovascular unit links areas with increased information processing to local increases in blood flow.

Reduction or cessation of blood flow to brain areas is caused by thrombosis or embolism. Other, less likely, causes are hemorrhage and cardiac arrest. A complete cessation of blood flow to a brain area for a few minutes causes neuronal death due to the limited energy resources. If blood flow is not stopped, but reduced in a way that over a longer time period the energy supply does not match energy demand, more subtle, but equally devastating consequences arise. In the vulnerable brain areas, chronic deterioration sets in. Depending on the affected area, it will often result in cognitive decline.

This chapter focuses on three scenarios. The first one and most important one deals with the results of an ischemic stroke, the most common neurovascular affliction. Then, the scenario of blood spilling into the parenchyma and its consequences as it happens during a hemorrhagic stroke is discussed. Finally, the consequences of the chronic deterioration of the neural tissue following a long-term reduction of blood supply and its resulting cognitive decline are discussed.

14.2 Ischemic Stroke

14.2.1 Introduction

Reductions of cerebral blood flow are usually the result of a deteriorated vascular structure. It affects the arterioles and capillaries by narrowing and thickening them, as well as increasing tortuosity. Aging, hypertension, diabetes, and hypercholesterolemia are risk factors for stiffening of arteries and atherosclerosis. These factors interfere with the neurovascular regulatory unit and cause a mismatch between energy demand and supply [2]. An acute stroke is usually caused by thrombosis due to the above conditions. Less frequently embolic occlusion by plaques of the carotid artery, aortic arch, or from the heart causes an ischemic stroke. So far, the effective remedy is injection of tissue plasminogen activator to lyse the clot and assist reperfusion. However, this therapeutic intervention has a narrow time window and is counterproductive in a hemorrhagic stroke [3]. In a small number of patients, it is possible to succeed with mechanical thrombectomy [4]. An ischemic stroke results in most cases in an ischemic core with necrotic cells. In the core, neurons die very fast as the infarct size is already at 3 mm^3 within 2 h of the onset of a photothrombotic stroke in mice [5]. It is surrounded by a penumbra consisting of live but distressed cells. Here, neurons stop functioning as signal processors, but they are alive. The energy mismatch is pronounced at the border with the core, but eases further from the border [6]. The core is likely to expand over time into the penumbra. Therefore, therapeutic efforts must focus on the penumbra area as the cells there are salvageable, but time is essential. Even if the neurons in the penumbra survive a stroke, they can succumb later to apoptotic death. Spreading depression waves radiate out of the extremely depolarized core into the penumbra and beyond. In the penumbra, these waves cause an increased energy demand. Thus, these spreading depression waves are instrumental in expanding the necrotic core into the penumbra. Interestingly, the waves continue into healthy tissue, where they have an opposite effect. There the waves strengthen the resistance of the tissue to subsequent strokes, a phenomenon called ischemic preconditioning [7, 8]. Cell death in the ischemic core is due to energy failure and breakdown of gradients. It is the result of a decrease of blood flow below 20% of normal distal of the clogged vessel. Cell death or damage in the penumbra is more complicated and changes over the time following the occlusion. The most important mechanisms are based on excitotoxicity, calcium overload, oxidative, and nitrosative stress [2]. The excess depolarization of cells leads to glutamate overload, which activates AMPA (α-amino-3-hydroxy-5-methyl-4-isooxazolepropionic acid) and NMDA (N-methyl-D-aspartic acid) ionotropic receptors. This leads to ion and water influx and therefore to cell swelling. Depolarization, glutamate overload, reverse sodium/calcium exchange, activation of acid-sensing ion, and volume-regulated anion channels, as well as transient receptor potential cation channels all combine to raise intracellular calcium levels. This increased cytoplasmic calcium causes release of stored calcium from mitochondria in a positive feedback loop. The calcium overload activates proteases, lipases, and nucleases and increases NO (nitric oxide) and superoxide production. Oxidative phosphorylation becomes uncoupled, which contributes to ATP (adenosine triphosphate) depletion and ROS (reactive oxygen species) production. Inflammatory processes play a role. These processes are discussed in connection with neuronal–glial/immune cell interactions.

14.2.2 Time Course of Cellular Interactions

Within seconds to minutes of an experimental stroke, neurons in the affected area become electrically silent but are still alive. Within another 30 min, microglia in the ischemic area show signs of an activated morphology. The microglia are activated by a reduction of neuronal OFF signals like chemokine fractalkine and the glycoproteins CD200 (Ox-2 membrane glycoprotein) and SIRPa (CD47 signal regulatory protein alpha) as well as an increased release of DAMPs (damage-associated molecular patterns) from stressed neurons and other cells in the afflicted area. Some prominent DAMPs are ATP (adenosine triphosphate), UTP (uridine triphosphate), heat shock protein, and high mobility group box 1. In this acute phase, microglia produce anti-inflammatory cytokines and growth factors. Inhibition of microglia activation during an experimental stroke results therefore in a worse outcome and an increased damage area [9]. However, after a day or later, neutrophils invade the core area which is now populated with necrotic neurons. The neutrophils replace the microglia as the microglia are highly susceptible to energy failure [10]. The invading neutrophils are from two sources. One is the skull bone marrow, where neutrophils are quickly activated after an experimental murine stroke and then migrate through small vascular channels that connect the skull bone marrow with the meninges. The neutrophils are probably activated by factors released from the damaged parenchyma and which diffuse through the channels to the skull bone marrow like SDF-1 (stromal cell-derived factor 1). The neutrophils can move against the flow into the CSF and the parenchyma to home into the ischemic core. A similar mechanism is probable in human stroke [11]. The blood–brain barrier is breached within 24–48 h of a stroke due to degradation of tight junctions, basement membrane, and astrocytic endfeet as well as loss of endothelial cells in the ischemic core. This provides an alternative, albeit delayed route for neutrophil invasion [12]. Neutrophils are mainly destructive for the tissue as they release proteases, reactive oxygen, and nitrogen species, inflammatory IL-1β (interleukin-1beta), and formation of NETs (neutrophil extracellular traps). The NETs are a meshwork of fibers composed of chromatin and serine proteases. The neutrophils phagocytose the dying and dead neurons. Neutrophil migration out of the core is prevented by microglia in the penumbra, which phagocytose neutrophils in turn and prevent the invasion of the penumbra by neutrophils. Within a few days microglia remove a significant part of the neutrophils. Inhibition of microglia, at this stage, enhances neutrophil activity and exacerbates stroke damage [13]. Until this stage, the impact of microglia is mostly beneficial for tissue survival. Border-associated macrophages seem to have only a minor role in ischemic stroke. In contrast, monocyte-derived macrophages enter the parenchyma from the vasculature after a stroke in large numbers. Inhibiting this macrophage invasion is detrimental to the stroke outcome [14]. The invasion of neutrophils and macrophages is facilitated by a disintegrating blood–brain barrier. There seem to be at least two subsets of invading macrophages. CCR2+ (C-C chemokine receptor 2 positive) macrophages are proinflammatory and dominate in the core, CX3CR1+ (C-X3-C motive chemokine receptor 1 or fractalkine receptor positive) macrophages have protective functions [15]. Within 48 h of a stroke, tight junctions are proteolytically degraded and endothelial cells and astrocyte endfeet are lost [16]. After entering the CX3CR1+ macrophages display a protective function and are very important for maintaining and re-establishing vascular integrity. If postischemic macrophage invasion is prevented in experimental stroke, the results are reduced angiogenesis, increased hemorrhagic leakage, and vascular disintegration [17]. The astrocytes in the ischemic core die in concert with the neurons and microglia. In the penumbra, within days the astrocytic JAK/STAT3 (Janus kinase/signal transducer and activator of transcription 3) pathway is activated by TGFα (transforming growth factor alpha), CNTF (ciliary neurotrophic factor), IL-6, and LIF (leukemia inhibitory factor). This pathway is one of the major mechanisms to transform astrocytes into reactive subtypes [18]. Some of these reactive astrocytes proliferate and migrate to the core-penumbra interface where they transform into border-forming reactive astrocytes. By excreting extracellular matrix molecules, they seal the core from the parenchyma and make sure the core is functionally isolated from the penumbra. Further migration of immune cells out of the core is prevented by the astrocyte barrier and this is an important protective mechanism. Within the penumbra, the reactive astrocytes fulfill a different role.

Fibroblasts play a role in the ischemic core. Normally these cells border the subarachnoid space as they line arachnoid and pial layer. Collagen fibers between these layers are covered with fibroblast-like cells and divide the subarachnoid space into compartments [19]. As the pial layer also surrounds the larger blood vessels (arteries, arterioles, veins, venules), perivascular fibroblasts accompany them into parts of the parenchyma. These fibroblasts proliferate and migrate into the core area after a stroke. Pericytes are probably a cell type consisting of different subtypes which lie within the basement membrane of capillaries. After injury, at least one subtype (tentatively called "type A pericyte") divides and transforms into fibroblasts [20]. The pericytes are thereafter quickly lost, whereas the newly generated fibroblasts organize the core area into a fibrotic scar. This fibrotic scarring is stimulated by proinflammatory cytokines like IL-1β, IL-6, and TNF-α (tumor necrosis factor alpha). These factors are secreted by microglia and CCR2+ macrophages. The fibroblasts also secrete laminins, collagens, and fibronectin. This creates a fibrous extracellular matrix, which together with the fibroblasts fills out the space of the core. This space was vacated by the original brain cells (neurons, microglia, astrocytes). It is surrounded by a tight border of reactive barrier astrocytes (see Fig. 14.1). The CX3CR1+ macrophages are active in reducing the fibrotic scar. They are now in the periphery of the core and are probably responsible for the fact that the fibrotic scar volume shrinks somewhat over time. As it ages, the fibrotic scar contains some cysts (fluid-filled spaces). Deletion of STAT3 from astrocytes, and therefore reducing the genesis of reactive astrocytes, leads to a diffuse rather than clearly marked core-penumbra border. It also increases the core space and neuronal death [21]. The repelling of the fibroblasts by astrocytes is accomplished with astrocytic ephrin-B2 and fibroblast EphB2 signaling. Semaphorin III is a chemorepellent that is secreted by fibroblasts and prevents neurite outgrowth [22].

What about the role of invading immune cells in these contexts? There are two main mechanisms to attract immune cells to the site of the infarct. As is the case during an infection (see previous chapter), antigens, not previously encountered in the healthy brain, leak out of damaged cells. Such antigens can be neuronally derived (e.g., microtubule-associated protein-2 and N-methyl D-aspartate receptor subunit NR-2A) or released by damaged oligodendrocytes and their myelin sheath (myelin basic protein and myelin oligodendrocyte glycoprotein). Pathways can be diffusion through the compromised blood–brain barrier or diffusion out of the brain via CSF (cerebrospinal fluid) and lymph. The antigens can also interact with dendritic cells at the brain surfaces, which then migrate to lymph nodes. The main target of the dendritic cells is the cervical lymph nodes, but others including the spleen can be involved [23]. The other main mechanism is phagocytosis of dead cells in the core infarct zone by the macrophages. These macrophages then present the antigen together with MHC II (major histocompatibility complex II) and migrate out of the

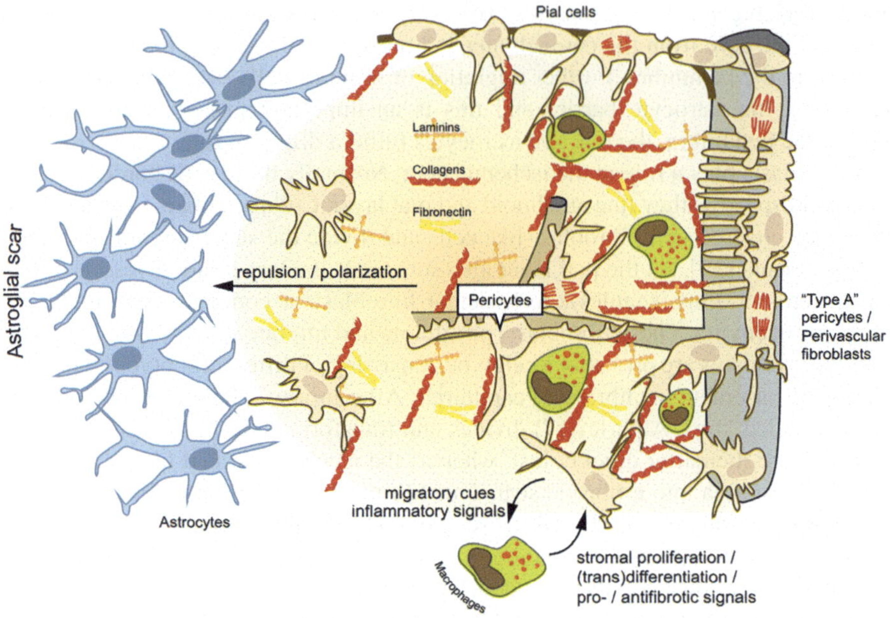

Fig. 14.1 Cellular and molecular interactions in central nervous system (CNS) scar formation. The fibrotic scar is characterized by the deposition of extracellular matrix molecules that are otherwise scarcely expressed in the neural parenchyma such as collagens, laminins, and fibronectin. These molecules are generated by stromal cells (myofibroblasts), which are normally absent from the CNS parenchyma. The stromal cells may originate from meningeal precursors (e.g., pial cells of the leptomeningeal lining), perivascular fibroblasts, or pericytes ("type A pericytes"). Blood-borne macrophages and microglia contribute to the proliferation and differentiation of stromal cells by producing profibrotic mediators, and they are also involved in the resolution of the fibrous scar. Conversely, stromal cells modulate neuroinflammation by producing cytokines, chemokines, and adhesion molecules. The astroglial scar is neatly separated from the fibrotic scar. Cellular and molecular constituents of the fibrotic scar induce the repulsion and polarization of astrocytes, and a new glia limitans is generated at the interphase of both cell populations. (From Ref. [20]; reprinted as part of Creative Common License Attribution 4.0 International from International Society of Neuropathology)

brain through the damaged blood–brain barrier. They reach the lymph nodes via blood circulation. B and T lymphocytes in the lymph nodes are stimulated and migrate into the brain through the compromised blood–brain barrier. This happens several days after the onset of the stroke. B lymphocytes seem to play only a minor role in stroke [12]. However, proinflammatory CD4+ and CD8+ T lymphocytes are in the penumbra area for up to 1 month after the infarct [24], where they proliferate and interact with astrocytes and contribute to a negative outcome. Another problem is that there is a risk that some of these T lymphocytes are causing autoimmune reactions to self-antigens, especially to myelin basic protein. These reactions can last over 90 days and be more intense than during a multiple sclerosis episode [25]. However, a large subclass of these T lymphocytes are regulatory T lymphocytes

(Tregs). Their density increases in the weeks after a stroke and is prevalent in the chronic phase afterward. They are attracted into the infarct area by the chemokines CCL1 and 20 (chemokine C-C motif ligand 1 and 20), released from macrophages, microglia, and other T lymphocyte populations. Their proliferation is stimulated by IL-33, released from reactive astrocytes. Tregs are a population of lymphocytes, which is active in preventing proliferation of other T lymphocytes, and they inhibit release of proinflammatory cytokines and prevent autoimmune reactions. Their activity supports neuronal recovery in the penumbra and their appearance correlates with improved neurological outcomes. The release of amphiregulin, an EGFR (epidermal growth factor receptor) ligand, by Tregs suppresses neurotoxic astrogliosis [26]. Thus, T cell invasion results in a delicate balance of proinflammatory and protective/restorative activity by different subgroups. In the chronic phase, the balance shifts normally toward the restorative activity.

Astrocytes in the necrotic core do not survive. Subsequently, the core is not repopulated by migrating reactive astrocytes or microglia. The core turns into a fibrotic scar, consisting of fibroblasts, pericytes, and macrophages. In the penumbra, astrocytes are turned into a reactive cell subtype by inflammatory factors released by activated microglia (TGF-α, IL-6, LIF, TNF-α) and distressed neurons (CNTF, IL-1, IL-6, ATP). Most of these factors activate the astrocytic JAK/STAT3 pathway. This is a key pathway to turn normal astrocytes into a reactive phenotype. Another important pathway in this context is the TGF-β/Smad (mothers against decapentaplegic homologs) signaling path. Fibrinogen is part of the blood coagulation system. It leaks out of the blood vessels due to the disintegration of the blood–brain barrier and diffuses into the surrounding tissue where it is deposited. It is a major signal for reactive astrocytes close to the core [27]. The reactive astrocytes interacting with fibrinogen proliferate and migrate to the border of the fibrotic scar. Here this subgroup of reactive astrocytes turns into border-forming palisading astrocytes. These newly formed, elongated, and polarized reactive astrocytes are necessary to contain the fibrotic scar and prevent macrophages and fibroblasts to invade into the penumbra. Some reports indicate these scar border-forming astrocytes might transform from a distinct astrocyte subpopulation [28]. In any event, blockade of the fibrinogen activity reduces the clear border between core and penumbra. It leads to influx of macrophages and fibrotic cells into the penumbra and increases neuronal death and demyelination in the penumbra [27]. The scar border-forming astrocyte is not the only cell type within this compact border. There are in addition activated microglia and NG2 (neuron-glial antigen 2) glia [29]. A tight extracellular matrix (ECM) is deposited between these cells. The major ECM constituent is CSPG (chondroitin sulphate proteoglycan). The further reactive astrocytes are away from this border, the less they exhibit a polarized structure and the more they appear with a stellate morphology [18], with the axis pointing toward the core area. Thus, there is a pronounced graded response with respect to the distance from the compact border. Closer to the core, the astrocytic reactivity is more pronounced than further away. Some reports point to a neurotoxic A1 and a neuroprotective A2 reactive subtype. However, this might be an oversimplification and the A1 and A2 subtypes are rather the two extremes of a highly flexible and plastic-reactive astrogliosis response.

Most evidence points toward an acute reactive astrocytic response in the early days after the infarct, when under the influence of microglial secretions astrocytes become inflammatory and damaging to neurons. Later, the astrocytes turn on more neuroprotective properties and assist neuronal survival and repair. This temporal pattern co-exists with the spatial gradient between core–penumbra barrier and the penumbra–healthy tissue interface.

The early neurotoxic reaction of astrocytes in the penumbra points to a close microglial–astrocytic interaction [30]. Under the influence of DAMPs, microglia release inflammatory cytokines (IL-1α, TNF-α) and the complement component subunit 1q (C1q) among other proinflammatory factors [18]. Inflammatory astrocytes in turn release proinflammatory mediators (IL-6, TNF-α, IL-1α and β, IFN-γ or interferon gamma) that act on both astrocytes and microglia in positive feedback loop, if only for a short time in the acute phase. These inflammatory astrocytes upregulate receptors, which are usually only found on cells that are part of the innate immune system (NOD-like receptors or nucleotide-binding oligomerization domain-containing protein-like receptors, mannose receptors, scavenger receptors like Megf10 or multiple EGF-like-proteins 10 and complement receptors). This inflammatory reaction is more intense closer to the core, contributing to the spatial reactivity gradient. Experimental inhibition of microglia reduces neurotoxic actions of astrocytes [31]. These actions consist of phagocytosis of synapses, which leads to neuronal apoptosis, glutamate release, causing excitotoxicity of neurons and release of ROS. Astrocyte endfeet, cell bodies, and processes swell. This swelling is dependent on the AQP4 (aquaporin 4) channels. However, astrocytes can reduce this swelling due to regulatory volume decrease, mainly mediated by taurine release [32]. Reactive astrocytes release exosomes (extracellular vesicles) which contain various miRNAs (micro ribonucleic acids). These miRNAs enter neurons and interact with their mRNA. However, there is a controversy if these miRNA actions are beneficial or deleterious for neuronal health. In the acute phase, astrocytes release several substances that attract neutrophils and CD8+ T lymphocytes (IL-15, -17). Astrocytes can act as a break to the inflammatory role of microglia by releasing TGF-β and thus re-establishing some form of a negative feedback system. Thus, in the chronic phase reactive astrocytes change their properties [33]. They downregulate their neurotoxic nature and upregulate neuroprotective features. They now upregulate their glutamate transporters. They release antioxidants and glutathione, which is used by neurons as a substrate for antioxidants. For retinal neurons, it was shown that they transfer damaged mitochondria to astrocytes for disposal to counteract apoptosis. Astrocytic EPO (erythropoietin), VEGF (vascular endothelial growth factor), and GDNF (glial cell line-derived neurotrophic factor) support neuroprotection in the chronic phase. A subset of astrocytes uses phagocytosis to clear cellular debris, but not functional synapses. In this phase, the astrocytes support angiogenesis and blood–brain barrier formation. One of the factors used by astrocytes for this task is Shh (sonic hedgehog protein). In rodents, reactive astrocytes form a scaffold and use CXCL12 as a chemoattractant to guide migrating neuroblasts from the SVZ (subventricular zone) to the ischemic penumbra. In addition, they release thrombospondin 1 and 2 and a sigma-1 receptor agonist to promote synaptogenesis and axon sprouting. They also promote the proliferation of Tregs by IL-33 and CCL1 release in the chronic phase, thereby contributing to a reduction in the inflammatory status of the penumbra.

White matter is more sensitive to stroke damage than gray matter [34]. The main reason is the demise of oligodendrocytes due to infarcts. In addition, myelin lipid metabolism is negatively affected by an infarct. The stroke, however, induces the proliferation of OPCs (oligodendrocyte precursors). This OPC proliferation is promoted by astrocytes and microglia. However, the problem is that most of these OPCs fail to mature into full oligodendrocytes, which could undertake the necessary remyelination. The reason is that whereas some OPCs differentiate into astrocytes, most of the differentiation process of OPCs into mature oligodendrocytes is arrested for reasons still unknown. Thus, repair and regeneration after a stroke leads only to partial remyelination.

14.2.3 Repair and Remodeling After Ischemic Stroke

Regeneration after stroke seems to superficially resemble developmental processes. However, the regenerative and developmental transcriptomes are different [35]. Remodeling or reorganization are not referred to as repair unless they lead to functional recovery [36]. In the gray matter, these processes begin within days of the infarct. They involve axonal sprouting and dendritic branching [37]. The density of dendritic spines may even increase to levels higher than normal [35]. There is angiogenesis in the penumbra and SVZ-derived neuroblasts are integrated into functional circuits [38]. In the white matter, the situation is somewhat more complicated as axons surviving the infarct succumb to degeneration days later. There is clear evidence that the glial barrier surrounding the fibrotic scar is inhibiting axonal growth. This is due to extracellular matrix components; the most important ones in this respect are chondroitin sulfate proteoglycans. However, it must be kept in mind that this barrier is shielding the penumbra from a fibrotic non-parenchymal microenvironment. This microenvironment contains not much glial support, even though it is supplied by new blood vessels. Thus, this space encapsulated by barrier astrocytes and some OPCs is not offering any reasonable prospects for the establishment of functional circuits. Indeed, some experimenters, using reactive astrogliosis ablation, found none or only minor detrimental effects for neural restoration due to the barrier in the cerebral cortex [39]. In mice, the first few days after a stroke are defined by hyperexcitability, most likely due to glutamate toxicity. Thereafter the circuits show a pronounced depression with increased GABA (gamma aminobutyric acid) signaling. It has been demonstrated that counteracting this inhibition with various strategies is highly beneficial [40]. In both rodents and humans, these active cellular and molecular processes lead to new connections and cortical map representations. The contralateral hemisphere is actively involved in this remapping. The larger the infarcted core area, the more the contralateral hemisphere is involved in these macro-level changes in cortical maps and networks [41]. These interactions are underlying modern rehabilitation approaches [42, 43]. Astrocytes have a role in these repair processes. Their impact on angiogenesis and neuroblast migration and incorporation was already pointed out. Astrocytes and microglia secrete neurotrophic factors which enhance neuronal plasticity including axonal sprouting and synaptogenesis (bFGF or basic fibroblast growth factor, NGF or nerve growth factor, CNTF, BDNF or brain-derived neurotrophic factor, EPO). Most of these factors are secreted during the acute phase after a stroke, when they support neuronal survival. In the restorative phase, these factors promote restoration of neuronal circuits [39]. Thrombospondins release by astrocytes is strongly increased in the recovery period and is an important factor in synaptogenesis [44].

14.2.4 *Ischemic Preconditioning*

Ischemic preconditioning is a very robust phenomenon that has been known for over 50 years and has been shown to exist in a variety of systems and situations. Stroke patients who suffered a previous TIA (transient ischemic attack) have a better recovery chance than those who do not [45]. This situation can be mimicked in in vivo and in vitro animal models of stroke or ischemia. In the CNS (central nervous system), if both the preconditioning stimulus and the following injury are an ischemic insult, the phenomenon is called ischemic preconditioning. If the first stimulus is of a different nature than the injury, the phenomenon is named cross tolerance [46]. Most cell types shift after a first and more mild ischemic challenge to a reduced energy metabolism reminiscent of cells in hibernating organisms [47]. Nevertheless, blood flow measurements and neuron-centric studies cannot easily explain the preconditioning phenomenon. However, most recent studies point toward a crucial involvement of microglia in this phenomenon. After a previous non-damaging, ischemic preconditioning stimulus, microglia are reprogrammed into a more protective subtype. The stimulus involves microglial TLR4 (Toll-like receptor 4) and IFNAR1 (interferon-alpha/beta receptor 1) as the knock-out of these receptors on microglia alone abolishes ischemic preconditioning [48]. Further studies showed that priming of microglia with repeated IFNβ injections inferred protection against a subsequent ischemic insult. Both TLR4 and IFNAR1 activation upregulated interferon-stimulated genes. The most important transcription factors involved are interferon regulatory factors [49]. This activation leads to strong microglial proliferation after the first (priming) stimulus. Thus, the picture that emerges points to a first mild ischemic stimulus releasing DAMPs from most parenchymal cells to engage microglial TLRs. Together with IFNAR1 stimulation this leads to a reprogramming of microglia toward a more anti-inflammatory and protective state. It skips the proinflammatory state that is upregulated in a more damaging ischemic infarct. It should be noted that microglia primed in this way will not stay in this protective mode for more than a few weeks.

14.3 Hemorrhagic Stroke

This type of stroke has two major forms, intracerebral hemorrhage (ICH) and subarachnoid hemorrhage. Here, the focus is on ICH. It is the rupture of a vessel with subsequent release of blood into the parenchyma or a ventricle. It can be caused by trauma, malformations, aneurism, tumor, or amyloid angiopathy. About 10–20% of strokes are of this nature and it has a far more detrimental prognosis than ischemic stroke. In pathophysiological terms, the difference of ICH is the blood extravasation leading to a hematoma [50]. There are two consequences of a rapid hematoma formation. There is a primary injury due to the mechanical impact of the blood mass formation. This is especially damaging if there is continued bleeding and the hematoma is expanding. Then there is secondary brain damage due to the action of extravasated blood ingredients like hemoglobin, hemin, thrombin, and iron [51]. The only possible treatment is surgical intervention in selected patients. Other than that, one must rely on supportive care.

The key cellular component in ICH secondary damage is the microglia. Within hours of an infarct activated microglia surround the hematomal brain region. This is the result of migration and proliferation. With a delay of several hours, additional blood-borne macrophages infiltrate the perihematomal region, probably recruited by the activated microglia. The microglia are proinflammatory, and their cytokine secretion recruits more microglia to the site. Other than cytokines (IL-1β, TNF-α), these microglia release chemokines, prostaglandins, proteases, ferrous iron, glutamate, and ROS [52]. These secretions lead to neuronal damage and death. Any experimental ICH model, which involves downregulation of microglia inflammation, reduces neuronal death and improves neurological outcome [53, 54]. In turn, the necrotic neurons stimulate microglia further by releasing ATP, neurotransmitters, heat shock proteins, and HMGB1 (high mobility group box 1 protein). A major factor shifting microglia toward a proinflammatory subtype is increased TLR4 expression and stimulation by hemin. This activates the NFƙB (nuclear factor kappa-light-chain-enhancer of activated B cells) signaling pathway that together the NLRP3 (NOD-like receptor family pyrin domain containing 3) pathway leads to increased inflammatory states. Release of chemokines (CXCL2) by microglia attracts even neutrophils, which can damage the blood–brain barrier further. Such a compromised blood–brain barrier is a major cause of brain edema. This swelling is associated with more severe forms of ICH [55]. Most studies agree that a large hematoma size and continuous expansion is the cause for the inflammatory status of microglia, macrophages, and neutrophils. Thus, hematoma resolution is key for circuit repair and recovery. Anti-inflammatory microglia subsets increase when inflammatory signals, acting, for example on Toll-like receptors, are reduced and anti-inflammatory signals (IL-10, IL-4, IL-13) take over. These anti-inflammatory microglia subtypes together with macrophages phagocytose cellular debris and hematoma components as well as catabolizes hemin and hemoglobin. These activities in turn reduce proinflammatory triggers [52]. Reactive astrocytes are involved in resolving the brain edema [56]. Reactive astrocytes are involved in hematoma expansion in that they actively secrete inflammatory cytokines and chemokines (IL-15, CCL2) which recruit proinflammatory microglia and macrophages to the perihematomal region [57]. Besides these mechanisms, the role of reactive astrocytes in ICH is still largely unknown [58].

14.4 Vascular Cognitive Impairment

As pointed out in a previous chapter, cognitive decline can have various causes, one of which may be vascular impairment. Cognitive decline or dementia due to mostly vascular causes is summarized as vascular cognitive impairment (VCI). The main VCI risk factor is increasing age, but genetic factors and the female gender [59] play some roles. Modifiable risk factors are diabetes, cholesterol/obesity, and late-life hypertension [60]. About 10% of patients with a first stroke develop dementia within a year of their stroke. Atherosclerosis, cerebral amyloid angiopathy, as well as smaller, asymptomatic infarcts like lacunar infarcts, microinfarcts, and hemorrhages all lead to hypoperfusion of the parenchyma for extended time periods [60]. This prolonged chronic under-supply leads to cortical atrophy, white matter disruption, and enlarged perivascular spaces. Thus, there is more accumulation of small, microscopic infarcts (less than 3 mm^3), which is causing the hypoperfusion in smaller vessels. Smaller hemorrhages and microbleeds also contribute to this chronic hypoperfusion [61]. Another important fact is that many dementia patients suffer from a mixed pathology. Here, VCI co-exists with Alzheimer, Lewy body dementia, frontotemporal dementia, or amyotrophic lateral sclerosis [62].

Analysis of a variety of VCI animal models suggests some key molecular mechanisms for brain damage [63]. Oxidative stress by increased ROS production and reduced antioxidants leads to excessive oxidation of proteins. A major contributor to ROS is nicotinamide adenine dinucleotide phosphate oxidase in endothelial cells. Its inhibition in animal models improved VCI-related behavior. Microglia seem to be the major cell type leading to VCI pathology. Microglia become chronically proinflammatory, releasing IL-6, TNF-α, ROS, and matrix metalloproteinases. The release damages the blood–brain barrier and myelin sheaths. As a result of this chronic microglia inflammation, the astrocytic endfeet–basement membrane anchoring is disrupted [64]. The main culprit is microglia-released matrix metalloproteinases. The result is a weakened neurovascular coupling, which in turn contributes further to the already reduced hypoperfusion. Nitric oxide (NO) is produced by endothelial cells to act on smooth muscle cells to control vascular tone. In VCI animal models, the reduction of NO release is seen as a major pathological factor. This may also be due to the additional role NO plays as a feedback signal in neuronal glutamate release [65]. The deterioration of the blood–brain barrier and neurovascular coupling further involves smooth muscle degeneration and loss of pericytes [66].

14.5 Conclusion

The role of the glial cells in infarcts is one of a delicate balance. The microglia and astrocytes must judge from a variety of signals the strength, localization, extent, and status of the parenchyma disturbance. In the acute phase, microglia take over and recruit external cells and try to limit the necrotic core. They use neutrophils for a brief period to remove damage, and then they use macrophages to build the core into an inert space, which will not act as a further thread to the cells struggling in the penumbra. Macrophages recruit pericytes, which transform into fibroblasts for this purpose. Astrocytes are used to limit the fibrotic core, assist neuronal survival, and then to rebuild neuronal connections. Both microglia and astrocytes go through an inflammatory stage, which seems detrimental to neurons. It results in a reorganization of the space and removal of damaging debris. In ischemic preconditioning, microglia seem to be able to skip this damaging phase, probably due to the milder first insult with reduced DAMPs. In infarcts there is not much of an aggressive involvement of T lymphocytes. Rather, Tregs dominate and contribute to an anti-inflammatory environment after the acute phase. The presence of blood components interferes with this picture and complicates the outlook somewhat.

References

1. Iadecola C. The neurovascular unit coming of age: a journey through neurovascular coupling in health and disease. Neuron. 2017;96(1):17–42.
2. Moskowitz MA, Lo EH, Iadecola C. The science of stroke: mechanisms in search of treatments. Neuron. 2010;67(2):181–98.
3. Stevenson TK, Moore SJ, Murphy GG, Lawrence DA. Tissue plasminogen activator in central nervous system physiology and pathology: from synaptic plasticity to Alzheimer's disease. Semin Thromb Hemost. 2022;48(3):288–300.
4. Dmytriw AA, Zhang Y, Mendes PV. Mechanical thrombectomy and the future of acute stroke treatment. Eur J Radiol. 2019;112:214–21.
5. Gorlamandala N, Parmar J, Craig AJ, Power JM, Moorhouse AJ, Krishnan AV, et al. Focal Ischaemic infarcts expand faster in cerebellar cortex than cerebral cortex in a mouse Photothrombotic stroke model. Transl Stroke Res. 2018;9(6):643–53.
6. Yang SH, Liu R. Four decades of ischemic penumbra and its implication for ischemic stroke. Transl Stroke Res. 2021;12(6):937–45.
7. Lauritzen M, Strong AJ. 'Spreading depression of Leão' and its emerging relevance to acute brain injury in humans. J Cereb Blood Flow Metab. 2017;37(5):1553–70.
8. Shen PP, Hou S, Ma D, Zhao MM, Zhu MQ, Zhang JD, et al. Cortical spreading depression-induced preconditioning in the brain. Neural Regen Res. 2016;11(11):1857–64.
9. Kawabori M, Kacimi R, Kauppinen T, Calosing C, Kim JY, Hsieh CL, et al. Triggering receptor expressed on myeloid cells 2 (TREM2) deficiency attenuates phagocytic activities of microglia and exacerbates ischemic damage in experimental stroke. J Neurosci Off J Soc Neurosci. 2015;35(8):3384–96.
10. Ritzel RM, Patel AR, Grenier JM, Crapser J, Verma R, Jellison ER, et al. Functional differences between microglia and monocytes after ischemic stroke. J Neuroinflammation. 2015;12:106.

11. Herisson F, Frodermann V, Courties G, Rohde D, Sun Y, Vandoorne K, et al. Direct vascular channels connect skull bone marrow and the brain surface enabling myeloid cell migration. Nat Neurosci. 2018;21(9):1209–17.
12. Iadecola C, Buckwalter MS, Anrather J. Immune responses to stroke: mechanisms, modulation, and therapeutic potential. J Clin Invest. 2020;130(6):2777–88.
13. Otxoa-de-Amezaga A, Miró-Mur F, Pedragosa J, Gallizioli M, Justicia C, Gaja-Capdevila N, et al. Microglial cell loss after ischemic stroke favors brain neutrophil accumulation. Acta Neuropathol. 2019;137(2):321–41.
14. Wattananit S, Tornero D, Graubardt N, Memanishvili T, Monni E, Tatarishvili J, et al. Monocyte-derived macrophages contribute to spontaneous long-term functional recovery after stroke in mice. J Neurosci Off J Soc Neurosci. 2016;36(15):4182–95.
15. Garcia-Bonilla L, Faraco G, Moore J, Murphy M, Racchumi G, Srinivasan J, et al. Spatiotemporal profile, phenotypic diversity, and fate of recruited monocytes into the post-ischemic brain. J Neuroinflammation. 2016;13(1):285.
16. Knowland D, Arac A, Sekiguchi KJ, Hsu M, Lutz SE, Perrino J, et al. Stepwise recruitment of transcellular and paracellular pathways underlies blood-brain barrier breakdown in stroke. Neuron. 2014;82(3):603–17.
17. Gliem M, Mausberg AK, Lee JI, Simiantonakis I, van Rooijen N, Hartung HP, et al. Macrophages prevent hemorrhagic infarct transformation in murine stroke models. Ann Neurol. 2012;71(6):743–52.
18. Pekny M, Wilhelmsson U, Tatlisumak T, Pekna M. Astrocyte activation and reactive gliosis-A new target in stroke? Neurosci Lett. 2019;689:45–55.
19. Xu L, Yao Y. Central nervous system fibroblast-like cells in stroke and other neurological disorders. Stroke. 2021;52(7):2456–64.
20. Fernández-Klett F, Priller J. The fibrotic scar in neurological disorders. Brain Pathol (Zurich, Switzerland). 2014;24(4):404–13.
21. Wanner IB, Anderson MA, Song B, Levine J, Fernandez A, Gray-Thompson Z, et al. Glial scar borders are formed by newly proliferated, elongated astrocytes that interact to corral inflammatory and fibrotic cells via STAT3-dependent mechanisms after spinal cord injury. J Neurosci. 2013;33(31):12870–86.
22. Pekcec A, Yigitkanli K, Jung JE, Pallast S, Xing C, Antipenko A, et al. Following experimental stroke, the recovering brain is vulnerable to lipoxygenase-dependent semaphorin signaling. FASEB J. 2013;27(2):437–45.
23. Planas AM, Gómez-Choco M, Urra X, Gorina R, Caballero M, Chamorro Á. Brain-derived antigens in lymphoid tissue of patients with acute stroke. J Immunol (Baltimore, Md: 1950). 2012;188(5):2156–63.
24. Xie L, Li W, Hersh J, Liu R, Yang SH. Experimental ischemic stroke induces long-term T cell activation in the brain. J Cereb Blood Flow Metab. 2019;39(11):2268–76.
25. Becker KJ, Kalil AJ, Tanzi P, Zierath DK, Savos AV, Gee JM, et al. Autoimmune responses to the brain after stroke are associated with worse outcome. Stroke. 2011;42(10):2763–9.
26. Ito M, Komai K, Mise-Omata S, Iizuka-Koga M, Noguchi Y, Kondo T, et al. Brain regulatory T cells suppress astrogliosis and potentiate neurological recovery. Nature. 2019;565(7738):246–50.
27. Conforti P, Mezey S, Nath S, Chu YH, Malik SC, Martínez Santamaría JC, et al. Fibrinogen regulates lesion border-forming reactive astrocyte properties after vascular damage. Glia. 2022;70(7):1251–66.
28. Martín-López E, García-Marques J, Núñez-Llaves R, López-Mascaraque L. Clonal astrocytic response to cortical injury. PLoS One. 2013;8(9):e74039.
29. Adams KL, Gallo V. The diversity and disparity of the glial scar. Nat Neurosci. 2018;21(1):9–15.
30. He T, Yang GY, Zhang Z. Crosstalk of astrocytes and other cells during ischemic stroke. Life (Basel, Switzerland). 2022;12(6):910.
31. Liddelow SA, Guttenplan KA, Clarke LE, Bennett FC, Bohlen CJ, Schirmer L, et al. Neurotoxic reactive astrocytes are induced by activated microglia. Nature. 2017;541(7638):481–7.

32. Shen XY, Gao ZK, Han Y, Yuan M, Guo YS, Bi X. Activation and role of astrocytes in ischemic stroke. Front Cell Neurosci. 2021;15:755955.
33. Li L, Zhou J, Han L, Wu X, Shi Y, Cui W, et al. The specific role of reactive astrocytes in stroke. Front Cell Neurosci. 2022;16:850866.
34. Chen D, Huang Y, Shi Z, Li J, Zhang Y, Wang K, et al. Demyelinating processes in aging and stroke in the central nervous system and the prospect of treatment strategy. CNS Neurosci Ther. 2020;26(12):1219–29.
35. Regenhardt RW, Takase H, Lo EH, Lin DJ. Translating concepts of neural repair after stroke: structural and functional targets for recovery. Restor Neurol Neurosci. 2020;38(1):67–92.
36. Bernhardt J, Hayward KS, Kwakkel G, Ward NS, Wolf SL, Borschmann K, et al. Agreed definitions and a shared vision for new standards in stroke recovery research: the stroke recovery and rehabilitation roundtable taskforce. Neurorehabil Neural Repair. 2017;31(9):793–9.
37. Carmichael ST, Kathirvelu B, Schweppe CA, Nie EH. Molecular, cellular and functional events in axonal sprouting after stroke. Exp Neurol. 2017;287(Pt 3):384–94.
38. Liang H, Zhao H, Gleichman A, Machnicki M, Telang S, Tang S, et al. Region-specific and activity-dependent regulation of SVZ neurogenesis and recovery after stroke. Proc Natl Acad Sci U S A. 2019;116(27):13621–30.
39. Liu Z, Chopp M. Astrocytes, therapeutic targets for neuroprotection and neurorestoration in ischemic stroke. Prog Neurobiol. 2016;144:103–20.
40. Alia C, Spalletti C, Lai S, Panarese A, Micera S, Caleo M. Reducing GABA(A)-mediated inhibition improves forelimb motor function after focal cortical stroke in mice. Sci Rep. 2016;6:37823.
41. Heiss WD, Thiel A. A proposed regional hierarchy in recovery of post-stroke aphasia. Brain Lang. 2006;98(1):118–23.
42. Wang L, Conner JM, Nagahara AH, Tuszynski MH. Rehabilitation drives enhancement of neuronal structure in functionally relevant neuronal subsets. Proc Natl Acad Sci U S A. 2016;113(10):2750–5.
43. Lu C, Wu X, Ma H, Wang Q, Wang Y, Luo Y, et al. Optogenetic stimulation enhanced neuronal Plasticities in motor recovery after ischemic stroke. Neural Plast. 2019;2019:5271573.
44. Liauw J, Hoang S, Choi M, Eroglu C, Choi M, Sun GH, et al. Thrombospondins 1 and 2 are necessary for synaptic plasticity and functional recovery after stroke. J Cereb Blood Flow Metab. 2008;28(10):1722–32.
45. Wegener S, Gottschalk B, Jovanovic V, Knab R, Fiebach JB, Schellinger PD, et al. Transient ischemic attacks before ischemic stroke: preconditioning the human brain? A multicenter magnetic resonance imaging study. Stroke. 2004;35(3):616–21.
46. McDonough A, Weinstein JR. The role of microglia in ischemic preconditioning. Glia. 2020;68(3):455–71.
47. Stenzel-Poore MP, Stevens SL, Simon RP. Genomics of preconditioning. Stroke. 2004;35(11 Suppl 1):2683–6.
48. Hamner MA, Ye Z, Lee RV, Colman JR, Le T, Gong DC, et al. Ischemic preconditioning in white matter: magnitude and mechanism. J Neurosci Off J Soc Neurosci. 2015;35(47):15599–611.
49. McDonough A, Lee RV, Noor S, Lee C, Le T, Iorga M, et al. Ischemia/reperfusion induces interferon-stimulated gene expression in microglia. J Neurosci. 2017;37(34):8292–308.
50. Qureshi AI, Mendelow AD, Hanley DF. Intracerebral haemorrhage. Lancet (London, England). 2009;373(9675):1632–44.
51. Lok J, Leung W, Murphy S, Butler W, Noviski N, Lo EH. Intracranial hemorrhage: mechanisms of secondary brain injury. Acta Neurochir Suppl. 2011;111:63–9.
52. Dasari R, Bonsack F, Sukumari-Ramesh S. Brain injury and repair after intracerebral hemorrhage: the role of microglia and brain-infiltrating macrophages. Neurochem Int. 2021;142:104923.
53. Li M, Li Z, Ren H, Jin WN, Wood K, Liu Q, et al. Colony stimulating factor 1 receptor inhibition eliminates microglia and attenuates brain injury after intracerebral hemorrhage. J Cereb Blood Flow Metab. 2017;37(7):2383–95.

54. Wasserman JK, Zhu X, Schlichter LC. Evolution of the inflammatory response in the brain following intracerebral hemorrhage and effects of delayed minocycline treatment. Brain Res. 2007;1180:140–54.
55. Appelboom G, Bruce SS, Hickman ZL, Zacharia BE, Carpenter AM, Vaughan KA, et al. Volume-dependent effect of perihaematomal oedema on outcome for spontaneous intracerebral haemorrhages. J Neurol Neurosurg Psychiatry. 2013;84(5):488–93.
56. Jeon H, Kim M, Park W, Lim JS, Lee E, Cha H, et al. Upregulation of AQP4 improves blood-brain barrier integrity and Perihematomal edema following intracerebral hemorrhage. Neurotherapeutics. 2021;18(4):2692–706.
57. Shi SX, Li YJ, Shi K, Wood K, Ducruet AF, Liu Q. IL (interleukin)-15 bridges astrocyte-microglia crosstalk and exacerbates brain injury following intracerebral hemorrhage. Stroke. 2020;51(3):967–74.
58. Scimemi A. Astrocytes and the warning signs of intracerebral hemorrhagic stroke. Neural Plast. 2018;2018:7301623.
59. Pendlebury ST, Rothwell PM. Incidence and prevalence of dementia associated with transient ischaemic attack and stroke: analysis of the population-based Oxford Vascular Study. Lancet Neurol. 2019;18(3):248–58.
60. Iadecola C, Duering M, Hachinski V, Joutel A, Pendlebury ST, Schneider JA, et al. Vascular cognitive impairment and dementia: JACC scientific expert panel. J Am Coll Cardiol. 2019;73(25):3326–44.
61. Yilmaz P, Ikram MK, Niessen WJ, Ikram MA, Vernooij MW. Practical small vessel disease score relates to stroke, dementia, and death. Stroke. 2018;49(12):2857–65.
62. Kapasi A, DeCarli C, Schneider JA. Impact of multiple pathologies on the threshold for clinically overt dementia. Acta Neuropathol. 2017;134(2):171–86.
63. Hort J, Vališ M, Kuča K, Angelucci F. Vascular cognitive impairment: information from animal models on the pathogenic mechanisms of cognitive deficits. Int J Mol Sci. 2019;20(10):2405.
64. Price BR, Norris CM, Sompol P, Wilcock DM. An emerging role of astrocytes in vascular contributions to cognitive impairment and dementia. J Neurochem. 2018;144(5):644–50.
65. Lourenço CF, Ledo A, Barbosa RM, Laranjinha J. Neurovascular-neuroenergetic coupling axis in the brain: master regulation by nitric oxide and consequences in aging and neurodegeneration. Free Radic Biol Med. 2017;108:668–82.
66. Montagne A, Nikolakopoulou AM, Zhao Z, Sagare AP, Si G, Lazic D, et al. Pericyte degeneration causes white matter dysfunction in the mouse central nervous system. Nat Med. 2018;24(3):326–37.

Chapter 15
Seizures

Abstract A brief introduction to the pathology of epilepsy is presented. Seizure-prone areas of the brain are always accompanied by reactive gliosis. These reactive astrocytes exhibit loss of domain organization, a phenomenon not seen in any other brain pathology. There is no evidence that astrocytic mutations are causing idiopathic seizures, although this remains a theoretical possibility. A different matter, however, is acquired epilepsy. The question is if the primary injury, which leads eventually to the seizures, is acting through reactive astrocytes to change neuronal excitability or if the primary injury directly affects neuronal excitation and this in turn causes reactive astrogliosis. The issue has not been addressed directly, but most (but not all) of the existing indirect evidence points to neuronal seizures causing gliosis rather than the other way around. In most animal models (but not all), initiating astrocytic calcium waves releases massive amounts of glutamate and this causes seizures as does any interference with astrocytic glutamate clearance. Another evidence, this time from human patients, points to genetic defects in the astrocytic potassium channel Kir4.1 interfering with potassium clearance and therefore lowering the neuronal firing threshold. There is evidence that, early in epileptogenesis, microglia secretions contribute to astrocyte transformation into a proconvulsive phenotype. The lost domain organization of reactive astrocytes in epileptic areas is an enigma. Not much research has focused on the functional properties of reactive astrocytes from epileptic brains versus those from brains afflicted with other pathologies.

Keywords Acquired epileptogenesis · Calcium waves · Domain organization · Glutamate release · Ictogenesis · Idiopathic epileptogenesis · Microglia · Potassium channel Kir4.1 · Proconvulsive astrocyte phenotype · Reactive astrocytes · Reactive gliosis

15.1 Introduction

Epilepsy as a disorder is characterized by recurrent seizures. There are two roots of epileptic pathologies: idiopathic and acquired epilepsies. An idiopathic epilepsy is due to unknown, probably genetic causes, as there are no known environmental risk

W. Walz, *The Gliocentric Brain*, https://doi.org/10.1007/978-3-031-48105-5_15

factors. Acquired epilepsies are caused by any previously existing brain diseases or conditions, like electrolyte disturbances, neurotoxins, medications, infections, trauma, stroke, cancer, or fever. Epilepsy affects 1–2% of the adult population and 4% of children. Seizures are paroxysmal events (meaning a sudden attack or outburst), which are transient and last only a short time (1–3 min). The exception to this short period is called status epilepticus (>5 min). Seizures are characterized by two features: abnormal neuronal excitation and synchronous discharges of groups of neurons. The active seizure period is called ictal, and the period between seizures is the interictal period. In this interictal period, the EEG (electroencephalogram) often exhibits interictal spiking without any symptoms. The discharges can affect any cortical or subcortical area. The affected brain area or epileptogenic zone determines the symptoms during the ictal discharge. Motor manifestations are the most prominent and may – in extreme cases – involve the complete voluntary musculature. These motor manifestations lead to convulsions (uncontrolled shaking). However, any brain activity like speech, cognition, perception, and consciousness can be affected. Not all seizures are focal or partial. Some can extend into both hemispheres by either expanding from a focal seizure (focal onset generalized seizures) or starting from the very beginning of the ictal discharges without going through a focal phase. These seizures are then called generalized onset seizures.

On the single neuron level, the ictal event often starts with calcium influx which affects channel openings and leads to a sustained depolarization (paroxysmal depolarizing shift). The depolarization increases toward a plateau, which normally reaches threshold and causes action potential bursts. The event is terminated by inhibitory actions of the GABA (γ-aminobutyric acid) transmitter system. Propagation to other neurons is due to the loss of surrounding inhibition and the spread of excitation by local networks. If strong enough, the excitation continues through the corpus callosum into the other hemisphere. The spread can happen through NMDA (N-methyl-D-aspartate) receptor activation, increased transmitter release due to presynaptic calcium overload, and accumulation of excess extracellular potassium, which will depolarize the membranes beyond action potential threshold.

There are around 30 different medications for epilepsy treatment, almost all targeting neuronal properties. The most prominent are lamotrigine, levetiracetam, topiramate, valproic acid, and zonisamide. About a third of the patients have residual seizures, which cannot be abolished by antiepileptic drugs. Alternative therapies include surgery to remove the focal point as well as corpus callosotomy and hemispherectomy. Other therapies involve neurostimulation and diet therapy (mostly ketogenic diet) [1].

A multitude of animal models for epilepsy exist, but only three are clinically validated [2]. These are the maximal electroshock seizure protocol, the subcutaneous pentylenetetrazol acute seizure test, and the kindled rodent model of chronic hyperexcitability.

Astrocytes play a significant role in epilepsy. However, one must distinguish between epileptogenesis and ictogenesis. Epileptogenesis is the process that causes the transformation of the normal brain into one with enduring epileptogenic potential. Ictogenesis is the generation of an ictal seizure in a brain with epileptogenic potential.

15.2 Reactive Gliosis and Epilepsy

The transformation of a brain area with normal neuronal output into an area with epileptogenic potential may be caused in some cases by glial cells. However, it appears that the epileptic area in human patients and animal models is almost always accompanied by gliosis [3]. In patients with acquired epilepsy, the severity of the astrogliosis at the focus is positively correlated with the strength of the epileptic discharges [4] with very few exceptions [5]. Temporal lobe epilepsy (TLE) is the most common epilepsy in adult humans. It is focused on the hippocampus, amygdala, and parahippocampal gyrus. It always involves hippocampal sclerosis, which is characterized by loss of neurons, axonal sprouting, and gliosis. The reactive astrocytes in hippocampal sclerosis have transformed homeostatic properties, which seem to actively support neuronal abnormal excitation [6].

A unique feature of these reactive astrocytes in epileptic tissue distinguishes these astrocytes from those in other neurological diseases. It is the loss of domain organization. The processes of a single astrocyte, despite encompassing two million synapses in humans and about 140,000 in rodents, stay within a single domain. They do not overlap with other astrocytic domains. In neurological diseases, reactive astrocytes show hypertrophy of their cellular processes, but they stay within their tiled domains and do not show overlap just as their normal counterparts [7]. The one exception, however, seems to be the epileptic brain. In each of three mouse models of epilepsy, the reactive astrocytes show a dramatic increase in overlap of processes [8]. Furthermore, if in these models the seizures were suppressed by antiepileptic drug treatment, the overlap of reactive processes was dramatically reduced. The models used were acquired epilepsy due to posttraumatic injury (ferrous chloride solution injection) as well as kainate-induced epilepsy and the inbred epileptic mouse strain SWLX-4. The last two are considered models for temporal lobe epilepsy. In focal epilepsy (ferrous chloride injection), the palisading astrocytes form a barrier around the lesion with processes radiating out into the periphery. This is similar to the barrier astrocyte in an ischemic lesion. The palisading astrocytes were surrounded by hypertrophic astrocytes. These palisading astrocytes were accompanied by decreased dendritic spine density, whereas the opposite occurred in the areas with hypertrophic astrocytes. It has been speculated that the loss of astrocytic domain organization occurs only in pathologies associated with EEG abnormalities [8]. To date there are no further systematic studies on this subject. It seems obvious, however, that loss of astrocytic domain organization is a hallmark of epilepsy and not of other prominent pathologies. It also seems clear that this loss is evident in both idiopathic and acquired epilepsies.

All together, these pathological analyses beg the question: are reactive astrocytes the cause or a consequence of seizures? The answer could be, of course, more complicated and context may determine the role of astrocytes. It may depend on the underlying pathology in acquired epilepsy as indicated in some forms of brain trauma and viral infections [9].

15.3 Causes of Idiopathic Epileptogenesis

In the normal brain, can astrocytic dysfunction cause neurons to turn epileptic? There are an increasing number of studies showing that optogenetic stimulation of astrocytes will affect excitation of adjacent neurons due to glutamate release [10, 11]. While astrocytes can be experimentally stimulated to cause epileptic discharges in normal brain tissue, this is an artificial construct. It indicates that normal astrocytes have the potential to induce abnormal neuronal discharges. Yet, this does not prove that in vivo astrocytes are at the root of some or all epileptogenic transformations of the brain. As pointed out in the following sections, the situation in acquired epilepsy is different. In that case, there is strong evidence that reactive astrocytes, transformed by the original insult, can contribute and sustain the switch from normal neurons into epileptic neurons. In contrast, idiopathic epilepsy seems to be more of an enigma. Recent genomic analysis points to strong genetic factors in idiopathic epilepsy [12]. There are about 140 epilepsy-related genes, many inherited within families. There are also many cases of de novo genetic mutations not inherited from either parent. In these cases, the contribution is mostly from post-zygotic mutations. Often these mutations lead to neurodevelopmental deformations, which are not prominent before the seizures start and then masked as idiopathic epilepsies. The majority is a combination of polymorphisms, where various mutations contribute unequally. Yet, there are single-gene epilepsies. They concern genes encoding for voltage- and ligand-gated ion channels and neuronal migration and maturation during development [13]. The astrocytic contributions in these cases have not yet been investigated. In animal models, astrocyte-specific deletion of the gene encoding for hamartin, which interacts with heat shock protein, leads to neuronal seizures [14]. Astrocyte-specific deletion of neogenin, which interacts with glutamate transporters, leads to increased neuronal excitability [15]. Also, deletion of the astrocyte potassium channel Kir4.1 (inward rectifier-type potassium channel 4.1) causes epileptiform discharges in neurons [16]. This deletion will affect potassium clearance in the extracellular space and therefore depolarize the neuronal membrane potential. The major astrocytic glutamate transporter is EAAT2 (excitatory amino acid transporter 2). A loss-of-function mutation of the gene encoding for EAAT2 is a risk factor for human epilepsy [17]. Studies in the drosophila cortex demonstrated that microdomain glial calcium oscillations require a glial-specific Na^+/Ca^{2+} and K^+ exchanger (NCKX) as part of a transmembrane calcium flux. Removal of NCKX interferes with glial calcium oscillations and makes the cortical neurons susceptible to seizures [18]. It appears that the genetic causes of idiopathic epilepsy do not overly focus on astrocytes. However, due to the role of astrocytes in homeostasis and control of neuronal excitability, some of the mutations, if occurring in strategically relevant astrocytic loci, are causing neuronal seizures. Yet, these might be exceptions. As pointed out earlier, once a part of the brain turns into an epileptic focus due to genetic mutations (or other causes), gliosis results irrespective of the cellular location of the mutations. This gliosis will in turn change the astrocytic (and microglial) properties. At this point, there are no recorded differences between reactive astrocytes in epileptic tissues from idiopathic and acquired pathologies [8].

15.4 Are Reactive Astrocytes a Cause of Acquired Epileptogenesis?

Acquired epileptogenesis can be a slow process, which starts before the first recurrent seizure and involves molecular and structural changes on the single neuron level, neuronal circuit, and glial cells. Recurrent epileptic seizure activity is almost always accompanied by reactive gliosis. The reactive astrocytes in a gliotic tissue have properties, which are altered to sustain recurrent neuronal hyperexcitability. The primary injury will involve gliosis with reactive astrocytes. The question is if this reactive gliosis in turn causes epileptic neurons. The alternative is that the primary event causes the hyperexcitable neurons, which in turn leads to reactive gliosis, which contributes to these abnormal discharges. If so, are there differences between the gliosis caused by the primary event and the one due to hyperactive neurons?

A traumatic injury like ferrous chloride injection to a prescribed area will cause reactive gliosis as a response to the trauma. The organization of the reactive tissue should reflect this, and the reactive astrocytes should be organized with restrictions due to domains. If eventually the injury causes epileptogenesis, at which point are the reactive astrocytes spreading into the domains of their neighbors? In other words, does the original injury cause the astrocytes to breach domain restrictions and this acts as epileptogenic stimulus, or is the original injury the cause of seizures, which in turn cause the epilepsy-specific organization of reactive astrocytes? The original research by Oberheim et al. [8] reports EEG abnormalities within hours of the injection. The astrocyte organization was investigated a week after the injection. A few hours are probably too short to allow reorganization of the astrocytes. The lost domain organization appeared in an area surrounding the injury for about 1 mm. The seizures stabilized over time but continued into month 6. At that time, the astrocytes still had a reactive morphology and a lost domain organization. Valproate treatment stopped the seizures and caused the astrocytes to reconvene in a domain organization. Valproate's mechanism of action is through enhancing GABA levels and blocking voltage-gated ion channels [19]. Valproate's main targets are most likely neuronally based systems and not astrocytic properties. Thus, so far these findings are an indication (but not a proof) that abnormal EEG activity and the underlying neuronal seizures are the triggers for the specific organization of reactive gliosis.

However, there are results that seem to contradict this conclusion. After astrocyte-specific deletion of β1-integrin, which contributes to integrin complexes needed for cell adhesion, widespread astrogliosis develops without any other pathologies or blood–brain barrier dysfunction. In this case, neuronal hyperexcitability develops within weeks [20]. Virus-induced reactive astrocytosis without microgliosis [21] did not change neuronal membrane properties. It disrupted the glutamate/glutamine cycle and reduced synaptic GABA availability. This in turn caused reductions in inhibitory synaptic transmission and led to hyperexcitability. In these two cases, it was demonstrated that reactive gliosis without significant accompanying inflammation or overt pathology causes hyperexcitability. However, the domain organization of the reactive astrocytes was not investigated. The question about astrogliosis as a cause of epileptogenesis has been so far not really addressed. The most likely possibility is that both concepts exist; reactive gliosis (caused by a primary pathology) can cause hyperexcitability and hyperexcitable neurons can cause gliosis. The complicated case of a primary pathology causing gliosis and both contributing to convulsive neurons, which in turn might feedback and change the nature of the gliosis (domain organization) has not been systematically investigated.

15.5 Ictogenesis due to Reactive Gliosis

It can be safely assumed that most epileptic neurons are surrounded by gliosis with reactive astrocytes not conforming to domain organization, irrespective of its genesis. What are the functional properties of reactive astrocytes in epileptic tissues? Are these astrocytes involved in triggering ictal discharges? There are several mechanisms by which astrocytes can trigger the periods of ictal discharges. There can be alterations in the coupling of the astrocytic syncytium, astrocytes can cause disruption of the blood–brain barrier, astrocytic-distorted calcium signaling can impact neurons, astrocytic glutamate transport can be altered, and, finally, potassium buffering can be reduced.

In the kainate model, astrocytic calcium signals precede neighboring hippocampal neuronal discharges [22]. In two different Zebrafish epileptic models, synchronous activity of the neuronal network is preceded by strong calcium waves in the astrocytic syncytium [23]. These waves can be blocked by gap junction inhibitors, and they involve massive astrocytic glutamate release. Optogenetic stimulation of these Zebrafish astrocytes leads to ictal neuronal discharges that are dependent on glial gap junctions and glutamate release. The authors hypothesize that the astrocytic network buffers neuronal glutamate release. If this homeostatic mechanism breaks down across the astrocytic syncytium, a massive glutamate release provokes neuronal ictal discharges. In a mouse model, TNFα (tumor necrosis factor alpha) was shown to initiate the calcium wave and concomitant glutamate release in the astrocytic syncytium to drive the neuronal discharges [24]. In several in vitro models, calcium elevation in astrocytes caused ictal discharges. Suppression of calcium increases prevented the discharges. In contrast, interictal discharges were not affected by astrocytic calcium [25]. However, it should be noted that one study using anaesthetized rats could not find astrocytic calcium transients preceding neuronal discharges [26].

GLT1 (glutamate transporter 1; the rodent form of EAAT2) knockout causes severe seizures as it is involved in extracellular glutamate clearance. To prevent backflow of accumulated glutamate from astrocytes, glutamate is turned mainly into glutamine by GS (glutamine synthetase). GLT1 is downregulated in reactive astrocytes [27] and pathological increases in extracellular glutamate have been measured in patients [28]. Reduced expression of GS is associated with the degree of gliosis in hippocampal sclerosis associated with TLE [29]. Inhibiting the glutamate-glutamine cycle causes epileptic seizures, which can be reversed by supplying exogenous glutamine [30].

Historically, the most persistent hypothesis connecting glia with epilepsy is the defect potassium homeostasis hypothesis. It indicates that reactive astrocytes in epileptic tissue are not effective enough in clearance of excess potassium. During neuronal activity, this would cause increased depolarization toward excitation threshold. During synchronous discharges, the astrocytic buffering capacity is exceeded as the concentration in the extracellular space reaches up to 12 mM [31]. This in turn could cause abnormal neuronal depolarization and therefore hyperexcitability. However, the defect potassium hypothesis is controversial. It was for a long time not clear if defective potassium homeostasis drove the discharges or if the potassium accumulation is a consequence of the paroxysmal activity. However, tissue samples from epileptic foci from patients who underwent surgery because of epilepsy have impaired potassium clearance and exhibited a downregulated Kir4.1 potassium channel, especially around astrocytic endfeet [32]. This channel is specific for astrocytes and is one of the main clearance mechanisms for extracellular potassium. Furthermore, astrocyte-specific deletion of this channel in mice leads to seizures and other pathological events [33]. Mutations in the gene encoding for Kir4.1 are a risk factor for seizures in humans [34]. Kir4.1 and the astrocyte-specific water channel AQP4 (aquaporin 4) work together in astrocytic endfeet to mediate ion and water flux for CNS volume regulation. Interestingly, deletion of AQP4, α-syntrophin, or dystrophin (both involved in anchoring AQP4 into the endfeet membrane) in mice lowers their seizure threshold substantially [35]. It has been argued that the changed astrocytic properties in the sclerotic hippocampus are generating seizure activity due to defunct AQP4 transporters and potassium buffering [36]. Gap junctions in astrocytes are involved in calcium signaling, glutamate homeostasis, ATP release, and energy metabolites shifts. Gap junction blockers suppress seizure activity, but they are not or only incompletely distinguishing between neuronal and astrocytic gap junctions. However, there is clear evidence that tissues taken from animal models and human patients exhibit increases of glial connexins, but not neuronal connexins [37].

In some forms of acquired epilepsies which involve stroke or trauma, the blood–brain barrier is compromised [3]. This leads to leakage of several blood-borne substances, which can interact with astrocytes and change properties. Most prominently is albumin which activates astrocytic TGFβ (transforming growth factor beta) receptors. This activation changes a whole range of astrocytic homeostatic mechanisms [38].

15.6 Microglia and Other Cells of the Immune System

Activated microglia are found in gliotic brain tissue samples from the foci of patients undergoing epilepsy-related surgery [39, 40]. The properties of microglia of TLE patients in the sclerotic and non-sclerotic areas were investigated and compared [41]. In the sclerotic areas, the microglia were more activated and expressed more proinflammatory cytokines, especially CXCL8 (interleukin 8 or chemokine [C-X-C motif] ligand 8, which has no rodent homologue), but also IL-1β (interleukin-1beta), IL-6, and TNFα (tumor necrosis factor alpha). Comparing the results with the time of the last seizure suggested that these cytokines are mainly released during seizure activity. Interestingly, IL-10, a cytokine associated with repair processes, was also strongly expressed. In patients with hippocampal sclerosis, presurgical memory tests were conducted and compared with intensity of microgliosis and astrogliosis of the tissue removed during the following surgery. Whereas there was no correlation between memory performance and astrogliosis, increased microgliosis correlated with reduced memory function [42]. Results from kainate animal models confirmed the release of the proinflammatory cytokines. Pretreatment with IL-1β prolonged the kainate-induced seizures [43]. Microglia engulf beaded dendrites in severe seizures and remain there for hours. The pouches are not involved in phagocytosis, and they seem to stabilize the dendrites and lead to shrinking of the beads [44]. In other models, depletion of microglia before seizure initiation increases the severity of the seizures [45, 46]. Thus, they are obviously involved in neuroprotection, at least in the early stages of epileptogenesis. In contrast, there is evidence that during prolonged periods of recurrent seizures, microglia might be drivers of neurodegeneration. Certainly, recurrent seizures are accompanied by neuroinflammatory processes, which contribute to seizure severity. Moreover, one study [47] suggests that in a mouse pilocarpine epilepsy model, there is a clear cellular sequence of events. At the onset microglia activate and release proinflammatory cytokines, for example, TNFα, IL-1β, and IL-6. This in turn activates astrocyte transformation into a reactive subset with large calcium signals. Reduction of microglia activation by various means reduced astrogliosis, astrocytic calcium signals, and severity of seizures. Deletion of IP_3R2 (inositol triphosphate 3 receptor type 2) reduced the aberrant astrocytic calcium signal and decreased seizure susceptibility after status epilepticus. This study demonstrates, at least for this model system, that the initial activation of microglia causes transformation of normal astrocytes into a proconvulsive phenotype (see Fig. 15.1). The inhibition of microglia activation is only effective in preventing further astrocyte reactivity if the inhibition occurs in the first 7 days after pilocarpine treatment. Thereafter microglial inhibition has no effect on the sequence of events. Various studies have demonstrated that microglia activation in early epileptogenesis includes release of proinflammatory cytokines [43]. In animal models of TLE, excess newborn neurons in the dentate gyrus cause the formation of hyperexcitable aberrant neuronal circuits. Microglia seem to attenuate this aberrant neurogenesis via activation of their TLR9 (toll-like receptor 9) receptors and release of TNFα. This mechanism could be suppressed by minocycline treatment [48]. Thus, in the long term, microglia contribute to circuitry maintenance. In the pilocarpine model, the transformation of proinflammatory microglia into an anti-inflammatory subtype by rosiglitazone (an antagonist of peroxisome proliferator-activated receptor γ) reduced neuronal loss [49].

Peripheral macrophages can contribute to the release of epileptogenic proinflammatory cytokines, especially if the blood–brain barrier is compromised [50]. In human TLE and hippocampal sclerosis, T lymphocytes are detected inside the parenchyma and around capillaries [50]. In other forms of human epilepsy, T cells were concentrated in the lesions and the more severe the epileptic outcome, the higher the density of these T cells [51]. Interestingly, regulatory T cells were also present, but their density correlated inversely with seizure severity.

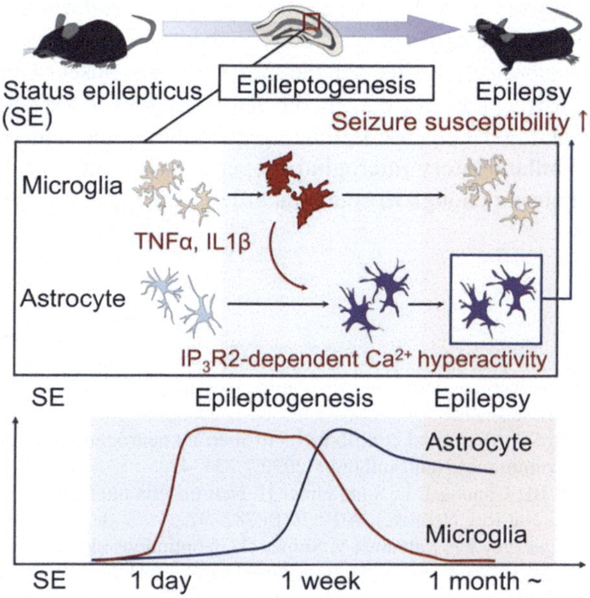

Fig. 15.1 Schematic diagram demonstrating the sequential activation of glial cells in the transformation from initiation of status epilepticus to epileptic seizure susceptibility. Microglia transform into a transient proinflammatory phenotype for about 1 week, whereas astrocytes are induced into a proconvulsive phenotype with a delay in response to these cytokines. (Reprinted with permission from Ref. [47]. Copyright American Association for Clinical Investigation)

15.7 Conclusion

There is not one cellular model that will fit all epileptic pathologies. Idiopathic epilepsies are caused by mutations that can affect any cell type. If such a loss-of-function mutation reduces the activity of a critical astrocytic homeostatic process, epileptogenesis will result. Acquired epilepsies are more complicated. The question if gliosis is the cause or consequence of epilepsy is too simplified. It seems that all activation possibilities exist, depending on the context. Microglia activation can transform astrocytes into a proconvulsive type. Neurons can become convulsive, and this will cause astrogliosis with lost domain organization, which will sustain the abnormal neuronal discharges. The opposite is possible too; reactive gliosis due to another primary event can cause convulsive neurons. The question is, why do not all or at least most reactive astrocytes, caused by a primary injury, result in hyperexcitable neurons? It appears that sustained neuronal hyperexcitability, accompanied by EEG abnormalities, is causing a peculiar reactive astrocyte subtype that does not conform to domain organization. Are there functional differences between the reactive astrocytes with and without domain organization? If astrogliosis is caused by a primary injury and involved neurons turning epileptogenic (with or without astrocytic involvement), does this change the organization and functional properties of these astrocytes? More experimental focus on these questions is urgently needed. Microglia can have protective effects in the short time after onset of epileptogenesis, despite directing astrocytes toward a proconvulsive subtype. In the long run, microglia as well as invading T cells are stabilizing astrogliosis and convulsive neuronal discharges. Anti-inflammatory microglial properties and regulatory T cells, while involved, are not strong enough to counteract these detrimental mechanisms.

References

1. Brigo F, Marson A. Approach to the medical treatment of epilepsy. Continuum (Minneapolis, Minn). 2022;28(2):483–99.
2. Victor TR, Tsirka SE. Microglial contributions to aberrant neurogenesis and pathophysiology of epilepsy. Neuroimmunol Neuroinflamm. 2020;7:234–47.
3. Patel DC, Tewari BP, Chaunsali L, Sontheimer H. Neuron-glia interactions in the pathophysiology of epilepsy. Nat Rev Neurosci. 2019;20(5):282–97.
4. Guerreiro MM, Quesney LF, Salanova V, Snipes GJ. Continuous electrocorticogram epileptiform discharges due to brain gliosis. J Clin Neurophysiol. 2003;20(4):239–42.
5. Rossini L, Garbelli R, Gnatkovsky V, Didato G, Villani F, Spreafico R, et al. Seizure activity per se does not induce tissue damage markers in human neocortical focal epilepsy. Ann Neurol. 2017;82(3):331–41.
6. Sendrowski K, Sobaniec W. Hippocampus, hippocampal sclerosis and epilepsy. Pharmacol Rep. 2013;65(3):555–65.
7. Hol EM, Pekny M. Glial fibrillary acidic protein (GFAP) and the astrocyte intermediate filament system in diseases of the central nervous system. Curr Opin Cell Biol. 2015;32:121–30.
8. Oberheim NA, Tian GF, Han X, Peng W, Takano T, Ransom B, et al. Loss of astrocytic domain organization in the epileptic brain. J Neurosci. 2008;28(13):3264–76.
9. Chen P, Chen F, Zhou B. Understanding the role of glia-neuron communication in the pathophysiology of epilepsy: a review. J Integr Neurosci. 2022;21(4):102.
10. Sasaki T, Beppu K, Tanaka KF, Fukazawa Y, Shigemoto R, Matsui K. Application of an optogenetic byway for perturbing neuronal activity via glial photostimulation. Proc Natl Acad Sci U S A. 2012;109(50):20720–5.
11. Perea G, Yang A, Boyden ES, Sur M. Optogenetic astrocyte activation modulates response selectivity of visual cortex neurons in vivo. Nat Commun. 2014;5:3262.
12. Ellis CA, Petrovski S, Berkovic SF. Epilepsy genetics: clinical impacts and biological insights. Lancet Neurol. 2020;19(1):93–100.
13. Sánchez-Carpintero Abad R, Sanmartí Vilaplana FX, Serratosa Fernández JM. Genetic causes of epilepsy. Neurologist. 2007;13(6 Suppl 1):S47–51.
14. Uhlmann EJ, Wong M, Baldwin RL, Bajenaru ML, Onda H, Kwiatkowski DJ, et al. Astrocyte-specific TSC1 conditional knockout mice exhibit abnormal neuronal organization and seizures. Ann Neurol. 2002;52(3):285–96.
15. Sun D, Tan ZB, Sun XD, Liu ZP, Chen WB, Milibari L, et al. Hippocampal astrocytic neogenin regulating glutamate uptake, a critical pathway for preventing epileptic response. Proc Natl Acad Sci U S A. 2021;118(16):e2022921118.
16. Steinhäuser C, Grunnet M, Carmignoto G. Crucial role of astrocytes in temporal lobe epilepsy. Neuroscience. 2016;323:157–69.
17. Guella I, McKenzie MB, Evans DM, Buerki SE, Toyota EB, Van Allen MI, et al. De novo mutations in YWHAG cause early-onset epilepsy. Am J Hum Genet. 2017;101(2):300–10.

18. Melom JE, Littleton JT. Mutation of a NCKX eliminates glial microdomain calcium oscillations and enhances seizure susceptibility. J Neurosci. 2013;33(3):1169–78.
19. Rosenberg G. The mechanisms of action of valproate in neuropsychiatric disorders: can we see the forest for the trees? Cell Mol Life Sci. 2007;64(16):2090–103.
20. Robel S, Buckingham SC, Boni JL, Campbell SL, Danbolt NC, Riedemann T, et al. Reactive astrogliosis causes the development of spontaneous seizures. J Neurosci. 2015;35(8):3330–45.
21. Ortinski PI, Dong J, Mungenast A, Yue C, Takano H, Watson DJ, et al. Selective induction of astrocytic gliosis generates deficits in neuronal inhibition. Nat Neurosci. 2010;13(5):584–91.
22. Heuser K, Nome CG, Pettersen KH, Åbjørsbråten KS, Jensen V, Tang W, et al. Ca2+ signals in astrocytes facilitate spread of Epileptiform activity. Cerebral Cortex (New York, NY: 1991). 2018;28(11):4036–48.
23. Diaz Verdugo C, Myren-Svelstad S, Aydin E, Van Hoeymissen E, Deneubourg C, Vanderhaeghe S, et al. Glia-neuron interactions underlie state transitions to generalized seizures. Nat Commun. 2019;10(1):3830.
24. Nikolic L, Shen W, Nobili P, Virenque A, Ulmann L, Audinat E. Blocking TNFα-driven astrocyte purinergic signaling restores normal synaptic activity during epileptogenesis. Glia. 2018;66(12):2673–83.
25. Gómez-Gonzalo M, Losi G, Chiavegato A, Zonta M, Cammarota M, Brondi M, et al. An excitatory loop with astrocytes contributes to drive neurons to seizure threshold. PLoS Biol. 2010;8(4):e1000352.
26. Baird-Daniel E, Daniel AGS, Wenzel M, Li D, Liou JY, Laffont P, et al. Glial calcium waves are triggered by seizure activity and not essential for initiating ictal onset or neurovascular coupling. Cerebral Cortex (New York, NY: 1991). 2017;27(6):3318–30.
27. Sha L, Wang X, Li J, Shi X, Wu L, Shen Y, et al. Pharmacologic inhibition of Hsp90 to prevent GLT-1 degradation as an effective therapy for epilepsy. J Exp Med. 2017;214(2):547–63.
28. During MJ, Spencer DD. Extracellular hippocampal glutamate and spontaneous seizure in the conscious human brain. Lancet (London, England). 1993;341(8861):1607–10.
29. van der Hel WS, Notenboom RG, Bos IW, van Rijen PC, van Veelen CW, de Graan PN. Reduced glutamine synthetase in hippocampal areas with neuron loss in temporal lobe epilepsy. Neurology. 2005;64(2):326–33.
30. Kaczor P, Rakus D, Mozrzymas JW. Neuron-astrocyte interaction enhance GABAergic synaptic transmission in a manner dependent on key metabolic enzymes. Front Cell Neurosci. 2015;9:120.
31. Heinemann U, Lux HD. Ceiling of stimulus induced rises in extracellular potassium concentration in the cerebral cortex of cat. Brain Res. 1977;120(2):231–49.
32. Steinhäuser C, Seifert G, Bedner P. Astrocyte dysfunction in temporal lobe epilepsy: K+ channels and gap junction coupling. Glia. 2012;60(8):1192–202.
33. Djukic B, Casper KB, Philpot BD, Chin LS, McCarthy KD. Conditional knock-out of Kir4.1 leads to glial membrane depolarization, inhibition of potassium and glutamate uptake, and enhanced short-term synaptic potentiation. J Neurosci. 2007;27(42):11354–65.
34. Buono RJ, Lohoff FW, Sander T, Sperling MR, O'Connor MJ, Dlugos DJ, et al. Association between variation in the human KCNJ10 potassium ion channel gene and seizure susceptibility. Epilepsy Res. 2004;58(2–3):175–83.
35. Binder DK, Nagelhus EA, Ottersen OP. Aquaporin-4 and epilepsy. Glia. 2012;60(8):1203–14.
36. de Lanerolle NC, Lee TS. New facets of the neuropathology and molecular profile of human temporal lobe epilepsy. Epilepsy Behav. 2005;7(2):190–203.
37. Mylvaganam S, Ramani M, Krawczyk M, Carlen PL. Roles of gap junctions, connexins, and pannexins in epilepsy. Front Physiol. 2014;5:172.
38. Weissberg I, Wood L, Kamintsky L, Vazquez O, Milikovsky DZ, Alexander A, et al. Albumin induces excitatory synaptogenesis through astrocytic TGF-β/ALK5 signaling in a model of acquired epilepsy following blood-brain barrier dysfunction. Neurobiol Dis. 2015;78:115–25.
39. Liu J, Reeves C, Michalak Z, Coppola A, Diehl B, Sisodiya SM, et al. Evidence for mTOR pathway activation in a spectrum of epilepsy-associated pathologies. Acta Neuropathol Commun. 2014;2:71.

40. Sosunov AA, Wu X, McGovern RA, Coughlin DG, Mikell CB, Goodman RR, et al. The mTOR pathway is activated in glial cells in mesial temporal sclerosis. Epilepsia. 2012;53(Suppl 1):78–86.
41. Morin-Brureau M, Milior G, Royer J, Chali F, Le Duigou C, Savary E, et al. Microglial phenotypes in the human epileptic temporal lobe. Brain J Neurol. 2018;141(12):3343–60.
42. Toscano ECB, Vieira ÉLM, Portela A, Caliari MV, Brant JAS, Giannetti AV, et al. Microgliosis is associated with visual memory decline in patients with temporal lobe epilepsy and hippocampal sclerosis: a clinicopathologic study. Epilepsy Behav. 2020;102:106643.
43. Kinoshita S, Koyama R. Pro- and anti-epileptic roles of microglia. Neural Regen Res. 2021;16(7):1369–71.
44. Eyo UB, Haruwaka K, Mo M, Campos-Salazar AB, Wang L, Speros XS 4th, et al. Microglia provide structural resolution to injured dendrites after severe seizures. Cell Rep. 2021;35(5):109080.
45. Mirrione MM, Konomos DK, Gravanis I, Dewey SL, Aguzzi A, Heppner FL, et al. Microglial ablation and lipopolysaccharide preconditioning affects pilocarpine-induced seizures in mice. Neurobiol Dis. 2010;39(1):85–97.
46. Liu M, Jiang L, Wen M, Ke Y, Tong X, Huang W, et al. Microglia depletion exacerbates acute seizures and hippocampal neuronal degeneration in mouse models of epilepsy. Am J Physiol Cell Physiol. 2020;319(3):C605–c10.
47. Sano F, Shigetomi E, Shinozaki Y, Tsuzukiyama H, Saito K, Mikoshiba K, et al. Reactive astrocyte-driven epileptogenesis is induced by microglia initially activated following status epilepticus. JCI Insight. 2021;6(9):e135391.
48. Luo C, Koyama R, Ikegaya Y. Microglia engulf viable newborn cells in the epileptic dentate gyrus. Glia. 2016;64(9):1508–17.
49. Peng J, Wang K, Xiang W, Li Y, Hao Y, Guan Y. Rosiglitazone polarizes microglia and protects against pilocarpine-induced status epilepticus. CNS Neurosci Ther. 2019;25(12):1363–72.
50. Zattoni M, Mura ML, Deprez F, Schwendener RA, Engelhardt B, Frei K, et al. Brain infiltration of leukocytes contributes to the pathophysiology of temporal lobe epilepsy. J Neurosci. 2011;31(11):4037–50.
51. Xu D, Robinson AP, Ishii T, Duncan DS, Alden TD, Goings GE, et al. Peripherally derived T regulatory and γδ T cells have opposing roles in the pathogenesis of intractable pediatric epilepsy. J Exp Med. 2018;215(4):1169–86.

Chapter 16
Traumatic Brain Injury

Abstract The most common traumatic brain injury type is mild form or concussion. This is especially a problem if these impacts happen frequently as in contact sports. The injury focus is white matter tracts due to mechanical shearing of axons. The resulting energy depletion and oxidative damage affects axons and oligodendrocyte precursors. Both axon degeneration and precursor depletion result in demyelination. This results in chronic inflammatory microglia. Repeated mild traumata increase the microglia inflammation and this is correlated with neurodegeneration and cognitive problems later in life. Repetitive mild traumata result in the appearance of an atypical reactive astrocyte phenotype, which lacks crucial homeostatic properties. The appearance of this atypical astrocyte is associated with acquired seizures. Interestingly, if there is damage to the meninges, macrophages with beneficial anti-inflammatory properties appear. In penetrating brain injury (PBI), oligodendrocyte precursors proliferate and migrate to the necrotic area. Their activity stabilizes the wound and is beneficial for the health of the surviving cells. A juxtavascular astrocyte phenotype appears around penetrating wounds. It prevents inflammatory macrophages from invading and, in doing so, stabilizes the situation. In explosive blast injury, microglia are the most resilient cell type and their degree of activation is correlated with blast intensity not neuronal damage.

Keywords Atypical reactive astrocyte · Chronic inflammation · Concussion · Explosive blast injury · Juxtavascular astrocyte · Macrophages · Mild traumatic brain injury · Mechanical shearing · Microglia · Microglia resilience · Oligodendrocyte precursor cells · Penetrating brain injury · Repetitive mild traumata

16.1 Introduction

Traumatic brain injury (TBI) concerns any mechanical impact on the brain from the outside as it might arise in traffic accidents, combat activity, gunshot wounds, or explosions. World incidence is around ten million cases each year and is increasing [1]. It can appear as closed injury when the brain is not exposed to the mechanical

W. Walz, *The Gliocentric Brain*, https://doi.org/10.1007/978-3-031-48105-5_16

impact. Another form is penetrating brain injury. In this case, at least the dura mater is breached and accompanying infections can be a problem. A third form is explosive blast TBI, where rapid pressure shockwaves cause injury [2]. The last one is frequently associated with posttraumatic stress disorder (PTSD), whereas a penetrating TBI is a risk factor for acquired epilepsy. Traumatic brain injury affects especially young people, and it is twice as common in males than females. The primary impact of penetrating and closed injury is a focal, necrotic injury at the location of the impact called coup (from "blow"). This coup is accompanied by a contrecoup, which is usually located on the opposite side of where the brain is struck. Apart from these focal injuries, there are always diffuse injuries. In explosive blast TBI, there is only diffuse injury. The diffuse injury causes extensive axonal damage. While these are all primary injuries, there are always secondary damages developing within hours and years of the primary insult. These include inflammation, degeneration, apoptosis, and acquired epilepsy. Many of these secondary injuries are related to the activities of glial cells and invading immune cells [2]. Mild TBI (mTBI) can appear in closed injury and in explosive blast injury. It is the most common form of TBI as it can also develop over time in contact sports. It can induce post-concussion syndrome with long-term neurological and psychological symptoms. Repeated mTBI is a risk factor for Parkinson's disease, chronic traumatic encephalopathy, and maybe others [3]. Mechanisms of neuronal and axonal damage include blood–brain barrier breakdown, excitatory neurotoxicity, mitochondrial dysfunction, oxygen radicals, and inflammation. Treatment options focus on secondary damage. However, clinical trials have been mostly disappointing, especially for glutamate receptor antagonists and calcium inhibitors (inhibitors against calcium channels and calcium-dependent enzymes). Other approaches are antioxidants, anti-inflammatory drugs, neurotrophic factors to support neuro-regeneration, and stem cell therapy.

16.2 Mild Traumatic Brain Injury/Concussion

An mTBI is often referred to as concussion. Close to 80% of head injuries fall under the category of mTBIs and concussions [4]. They are likely underreported. There are at this point no clear definitions which would distinguish between mTBI and concussion [5]. Concussions/mTBIs are caused by accidents, war-related injuries, and sporting injuries. In mTBI, there is no visible injury to the outside of the head other than maybe bruising or cuts and there is no hemorrhage. The affected person does not necessarily lose consciousness, even for a brief time. The resultant disturbance in brain function in mTBI is brief, lasting at most a few weeks. The symptoms in this brief period are often headaches, loss of balance control, confusion, nausea, and visual disturbances. In an unknown, but significant number of patients, the symptoms persist longer than 6 months and often for years. These symptoms are fatigue, depression, chronic pain, anxiety, and cognitive dysfunction [6]. It is estimated that in the USA alone, about 5.3 million people live with a permanent disability related to mTBI [7]. There is also a higher mortality [8]. This is especially a problem if the mTBI is a repeat event such as with military personal, athletes, elderly, and survivors of domestic violence. Repeated mTBI is a risk factor for PTSD and neurodegenerative diseases. Retired athletes who were at a younger age exposed to repeated mild TBI have an increased mortality and risk of neurodegenerative diseases than matched controls [9]. It has therefore been suggested that the use of the "mild" terminology may be a misnomer [10]. There are no generally agreed upon and standardized animal models for mild and closed head injury. The most used ones involve weight drop models, but piston-driven models are also widely applied [7]. The use of rodents can be problematic for mTBI studies, as rodents have substantially less white matter tracts than humans as a proportion of the brain volume. The major damage reported is axonal shearing injury in the white matter tracts [7, 11]. Clinical studies using imaging found white matter microstructure disturbances, alterations in cortical thickness, neurotransmitter distribution, and metabolism. These findings were generally replicated in the animal models [4]. Presently, it is not possible to specifically treat mTBI other than comforting and supporting the patient [12].

One of the pathological hallmarks of mTBI is the damage to white matter tracts. Combined results from human patients' imaging analysis and animal models elucidated the following pathophysiological mechanisms [13]. The primary impact leads to axonal shearing and stretching, but the consequences of the primary impact are not noticeable for several weeks as the axons react with increased Na,K-ATPase (sodium-potassium adenosine 5′-triphosphatase) activity to compensate for the decrease in ionic gradients. This enhanced activity increases both the energy metabolism and the blood supply. The increase in blood supply lasts several days before it diminishes again. The reason for this delayed reduction in blood flow seems to be oxidative stress. However, the reduction in the energy supply alone causes an imbalance of energy demand and supply with concomitant problems [14]. The oxidative stress interferes negatively with OPC (oligodendrocyte progenitor cell) differentiation and therefore within weeks decreases myelin renewal. The net result is a reduction in the number of differentiated, mature oligodendrocytes, and this reduction in turn contributes to demyelination. A second mTBI during this vulnerable period will potentiate the damage [15, 16]. Now mitochondrial dysfunction and cytoskeletal deterioration will normally occur. The lack of compensation for the loss of ion gradients leads to swelling of the axons. Within months, this can lead to axonal disconnection and Wallerian degeneration. Myelin sheaths start to fragment and degenerate. This in turn will cause the invasion of phagocytosing microglia to the affected white matter area. The myelin sheaths are especially vulnerable to secondary mTBI after a first mild impact. The myelin lamellae are held together by proteins that seem to be very sensitive to the actions of ROS (reactive oxygen species) and lipid peroxidation. Both ROS production and lipid peroxidation are potentiated after such a second mTBI in the vulnerable period.

Activated microglia are surrounding the damaged myelin sheaths and are involved in phagocytosing and inflammation inside the damaged area. Clinical evidence is mostly based on PET (positron emission tomography) imaging. This form of imaging uses ligands that bind to TSPO (mitochondrial 18 kDa translocator protein). During neuroinflammatory processes, TSPO is not only newly expressed in microglia but also to a lesser extent in astrocytes. It is therefore used in clinical imaging as a biomarker. Former NFL (National Football League) players had increased de novo TSPO expression compared with age-matched controls. They also showed hippocampal atrophy and reduced performance in verbal learning and memory tests [17]. Younger players who reported mTBI had increased TSPO expression, but no reduced performance level in the psychological tests [18]. Another study [19] investigated the postmortem brains of former American football players and compared them with matched controls. The appearance of activated microglia in parts of the cortex correlated with the players' history of repetitive head impacts (indicated by the length of the players' active career). The extent of chronically activated microglia correlated with dementia diagnosis of the players after their active career. This connection between activated microglia and dementia seems to be mediated by p-tau (phosphorylated tubulin associated unit), as the appearance of p-tau in addition to activated microglia is a prerequisite for a dementia diagnosis of such an individuum. In another study [20], CSF (cerebrospinal fluid) samples of a group of former NFL football players were compared with matched controls. The samples were analyzed for biomarkers. One, sTREM2 (soluble triggering receptor expressed on myeloid cells 2), is an in vivo indicator of chronic microglial activation and therefore neuroinflammation. It is a biomarker for Alzheimer's disease [21]. The sTREM2 and t-tau (total tau protein) concentrations were higher in the CSF of the former players. A higher concentration of t-tau in the CSF is regarded as a biomarker for neurodegeneration. A similar result was found in another group of patients with a recent single episode of mTBI [22]. Indeed, in a matched cohort study of former Scottish soccer players, mortality from neurodegenerative diseases was significantly higher in the players' group, although their mortality from cardiovascular diseases was lower [9].

The conclusions from animal models are not quite as straightforward. Rodents have a lower white to gray matter ratio than humans and this must be kept in mind when drawing conclusions. Pigs are an example of an animal model with a similar ratio to humans, but pigs are not as accessible as rodents for experimental medicine. Another complication is that certain anesthetics interfere with microglia activation [23]. The alternative experiments with awake animals (if permitted) cause undue stress and might therefore interfere with the results. In a pig model of mTBI (rapid head rotation), it appears that microglia activate within days of an mTBI insult. Within weeks, microgliosis develops and these microglia remain activated for up to 1 year [24]. The microglia changes appear at the same time points as synaptic and morphological neuronal changes. From various rodent models of mTBI, it appears that microglia are activated very early on, possibly with beneficial effects for neuronal health. However, after 30 days or longer microgliosis is noticeable around axonal damage. Both go together with behavioral deficits as reviewed in ref. [25]. If LPS (lipopolysaccharide; a toll-like receptor 4 [TLR4] agonist) was given a day after the last injury of three repeated mTBIs in rats, injuries to neurons and myelin as well as behavioral deficits were much abolished as measured 3 months after the last event due to the LPS administration. However, if LPS was given 5 days after the last repetitive mTBI, the results after 3 months indicated heavy release of proinflammatory cytokines, increased neuronal damage, accumulated p-tau, and increased behavioral deficits, including depressive behavior [26]. On day 1, the LPS injection led to increased microglia activation, but at the same time to a reduction of proinflammatory cytokines [26]. This is another indication that microglia activation early in mTBI, even in repeated mTBI, has in balance a more protective effect. In contrast chronic microglia inflammatory activity is detrimental and is involved in behavioral deficits. This might be the reason why a study did not report any microglial impact on repetitive mTBI outcome, as reducing microglia in the mouse brain before initiating repetitive mTBI had no effect on axon injury [27].

Interestingly, in animal models of repetitive mTBI there is little evidence of reactive astrogliosis, only mild upregulation of GFAP and no proliferation. These mildly reactive astrocytes are organized into small groups, scattered across the cerebral cortex [28]. Usually, mTBIs do not cause ruptured blood vessels with hemorrhage. Nevertheless, the blood–brain barrier can be damaged and cause leakage of blood constituents into the parenchyma. In a mouse model of repetitive mTBI, a new type of atypical reactive astrocyte was found [29]. Groups of this atypical astrocyte subtype are associated with areas of blood–brain barrier leakage. These astrocytes are GFAP negative, they lack Kir4.1 (ATP-sensitive inward rectifier potassium channel 4.1) and glutamate transporter 1 (GLT1). They are also not coupled to each other by connexins. This means they lack crucial homeostatic elements, which must impact neighboring neurons. Lack of coupling, of glutamate and potassium homeostasis in reactive astrocytes, is associated with epileptic seizures. Indeed, it was shown that whenever these groups of atypical astrocytes appear, spontaneous convulsive and nonconvulsive seizures appeared in these animals. There was a positive correlation between the brain volume, which was infested with these atypical astrocytes and the likelihood to develop seizures [30]. Epilepsy incidence of patients exposed to mTBI is at least twice as high as in the general population and this is probably an underestimation [31]. Interestingly, in these mice exposed to repetitive mTBI, the damaged blood–brain barrier was not repaired months after cessation of the insults. Atypical astrocytes and seizures continued to be present.

In about half of patients with mTBI, there is damage to the meninges, most likely related to the site of the coup, if there is one [32]. Such an injury leads to a recruitment of blood-borne immune cells to the meningeal vasculature. A thorough study involving both, mTBI patients and a mice model with meningeal compression injury, evaluated the interactions at the meningeal barrier [33]. At first neutrophils and inflammatory macrophages swarm the injury sites and remove dead cells and debris. This process usually lasts 2 days. In a second step, there is thereafter a second invasion of another subtype of macrophages, CD11b+ and CD206+ (cluster of differentiation molecules 11b and 206) with wound-healing properties. These macrophages proliferate, focusing on repair of the meninges and angiogenesis together with clearance of extravascular fibrin deposits. In most patients, this restoration is accomplished within 1–3 weeks. In these patients and animal models, this leads to angiogenesis and repair of the meninges. These macrophages proliferated and secreted tissue-remodeling enzymes like MMP-2 (matrix metalloproteinase 2). These processes are less efficient in more severe forms of TBI. In the case of mTBI, if a second impact arrives in the first phase of debris removal, a new wave of invasive inflammatory monocytes arrives and impairs the second anti-inflammatory repair process. This leads in turn to chronic inflammation and secondary cell damage. In contrast, when a second impact occurs during the anti-inflammatory repair phase, the tissue is resistant to secondary damage and inflammation.

16.3 Penetrating Brain Injury

Penetrating brain injury (PBI) is the most severe of the TBIs. It has a high mortality and affects mainly younger people. High-velocity penetrations (gunshots, shell fragments) have the worst prognosis. The prognosis is worse with increased object size and velocity. Increases in both parameters escalate the energy the object is transferring to the skull and brain tissue. The penetration pathway is also an issue. The crossing of the object of the midline and damage to the ventricles is highly detrimental to the outcome [34]. The presence of CSF leakage is a bad sign as are hemorrhages accompanied by edema. The inelasticity of the brain, due to the skull encasing, is another critical factor. Along the leading edge of the penetrating tract, there will be crush injury with necrosis. Ultimately, a cavity will be formed with a size larger than the original penetrating tract. This is due to cell death following stretching, shearing, and compression [35, 36]. The clinical outcome then depends not only on these circumstances but also on the secondary injury in the days and weeks after the impact. Complications can occur due to ischemia, bleeding, infection, and early or late onset epilepsy. Secondary neuronal injury is mainly due to calcium overload, excitotoxicity, mitochondrial dysfunction, oxidative stress, and inflammation. Treatment focuses on the prevention of secondary damage by stabilizing the patient and the removal of a deeply located object might not be desirable. In contrast, hematomas might need surgical intervention. Antibiotics and anticonvulsants are given routinely [37, 38].

The most common animal model to study penetrating brain injury uses a calibrated stab wound. One might assume that the responses of astrocytes and microglia are like those in focal ischemia with its necrotic core and a surrounding penumbra. Yet, as will be shown below, there are differences. One difference pertains to OPCs. In white matter, they transform into oligodendrocytes; in gray matter, they survive as OPCs into adulthood. Interestingly, OPCs transform after focal ischemia into astrocytes. However, such a transformation is not observed after a stab wound [39]. The OPCs accumulate within 2 days of a stab wound in the scar by migration and proliferation. Gradients of VEGF (vascular endothelial growth factor) and bFGF (basic fibroblast growth factor) are responsible for this directional migration of the OPCs. The signals for OPC proliferation are unknown. However, the OPC proliferation is correlated with blood–brain barrier disruption, rather than neuronal death. OPC ablation leads to a delayed wound closure [40]. The role of OPCs seems to be in support of restoration and stabilization of the tissue. After about 4 weeks, the OPCs normalize and reorient away from the scar. Apoptosis normalizes their cell number around this time point. NG2 (neuron-glial antigen 2) knockout mice exhibit increased astrogliosis, prolonged blood–brain barrier disruption, and a slower healing of the lesion after a stab wound [41]. The OPCs seem to go through only one cell division within the lesion. By contrast, mature oligodendrocytes seem not to be involved in scar formation or subsequent wound healing.

Astrocytes play a major role in the stab wound-healing process. There seem to be at least three major subsets of astrocytes after the injury. To start with, juxtavascular astrocytes are astrocytes that in the healthy brain have the soma attached to the basement membrane but are inside the parenchyma. This distinguishes them from perivascular astrocytes, which are located between the basement membranes [42]. After a stab wound, these astrocytes change their physiological properties from a more passive, homeostatic current pattern with Kir4.1 to a dominating pattern where voltage-gated currents dominate. This resembles the pattern of many immature and proliferating cells [43]. Indeed, the juxtavascular astrocytes are the only astrocyte subtype that proliferates after a stab wound, accounting for the 20% increase in astrocyte numbers after this type of injury. This proliferation rate is independent of the size of the stab wound lesion [42]. Another population of astrocytes polarizes with respect to the lesion site, elongates processes, and extends them toward the lesion. A third astrocyte population retains its bushy morphology and hypertrophies.

At this point, the most interesting fact relates to the proliferation of the juxtavascular astrocytes. The daughter cells of these astrocytes remain compact and stay close to the blood vessels. If their proliferation is prevented, macrophages invade the parenchyma, and this invasion is correlated with increased scar formation and reduced healing [44]. Thus, these proliferating juxtavascular astrocytes play a major role in preventing macrophage infiltration toward the stab wound lesion. The factors responsible might be tight junctions or the expression of AhR (aryl hydrocarbon receptor) in the astrocytes. AhR expression is known to downregulate chemokines and proinflammatory cytokines [45]. This means that in a stab wound a certain macrophage invasion takes place, which is detrimental to the wound healing. One of the main tasks of these juxtavascular astrocytes is to prevent the invasion of proinflammatory macrophages. There are no significant intrusion of T or B lymphocytes into the parenchyma after a stab wound injury [46].

In the absence of macrophage infiltration, the microglia activation is enhanced although their proliferation is not changed. This indicates that in the case of a stab wound reduced macrophage invasion due to juxtavascular astrocyte proliferation and resulting increased microglia activation has a pronounced impact on the scar formation. These anti-inflammatory actions of the microglia together with the reactive astrocytes and OPCs reduce scar size and improve the outcome [44]. Preventing macrophage invasion not only reduces scar formation but also several components of the extracellular matrix (collagen IV, tenascin C, chondroitin sulfate proteoglycan) are less prominent.

16.4 Explosive Blast Injury

The last decades saw an increased use of improvised explosive devices in military combat, and also in guerilla warfare and terrorist attacks on civilian targets. The effects of such explosive blasts are well investigated for the pulmonary and auditory system, but its consequences for brain tissue are less known. Blast physics and impact are complicated [47]. In brief, there is a positive pressure wave (overpressure), then a negative pressure wave, in turn followed by blast winds, which can be excessive and cause amputation of body parts. This picture is complicated by reflected pressure waves. The effects on the body start with primary blast injury followed by a secondary one consisting mainly of debris and shrapnel, which might pierce the skull. Then follows injury due to the acceleration/deceleration of the head, followed by a heatwave, that may cause burns. Finally, contamination due to the blast source may play a role (like bacteria, radiation, fuel, and metal residues). Another type of injury is cavitation. Cavitation occurs when the blast wave in the tissue compresses dissolves gases in the different fluid compartments. This results in compressed gas bubbles. After the wave passes, the bubbles re-expand and burst. The resulting tissue injury is known as cavitation. In humans, hemorrhage, hematomas, and vasospasms occur. In neurons, the strain activates proteolytic enzymes, whose actions can initiate neurodegenerative processes [48]. The pathophysiology is complicated in that normally the brain is not the only organ affected. Thus, damage to the lungs and damage/overstimulation of the autonomic nervous system cause changes in the blood biochemistry, like pH ("potential of hydrogen"). Animal models employ mostly shock tubes in various forms [49].

As with all other forms of TBI there is a strong glial and neuroinflammatory reaction days after the blast. In contrast to other TBI forms, not much research has focused on glial cells or the role of invasive immune cells. Both astrocytes and microglia show early reactivity to a blast. Microglia tolerate and survive pressure waves that are far higher than the those of neurons. Interestingly their reactivity is not correlated with neuronal loss or axonal damage [50]. Microglia reactivity seem to be more tuned to pressure amplitude than duration [51]. Reactive microglia release ROS and NO (nitric oxide) as well as inflammatory cytokines (most prominently TNFα, tumor necrosis factor alpha) [52]. The microglial expression of cytokines correlates with pressure amplitude, but in a nonlinear relationship [53].

16.5 Conclusion

The different injury types caused by the various TBI forms cause different response patterns by nonneuronal cells. Interestingly, they do not recapitulate the patterns seen in focal or global ischemic injury. In mTBI, microglia take the lead due to myelin debris and damage to OPCs. In explosive blast injury, they appear most resilient and prominent. As in other injury types, microglia at first display anti-inflammatory properties, which, within days, transform to proinflammatory behavior. This is a major determining event for damage to repeated TBIs. The secondary TBI's impact is determined by its delay with the first impact. This delay determines if the impact appears in the pro- or anti-inflammatory phase. In mTBI, astrocytes express a mild reactivity and take a back seat to microglial reactions. In penetrating brain injury, however, if blood–brain barrier leakage is present, the normally low-key juxtavascular astrocyte type activates and proliferates to prevent macrophage invasion – a key mechanism in the turn of events. OPCs play a major role in the first month following penetrating TBI. Their behavior is different in this focal injury from focal ischemia, where they can transform into astrocytes. In penetrating TBI, there is no sign of this transformation. Instead, OPCs proliferate and migrate toward the injury site and prevent macrophage accumulation. This has a major restorative effect in the first month after injury. Macrophage invasion has in these injuries a proinflammatory and damaging role. This role is antagonized by both microglia and astrocytes. However, in injuries damaging the meninges, the macrophages have an opposite role as they terminate the action of debris removing and phagocytosing neutrophils. Comparing these forms of TBIs and ischemic injuries with each other, a surprising varying pattern of interactions and roles appear within the astrocyte–microglia–macrophage–OPC plasticity. It seems no clear general and stereotypical rule of these interactions can be deduced. The context of the various damage-induced tissue factors and cues are most decisive. This points therefore not to stereotypical mechanisms but an extremely context dependent plasticity.

References

1. Mira RG, Lira M, Cerpa W. Traumatic brain injury: mechanisms of glial response. Front Physiol. 2021;12:740939.
2. Ng SY, Lee AYW. Traumatic brain injuries: pathophysiology and potential therapeutic targets. Front Cell Neurosci. 2019;13:528.
3. Kamins J, Giza CC. Concussion-mild traumatic brain injury: recoverable injury with potential for serious sequelae. Neurosurg Clin N Am. 2016;27(4):441–52.
4. Wiegand TLT, Sollmann N, Bonke EM, Umeasalugo KE, Sobolewski KR, Plesnila N, et al. Translational neuroimaging in mild traumatic brain injury. J Neurosci Res. 2022;100(5):1201–17.
5. Abbott NJ, Pizzo ME, Preston JE, Janigro D, Thorne RG. The role of brain barriers in fluid movement in the CNS: is there a 'glymphatic' system? Acta Neuropathol. 2018;135(3):387–407.
6. Dean PJ, Sterr A. Long-term effects of mild traumatic brain injury on cognitive performance. Front Hum Neurosci. 2013;7:30.
7. Bodnar CN, Roberts KN, Higgins EK, Bachstetter AD. A systematic review of closed head injury models of mild traumatic brain injury in mice and rats. J Neurotrauma. 2019;36(11):1683–706.
8. Carroll LJ, Cassidy JD, Cancelliere C, Côté P, Hincapié CA, Kristman VL, et al. Systematic review of the prognosis after mild traumatic brain injury in adults: cognitive, psychiatric, and mortality outcomes: results of the International Collaboration on Mild Traumatic Brain Injury Prognosis. Arch Phys Med Rehabil. 2014;95(3 Suppl):S152–73.

9. Mackay DF, Russell ER, Stewart K, MacLean JA, Pell JP, Stewart W. Neurodegenerative disease mortality among former professional soccer players. N Engl J Med. 2019;381(19):1801–8.
10. Palacios EM, Yuh EL, Mac Donald CL, Bourla I, Wren-Jarvis J, Sun X, et al. Diffusion tensor imaging reveals elevated diffusivity of white matter microstructure that is independently associated with long-term outcome after mild traumatic brain injury: a TRACK-TBI study. J Neurotrauma. 2022;39(19–20):1318–28.
11. Hellstrøm T, Westlye LT, Kaufmann T, Trung Doan N, Søberg HL, Sigurdardottir S, et al. White matter microstructure is associated with functional, cognitive and emotional symptoms 12 months after mild traumatic brain injury. Sci Rep. 2017;7(1):13795.
12. Silverberg ND, Duhaime AC, Iaccarino MA. Mild traumatic brain injury in 2019-2020. JAMA. 2020;323(2):177–8.
13. Weber AM, Torres C, Rauscher A. Imaging the role of myelin in concussion. Neuroimaging Clin N Am. 2018;28(1):83–90.
14. Kochanek PM, Dixon CE, Shellington DK, Shin SS, Bayır H, Jackson EK, et al. Screening of biochemical and molecular mechanisms of secondary injury and repair in the brain after experimental blast-induced traumatic brain injury in rats. J Neurotrauma. 2013;30(11):920–37.
15. Bailes JE, Dashnaw ML, Petraglia AL, Turner RC. Cumulative effects of repetitive mild traumatic brain injury. Prog Neurol Surg. 2014;28:50–62.
16. Donovan V, Kim C, Anugerah AK, Coats JS, Oyoyo U, Pardo AC, et al. Repeated mild traumatic brain injury results in long-term white-matter disruption. J Cereb Blood Flow Metab. 2014;34(4):715–23.
17. Coughlin JM, Wang Y, Munro CA, Ma S, Yue C, Chen S, et al. Neuroinflammation and brain atrophy in former NFL players: an in vivo multimodal imaging pilot study. Neurobiol Dis. 2015;74:58–65.
18. Coughlin JM, Wang Y, Minn I, Bienko N, Ambinder EB, Xu X, et al. Imaging of glial cell activation and white matter integrity in brains of active and recently retired national football league players. JAMA Neurol. 2017;74(1):67–74.
19. Cherry JD, Tripodis Y, Alvarez VE, Huber B, Kiernan PT, Daneshvar DH, et al. Microglial neuroinflammation contributes to tau accumulation in chronic traumatic encephalopathy. Acta Neuropathol Commun. 2016;4(1):112.
20. Alosco ML, Tripodis Y, Fritts NG, Heslegrave A, Baugh CM, Conneely S, et al. Cerebrospinal fluid tau, Aβ, and sTREM2 in Former National Football League Players: modeling the relationship between repetitive head impacts, microglial activation, and neurodegeneration. Alzheimers Dement. 2018;14(9):1159–70.
21. Suárez-Calvet M, Kleinberger G, Araque Caballero M, Brendel M, Rominger A, Alcolea D, et al. sTREM2 cerebrospinal fluid levels are a potential biomarker for microglia activity in early-stage Alzheimer's disease and associate with neuronal injury markers. EMBO Mol Med. 2016;8(5):466–76.
22. Ebert SE, Jensen P, Ozenne B, Armand S, Svarer C, Stenbaek DS, et al. Molecular imaging of neuroinflammation in patients after mild traumatic brain injury: a longitudinal (123) I-CLINDE single photon emission computed tomography study. Eur J Neurol. 2019;26(12):1426–32.
23. Madry C, Kyrargyri V, Arancibia-Cárcamo IL, Jolivet R, Kohsaka S, Bryan RM, et al. Microglial ramification, surveillance, and interleukin-1β release are regulated by the two-pore domain K(+) channel THIK-1. Neuron. 2018;97(2):299–312.e6.
24. Grovola MR, Paleologos N, Wofford KL, Harris JP, Browne KD, Johnson V, et al. Mossy cell hypertrophy and synaptic changes in the hilus following mild diffuse traumatic brain injury in pigs. J Neuroinflammation. 2020;17(1):44.
25. Verboon LN, Patel HC, Greenhalgh AD. The immune system's role in the consequences of mild traumatic brain injury (concussion). Front Immunol. 2021;12:620698.
26. Corrigan F, Arulsamy A, Collins-Praino LE, Holmes JL, Vink R. Toll like receptor 4 activation can be either detrimental or beneficial following mild repetitive traumatic brain injury depending on timing of activation. Brain Behav Immun. 2017;64:124–39.

27. Bennett RE, Brody DL. Acute reduction of microglia does not alter axonal injury in a mouse model of repetitive concussive traumatic brain injury. J Neurotrauma. 2014;31(19):1647–63.
28. Shandra O, Winemiller AR, Heithoff BP, Munoz-Ballester C, George KK, Benko MJ, et al. Repetitive diffuse mild traumatic brain injury causes an atypical astrocyte response and spontaneous recurrent seizures. J Neurosci. 2019;39(10):1944–63.
29. George KK, Heithoff BP, Shandra O, Robel S. Mild traumatic brain injury/concussion initiates an atypical astrocyte response caused by blood-brain barrier dysfunction. J Neurotrauma. 2022;39(1–2):211–26.
30. Shandra O, Robel S. Inducing post-traumatic epilepsy in a mouse model of repetitive diffuse traumatic brain injury. J Vis Exp. 2020;156:10.3791/60360.
31. Christensen J. Traumatic brain injury: risks of epilepsy and implications for medicolegal assessment. Epilepsia. 2012;53(Suppl 4):43–7.
32. Roth TL, Nayak D, Atanasijevic T, Koretsky AP, Latour LL, McGavern DB. Transcranial amelioration of inflammation and cell death after brain injury. Nature. 2014;505(7482):223–8.
33. Russo MV, Latour LL, McGavern DB. Distinct myeloid cell subsets promote meningeal remodeling and vascular repair after mild traumatic brain injury. Nat Immunol. 2018;19(5):442–52.
34. Turco L, Cornell DL, Phillips B. Penetrating bihemispheric traumatic brain injury: a collective review of gunshot wounds to the head. World Neurosurg. 2017;104:653–9.
35. Vakil MT, Singh AK. A review of penetrating brain trauma: epidemiology, pathophysiology, imaging assessment, complications, and treatment. Emerg Radiol. 2017;24(3):301–9.
36. Young L, Rule GT, Bocchieri RT, Walilko TJ, Burns JM, Ling G. When physics meets biology: low and high-velocity penetration, blunt impact, and blast injuries to the brain. Front Neurol. 2015;6:89.
37. Takahashi CE, Virmani D, Chung DY, Ong C, Cervantes-Arslanian AM. Blunt and penetrating severe traumatic brain injury. Neurol Clin. 2021;39(2):443–69.
38. D'Agostino R, Kursinskis A, Parikh P, Letarte P, Harmon L, Semon G. Management of penetrating traumatic brain injury: operative versus non-operative intervention. J Surg Res. 2021;257:101–6.
39. Kirdajova D, Valihrach L, Valny M, Kriska J, Krocianova D, Benesova S, et al. Transient astrocyte-like NG2 glia subpopulation emerges solely following permanent brain ischemia. Glia. 2021;69(11):2658–81.
40. von Streitberg A, Jäkel S, Eugenin von Bernhardi J, Straube C, Buggenthin F, Marr C, et al. NG2-glia transiently overcome their homeostatic network and contribute to wound closure after brain injury. Front Cell Dev Biol. 2021;9:662056.
41. Huang C, Sakry D, Menzel L, Dangel L, Sebastiani A, Krämer T, et al. Lack of NG2 exacerbates neurological outcome and modulates glial responses after traumatic brain injury. Glia. 2016;64(4):507–23.
42. Bardehle S, Krüger M, Buggenthin F, Schwausch J, Ninkovic J, Clevers H, et al. Live imaging of astrocyte responses to acute injury reveals selective juxtavascular proliferation. Nat Neurosci. 2013;16(5):580–6.
43. Götz S, Bribian A, López-Mascaraque L, Götz M, Grothe B, Kunz L. Heterogeneity of astrocytes: electrophysiological properties of juxtavascular astrocytes before and after brain injury. Glia. 2021;69(2):346–61.
44. Frik J, Merl-Pham J, Plesnila N, Mattugini N, Kjell J, Kraska J, et al. Cross-talk between monocyte invasion and astrocyte proliferation regulates scarring in brain injury. EMBO Rep. 2018;19(5):e45294.
45. Rothhammer V, Mascanfroni ID, Bunse L, Takenaka MC, Kenison JE, Mayo L, et al. Type I interferons and microbial metabolites of tryptophan modulate astrocyte activity and central nervous system inflammation via the aryl hydrocarbon receptor. Nat Med. 2016;22(6):586–97.
46. McKee CA, Lukens JR. Emerging roles for the immune system in traumatic brain injury. Front Immunol. 2016;7:556.
47. Cernak I. Understanding blast-induced neurotrauma: how far have we come? Concussion (London, England). 2017;2(3):Cnc42.

48. Valiyaveettil M, Alamneh YA, Wang Y, Arun P, Oguntayo S, Wei Y, et al. Cytoskeletal protein α-II spectrin degradation in the brain of repeated blast exposed mice. Brain Res. 2014;1549:32–41.
49. Needham CE, Ritzel D, Rule GT, Wiri S, Young L. Blast testing issues and TBI: experimental models that lead to wrong conclusions. Front Neurol. 2015;6:72.
50. Goodrich JA, Kim JH, Situ R, Taylor W, Westmoreland T, Du F, et al. Neuronal and glial changes in the brain resulting from explosive blast in an experimental model. Acta Neuropathol Commun. 2016;4(1):124.
51. Kane MJ, Angoa-Pérez M, Francescutti DM, Sykes CE, Briggs DI, Leung LY, et al. Altered gene expression in cultured microglia in response to simulated blast overpressure: possible role of pulse duration. Neurosci Lett. 2012;522(1):47–51.
52. Xu D, Zhang N, Wang S, Yu Y, Zhang P, Li Y, et al. A novel in vitro platform development in the lab for modeling blast injury to microglia. Front Bioeng Biotechnol. 2022;10:883545.
53. Sawyer TW, Lee JJ, Villanueva M, Wang Y, Nelson P, Song Y, et al. The effect of underwater blast on aggregating brain cell cultures. J Neurotrauma. 2017;34(2):517–28.

Chapter 17
Major Psychiatric Disorders

Abstract Glial and immune cells are involved to various degrees in initiation and progression of mental diseases. However, one must judge each disease separately for glial involvement as there is no common guiding principle. In schizophrenia, the most pronounced finding is that during neurodevelopment the differentiation of progenitors into oligodendrocytes and astrocytes is arrested. The lack of new oligodendrocytes is the major factor in myelin and oligodendrocyte loss. The homeostatic properties of astrocytes are reduced. Both factors – if reproduced in animal models – cause behavior patterns reminiscent of schizophrenia. Reactive astrogliosis and microgliosis are not pronounced and mostly visible in a later stage. In major depressive disorder, the only consistent result points to decreased astrocytic homeostatic properties, some of which can be restored by medication. In bipolar disorder, the variations of glial properties in human studies do not allow any conclusions about glial participation. Studies report consistently high levels of inflammatory cytokines in the cerebrospinal fluid of patients attempting or committing suicide. This would suggest the participation of microglia. In anxiety disorders, many animal models point to an active if not causal involvement of astrocytes, microglia, and oligodendrocyte precursors. The problem is that there are no human studies and animal models for anxiety are highly controversial. In alcohol use disorder, all glial cells are affected due to inflammatory responses and impaired myelin sheets. This seems due to the toxic effect of ethanol. Both astrocytes and microglia have opioid receptors. In opioid use disorder, the stimulation of these astrocytic receptors leads to strong disturbances of their role in glutamate homeostasis. The main actions of nicotine are as agonist of the nicotinic acetylcholine receptor family. Astrocytes possess these receptors, and their stimulation causes release of D-serine which interacts with neuronal NMDA (N-methyl-D-aspartic acid) glutamate receptors and contributes to addictive behavior.

Keywords Addiction · Alcohol use disorde · Anxiety disorders · Bipolar disorder · Brain reward system · D-serine · Glutamate homeostasis · Inflammatory cytokines · Major depressive disorder · Mood disorders · Nicotine dependence · NMDA receptors · Oligodendrocyte loss · Opioid use disorder · Progenitor cell differentiation · Schizophrenia · Substance use disorders · Suicide

W. Walz, *The Gliocentric Brain*, https://doi.org/10.1007/978-3-031-48105-5_17

17.1 Introduction

An increasing number of observations link psychiatric disorders with glial cell reactions. Neuroinflammation and therefore microglia are certainly involved. However, astrocytes are important for synaptic connections and these connections are affected in many if not all disorders. Oligodendrocytes due to their importance for white matter tracts are also involved. This chapter introduces each major disorder and focuses as much as possible on postmortem or patient findings. Animal models of psychiatric diseases reproduce normally only part of the pathology with various strengths and weaknesses. The models are a problem when it comes to understanding the pathogenesis of the diseases. Partly this is because most psychiatric diseases can have different origins which then convergence into similar symptoms. These symptoms are then – for simplicity's sake – grouped as a single disease. There is no question, however, that animal models have their strengths in elucidating some of the pathophysiological mechanisms and it is in this context that this chapter refers to them.

17.2 Schizophrenia

Schizophrenia is one of the 20 major causes of global disability, affecting about 1% of the population [1]. According to the DSM-5 (*Diagnostic and Statistical Manual of Mental Disorders*, 5th edition), individuals must manifest at least two of the following criteria: delusions, hallucinations, disorganized speech and behavior, avolition, and anhedonia. Schizophrenia manifests itself in a large heterogeneity of behaviors, but in the end, it leads to highly dysfunctional individuals with a life expectancy reduced by about 15 years [2]. The strongest risk factor is a family history, pointing to dominating genetic factors, although environmental factors play a role [3]. Studies have shown that patients on antipsychotic drugs (e.g., risperidone) have lower mortality than untreated patients, despite some side effects especially of the older drugs [1]. The dopamine hypothesis has been very influential. It states that schizophrenia sufferers have an increased striatal presynaptic dopamine synthesis rate. More recent investigations point to the decreased activity of NMDA (N-methyl-D-aspartic acid) receptors on GABA (gamma aminobutyric acid) interneurons. However, the dopamine hyperactivity could easily be explained by the downstream effect of the glutamate dysfunction. Similarly, excess serotonin could via the glutamatergic system cause further downstream dopamine hyperactivity [4]. Structural abnormalities are present in schizophrenia and may be there since early neurodevelopment; however, they are diffuse and widespread [5]. A very influential theory is that genetic and environmental factors combine in embryonic and early postnatal development to derail some developmental processes. This leads during the transition from adolescence to early adulthood to psychosis [6]. Many candidate genes encode for synaptic proteins and, indeed, the major pathophysiological hallmark of schizophrenia is synapse loss [7]. Such an increased synaptic stripping into adulthood will be accompanied by neuroinflammation.

Many animal models, like the chronic ketamine treatment model, find increased astrogliosis and relatively little microgliosis [8]. It must be kept in mind that such models use normal, juvenile rodents. This will not take into account the newer findings on the impact of a combination of detrimental genetic and environmental factors in embryogenesis on cell differentiation and subsequent manifestation of these aberrant neurodevelopmental processes as schizophrenic symptoms later in life [9]. There are environmental risk factors which detrimentally affect people with a genetic predisposition to schizophrenia. These are physical or emotional stress in the first trimester of pregnancy and later during pregnancy, maternal infections and obstetrical complications as well as inflammation for any reason [10]. At the onset of symptoms, MRI (magnetic resonance imaging) images of patients reveal reduced white matter tract integrity of frontoparietal circuitry, which is important for working memory. In drug-naïve patients, the more severe the psychotic symptoms, the more pronounced the white matter tract abnormalities [11]. Chronic administration of clozapine (a catecholamine antagonist), which improves the symptoms, reverses most of the abnormalities [12]. It also appears that schizophrenia might lead to loss of mature oligodendrocytes but not of OPCs (oligodendrocyte precursor cells). This is accompanied by a reduction of myelin-associated gene activities [13]. Hippocampal volume is reduced in schizophrenic patients compared to controls. A recent careful study found this to be exclusively due to loss of oligodendrocytes, which may lead to altered functional connections [14].

The question is then, is white matter pathology at the core of the genesis of this disease early during embryogenesis? Divisions of glial progenitor cells in the embryo result in both new oligodendrocytes and astrocytes. The path of this differentiation depends on Olig2. This transcription factor, if overexpressed, leads to increased oligodendrocyte production and decreased astrocyte numbers and vice versa. Postmortem analysis of schizophrenic patients suggests that glial progenitor cells remain far more proliferative then matched controls. Further imaging studies in patients reveal arrested differentiation into mature oligodendrocytes and astrocytes [15]. This observation would explain the white matter pathology. A reduction in the number of mature astrocytes would negatively affect homeostatic mechanisms. Elimination or reduction of astrocytes in animal models produces behavioral patterns consistent with schizophrenia like deficits in working memory and reduced prepulse inhibition, a litmus test for schizophrenia [15]. Postmortem studies tend to find increased astrogliosis. However, when antipsychotic drug treatment is considered as a factor, the increased astrogliosis seems to be caused by the drug treatment and not by the schizophrenia [16]. This indicates that reactive astrogliosis is not a prerequisite for the development of the disease, but probably a secondary phenomenon arriving later. Astrocyte-specific enzymes and transporters related to glutamate metabolism are reduced in these postmortem samples and this might contribute to the symptoms. There are also indications that Kir4.1 is downregulated in astrocytes and oligodendrocytes in postmortem samples of patients. In the astrocytes, Kir4.1 is important for potassium homeostasis, whereas its expression in oligodendrocytes is a prerequisite for normal differentiation. Most studies found an increase in microglia density, some found no changes, and a few reported even reductions. Suicide rates are high among schizophrenia sufferers and if the brain tissue of schizophrenia patients, who committed suicide, is analyzed separately, the following picture emerges. Suicide victims have a higher level of proinflammatory cytokines and therefore a higher density and activation of microglia. If these suicide cases are considered, the microglia show no change in postmortem schizophrenia brains compared with controls [17]. This conclusion seems to be confirmed in PET (positron emission tomography) studies of schizophrenia patients, which, despite variations in the results, found no convincing evidence for microglial activation [18]. Furthermore, genes whose activity is crucial for microglial activation are not contributing to schizophrenia susceptibility [19]. However, the PFC (prefrontal cortex) of schizophrenic patients shows a dendritic spine deficit. Complement and phagocytic microglia markers are increased in these patients, indicating increased phagocytosis of spines by microglia as compared to controls [20]. A genome-wide association study found, however, no association of schizophrenia with microglia-defining genes. At the same time, six astrocytic and three oligodendrocyte-specific gene sets were associated with schizophrenia risk [19].

Some crucial experiments were performed by Windrem et al. [21, 22] using a new model of human glial chimeric mice. Glial progenitor cells were prepared from iPSCs (induced pluripotent stem cells) derived from fibroblasts from either juvenile-onset schizophrenic patients or their normal controls. The glial progenitors from patients had downregulated expression of genes important for glial differentiation compared to the controls. These cells were then transplanted into immunodeficient neonatal mice to produce these patient-specific human glial chimeric mice. The human glial progenitor cells overwhelmed the host counterparts and colonized the mouse brains. They differentiated into myelinating oligodendrocytes and astrocytes. There were pronounced differences between the schizophrenia-derived human glial chimeras and control chimeras. In the former, both oligodendrocyte and astrocyte maturation were impaired. This led to decreased myelination, less astrocytic branching, and reduced coherent domain organization, reminiscent of a reduced homeostatic role. Indeed, the potassium uptake of both cell types was reduced. The behavior of the schizophrenia-derived human glial chimeras was different from the control chimeras. The schizophrenia-related chimeras exhibited reduced prepulse inhibition, reduced working memory, and other schizophrenia-related behavioral deficits.

These results compelled Dietz et al. [15] to develop a hypothesis for the genesis of juvenile-onset if not other forms of schizophrenia. According to this hypothesis, aberrant glial functions are the most important mechanism. It postulates microglial activation leading to neuroinflammation at one point in the embryogenesis either by infections or by other kind of risk factors. Infections occurring late in embryogenesis would have a greater effect on oligodendrocytes than neuronal differentiation because gliogenesis occurs in later stages of fetal development than neurogenesis. The inflammation derails normal differentiation of the glial progenitors into mature oligodendrocytes and astrocytes. Key would be the transcription factor Olig2, which has a crucial role in the balance of oligodendrocyte versus astrocyte differentiation. The activity of Olig2 and other associated processes are highly sensitive to environmental factors like neuroinflammation, neurotransmitters, or growth factors as well as neuronal activity [23, 24]. The resulting reduction in myelination will affect the connectivity of cortical and subcortical tracts. In addition, reduced astrocyte glutamate and potassium homeostasis disrupts normal function further. Secondary effects in the development of the disease in adult patients could exhibit quite a large disparity. This would explain the observed heterogeneity in reactive astrogliosis and microgliosis in patients and postmortem tissue.

17.3 Mood Disorders/Suicide

17.3.1 Introduction

The overwhelming manifestation of mood disorders is via two disorders: major depressive disorder (MDD) and bipolar disorder. Both are related but have different underlying causes, symptoms, and treatments. Bipolar disorder has often components of depression. MDD has a lifetime occurrence of 16%, is the second leading cause of disability, and has a median onset of 32 years of age. Bipolar disorder is more complicated as it consists of a more heterogeneous picture than MDD. Lifetime incidence is 5%, but it has a relatively high mortality [25].

17.3.2 Major Depressive Disorder

MDD appears usually in recurrent episodes each lasting about 20 weeks. There is a genetic component/family history risk as individuals with a first-degree relative suffering from MDD have an almost three-time increased risk. In addition, women have almost double the risk of developing MDD. It is assumed this is due to fluctuating hormonal levels and the frequency of postpartum depression. The start of the disorder and the single episodes can be triggered by external stress or appear spontaneously. Two-thirds of patients respond to therapy, which is a choice of either one or a combination of psychotherapy, bright light therapy, and medication. First-line medications are those that increase monoamine levels in the synaptic cleft (e.g., fluoxetine) and atypical antidepressants (e.g., opioid receptor antagonists). Other drugs used are tricyclic antidepressants (imipramine) and monoamine oxidase inhibitors (iproniazid). This last group has the potential for significant side effects.

Several pathophysiological mechanisms are driving the current research directions. The monoamine hypothesis was very influential and was based on reduced serotonin and norepinephrine availability at the synapse. This deficit would be addressed by most of the first-line antidepressants. This hypothesis is now challenged by two more recent approaches. The neurotrophin hypothesis claims a reduction in neurotrophins, like BDNF (brain-derived neurotrophic factor), is at the core of the disease, and leads to neuronal atrophy, reduced neurogenesis, and loss of glia. Together these effects lead to an impaired brain circuitry. Antidepressants elevate

BDNF and reverse atrophy. One of these, ketamine, which at subanesthetic doses has an antidepressant effect, increases dendritic spine function and number in the PFC. The neuroinflammation hypothesis states that the anti- versus proinflammatory balance is shifted to the latter, as increased serum levels of proinflammatory cytokines are found. The proinflammatory cytokines will shift serotonin metabolism toward excitotoxic compounds and interfere with the HPA (hypothalamus–pituitary–adrenal) axis. This last interference would cause chronic cortisol release with resulting hippocampal damage [26].

Of the major glial cell populations, most efforts so far focused on the astrocytes. Young and middle-aged MDD patients in imaging and postmortem studies show consistently reduced densities of GFAP+ (glial fibrillary acidic protein positive) astrocytes compared with controls. In contrast, late-onset MDD patients show an increase in these astrocytes [27]. This could be due to age-related increases in GFAP as it is not clear if age-matched controls were used. Medication history could also play a role. Patients show consistently reduced levels of the astrocyte glutamate homeostatic system (glutamate transporter 1 GLT-1 also known as excitatory amino acid transporter 2 EAAT2, glutamate aspartate transporter GLAST and glutamine synthetase GS) as well as decreased coverage of blood vessels by endfeet including AQP4 (aquaporin 4) expression. Blocking GLT-1 in the rodent PFC leads to depressive-like behavior. Selective ablation of cortical astrocytes also produces depression-like rodent behavior. Various rodent models of depression exhibit reduced PFC astrocyte density and reduced endfeet covering of blood vessels. If these rodents are then treated with fluoxetine, most of these pathologies reverse [26]. In general, fluoxetine treatment increases gliogenesis in animals [28]. In a mouse model of chronic social defeat, the chemogenetic activation of astrocytes releases ATP (adenosine triphosphate) which via P2X2 (purinergic P2X2) receptors mediate antidepressant effects in these mice [29]. Interestingly, selective serotonin reuptake inhibitors upregulate selective serotonin receptors in astrocytes, but not neurons. The stimulation of these receptors also increases ATP release from the astrocytes [30].

Patients with MDD express increased proinflammatory cytokines in their peripheral blood (interleukin IL-1, IL-6, tumor necrosis factor alpha [TNFα]) and are regarded as a biomarker for the disorder. As peripheral blood monocytes of MDD patients express increased cytokines and mRNA (messenger ribonucleic acid) for inflammatory responses, the question arises, if these cytokines originate from peripheral organs or brain parenchyma [31]. However, postmortem studies were contradictory and, in the end, showed no clear evidence of microglia activation in any part of the brain. TSPO (translocator protein 18 kDa) binding studies showed a correlation of the binding in some brain areas with the degree of MDD [31]. However, TSPO is not universally acknowledged as a good indicator of microglia activation in humans [32]. Thus, not a clear picture of microglia involvement in MDD emerges [33]. A single-cell study investigated microglia from postmortem brain tissue of MDD patients [31]. The study revealed a non-inflammatory and homeostatic type of microglia in the brain of MDD patients but not in matched controls. The type appeared in clusters, had downregulated inflammatory and immune molecules, and upregulated TMEM119 (transmembrane protein 119) and P2Y12 (purinergic receptor for adenosine diphosphate). Both markers are associated with an anti-inflammatory and homeostatic stage in microglia. The study did not consider the effect of medication. In a chronic despair mouse model of depression, a microglia cluster is found, that showed hyper-ramified microglia [34]. The two clusters of microglia might be functionally related. All in all, the various results do not point to microglial-mediated inflammation as a major player in depression.

Imaging of patient's brains and postmortem analysis show inconsistent results for white matter tracts. Partly this is due to the white matter integrity affected in different tracts depending on the type of depression, the age of onset, the patient at the time of the scan or death, and medication used, if any. White matter integrity seems to be more affected in schizophrenia and bipolar disorder than in MDD [35]. When maturity and density of oligodendrocytes are assessed, there seems to be no straightforward link with MDD. However, social stress, if connected to MDD or not, seems to have an impact [36]. MDD patients with childhood abuse had a decreased oligodendrocyte density with a shift toward more mature oligodendrocytes as compared to MDD patients without childhood abuse [37].

17.3.3 Bipolar Disorder

Bipolar disorder is a complicated disorder as patients experience long-lasting cyclical mood shifts. In general, they spent more time in depressive than in manic, hypomanic, or mixed episodes. Average age of onset is 18–20 years. There is a very strong genetic component, but this does not rule out environmental influences [38]. Treatment is medication or medication combined with psychotherapy, but not psychotherapy alone. Medications include mood stabilizers (lithium) and anticonvulsants. Antidepressants are only recommended together with a mood stabilizer. Antipsychotics (dopamine receptor antagonists, e.g., chlorpromazine) can be used on their own. Underlying neurobiological mechanisms of the disorder are desynchronized circadian rhythms, including mutations of genes that affect the circadian timing system and disturbances of the melatonin system [25].

Interestingly, bipolar disorder patients exhibit chronic and mild systemic inflammation [39]. This fact makes the inflammatory properties of microglia and astrocytes a key focus of research. Yet, postmortem studies of bipolar disorder patients, although exhibiting large variations, revealed no clear and unambiguous change in microglia or astrocyte appearance or density [16, 40]. A thorough study of inflammatory properties of microglia from postmortem brain samples of bipolar disorder patients found no evidence of proinflammatory upregulation [41]. There might still be a possibility for microglia involvement. Studies found increased serum levels of the proinflammatory cytokines IL-2 (interleukin 2), IL-4, and IL-6 in bipolar disorder patients during manic phases. The same patients in the depressive phase had only increased IL-6 levels; in remission only IL-4 was increased compared with matched controls [41]. Thus, the possibility exists that there are cyclic swings in the microglial activation, depending on the bipolar cycle including medication. This would explain the large variations and non-significant overall results.

Studies targeting the involvement of astrocytes in bipolar disorder reflect the microglia results. There are large fluctuations, especially in inflammatory markers, but no consistent picture [42].

17.3.4 Suicide Behavior

The suicidal risk of patients with mood disorders is up to 20 times higher compared with the one in the general population [43]. Worldwide, it is the second leading cause of death among people 15–29 years of age [44]. There is a clear family history of suicidal behavior. Patients diagnosed with bipolar disorder with mixed symptoms are more likely to attempt suicide than either those with depressed or hypomanic symptoms alone [45]. There are other risk factors like schizophrenia, PTSD (post-traumatic stress disorder), and childhood abuse. Ninety percent of suicidal victims are diagnosed with a psychiatric illness or substance abuse [46]. Are there any unique functional glial properties, which make the decision process so vulnerable to self-destruction? To answer this question, one must investigate if there are any unique parameters in patients with suicidal behavior which are not found in existing mental disorders. This is a problem, as most studies compared findings from suicide victims with matched healthy controls and not with patients suffering from depression without being suicidal. There are animal models, but they are not very satisfactory. These models are mostly based on aggression, impulsivity, irritability, hopelessness, and social isolation [47]. It is difficult to pinpoint neurobiological mechanisms that may underlie suicidal behavior. There are two major mechanisms discussed, but neither is without its critics [48]. Abnormalities of the serotonergic system frequently correlate with suicidal behavior irrespective of underlying disease. Similarly, chronic stimulation of the HPA (hypothalamic–pituitary–adrenal) axis might cause cortisol overproduction and – as it crosses the blood–brain barrier unhindered – might create dysfunction, maybe even in conjunction with a faulty serotonin system.

Inflammatory events are also considered as a factor; nevertheless, they are no less controversial than the serotonin system and the HPA axis. Serum levels of IL-2, IL-4, and TGFβ (transforming growth factor beta) are higher in suicidal subjects compared to healthy matched controls and non-suicidal patients [44]. Within the PFC and the cerebrospinal fluid, IL-1β, IL-6, and TNFα are elevated in the postmortem brain of suicide victims irrespective of the psychiatric diagnosis [49]. A majority, but not all, of studies report a connection between suicide and increased inflammatory cytokines in the brain [44]. This suggests that suicidal individual may have a unique proinflammatory cytokine profile. Microglia densities are elevated in suicide victims in the anterior cingulate and PFC [50, 51]. However, earlier suggestions of a compromised blood–brain barrier and invading macrophages in the PFC were found to represent increased microglia population around the blood vessels rather than macrophages [52]. TLR3 (toll-like receptor 3), TLR4, NLRPs (nucleotide-binding oligomerization domain, leucin-rich repeat, and pyrin domain containing), and inflammasomes were found to be upregulated in microglia of suicide victims irrespective of underlying psychiatric disease [53, 54]. This suggests that activation of microglial damage/danger-associated molecular patterns is involved in these mechanisms. No studies addressed the possible ligands and sources of this activation. Some studies, however, investigated the possible action of this increased cytokine production with respect to suicidal tendencies. Inflammatory cytokines stimulate the breakdown of tryptophan into kynurenine and then further quinolinic acid and kynurenic acid in microglia. Quinolinic acid levels are 2–3 times higher in the CSF of patients who attempted suicide than in controls. The increased CSF levels correlated with increases in IL-6 [55]. Quinolinic acid is an NMDA antagonist and may therefore dysregulate glutamate signaling in neurons. Follow-up studies with these patients indicated a correlation between the levels of quinolinic acid and IL-6 with the severity of suicide ideation [56]. In addition to this kynurenine pathway, inflammatory cytokines can change serotonin metabolism and the HPA axis, thereby providing other pathways for promoting suicidal behavior [57]. Patients with multiple sclerosis are often treated with IFN-γ (interferon gamma). A study investigated such a cohort further [58]. In a subgroup of these patients with no previous history of psychiatric disorders, IFN-γ treatment induced suicidal ideation or attempts.

There are postmortem studies reporting a decrease in astrocyte density in depressed suicides, but these brains were compared with those of healthy individuals and not postmortem brains of depressive patients who had no suicide attempts during their lifetime [59, 60]. Nevertheless, there is a strong possibility of changes in the astrocyte–neuronal interaction in dorsomedial prefrontal cortex and anterior cingulate cortex [61]. The same criticism applies to studies investigating the role of oligodendrocytes in suicide [62].

17.4 Anxiety Disorders

Vigilance is a basic behavioral response of all mammals to possible threats from the environment and to a large extent based on circuits involving the amygdala. However, this fear system can under certain circumstances transform into a system sensing a chronic threat, irrespective of external circumstances. In humans, if this behavior pattern manifests itself and interferes with daily life, the individual is regarded as suffering from anxiety disorders. These disorders are the most common group of mental disturbances [63]. Anxiety disorders are a heterogeneous group of disorders with generalized anxiety disorder (GAD), the most basic component. GAD is underlying most other anxieties (agoraphobia, social phobia, and other special phobias). It is a chronic disorder which is present on most days for long periods (months or longer). According to the *Diagnostic and Statistical Manual of Mental Disorders*, 5th edition, the sufferer is not able to control his/her worries, and this interferes with the basic functioning of life tasks. Symptoms can be restlessness, fatigue, loss of concentration, irritability, muscle tension, and sleep disturbances [64]. Family history is a strong risk factor due to both genetic and environmental factors [65, 66]. GAD occurs often with MDD, panic disorder, and substance use disorder. It was proposed that GAD, MDD, and PTSD are in a category by themselves called "distress disorders." However, this is controversial [67]. MDD is dealt separately (discussed earlier) and PTSD is incorporated into the chapter on traumatic brain injuries.

Not surprisingly, MRI studies revealed that the anxiety neurocircuitry in patients (amygdala, anterior cingulate cortex ACC, PFC, hippocampus among others) has abnormal activities [64, 68]. The HPA axis is unchanged in GAD, however. In addition, dopamine reuptake sites and GABA receptors were decreased, but serotonin reuptake was unchanged. Animal models usually involve the application of stressors to induce anxiety. This has been criticized [69]. For example, the stressors change the HPA axis whereas in GAD this is not observed. Thus, there is a large overlap in animal models between depression, anxiety, and trauma-related disorders [70]. Another criticism has been the focus in animal models on the male sex, whereas in human anxiety there are clearly more women in the patient cohort [71]. There are no new developments in medication treatment of anxiety disorders, even though only around 70% of patients respond to the current treatments. The recurrence is especially high when patients at the same time suffer from MDD. Selective serotonin reuptake inhibitors and serotonin–norepinephrine reuptake inhibitors (escitalopram, duloxetine) are first-line treatments. Tricyclic antidepressants are second options. Buspirone (5-HT_{1A} partial agonist) is another option [72]. Psychotherapy, especially cognitive-behavioral therapy (CBT), is being used with mixed success. About half of the patients respond to it and for patients with GAD CBT is the least effective treatment [73].

There are several in vivo studies in mice, which indicate that astrocytes within the fear/anxiety circuits are actively involved in manipulating neuronal output. This astrocytic impact can induce or reduce anxiolytic behavior, depending on the location of the astrocytes. Most of these interactions are in the amygdala, and the hippocampus is also involved [74]. In the medial region of the central amygdala (CeM), the activation of astrocytes by DREADD (designer receptor exclusively activated by designer drugs) increases inhibitory synaptic transmission via adenosine A1 receptors and reduces excitatory transmission via A2A receptors. The overall effect is to reduce neuronal output of the CeM with decreased fear expression in a mouse fear conditioning model [75]. If in the lateral region (CeL) the oxytocin receptors of a subgroup of astrocytes are activated, a calcium wave propagates through the syncytium. This wave releases D-serine, which in turn activates CeL neurons. These activated neurons then in turn inhibit CeM neurons. The result is an increased anxiolytic behavior of these mice in a behavioral paradigm. Knockout of the oxytocin receptors in the same CeL astrocyte population has the opposite effect and decreases anxiolytic behavior. Thus, activation of oxytocin receptors on CeL astrocytes causes anxiolytic behavior [76]. There are two more conclusions from these experiments. First, increases in calcium of astrocytes in the central amygdala cause decreases in fear/anxiety. Second, an oxytocin-positive astrocyte subpopulation can recruit an oxytocin-negative population of astrocytes to manipulate the neuronal output with consequences for the anxiolytic behavior of the animal [74]. If astrocytes in the basolateral amygdala (BLA) are chronically stimulated, the anxiolytic behavior of mice in a chronic unpredictable mild stress model is reduced. However, if BLA neurons are stimulated, there is no effect in this model. The connection is not yet clear [77]. In hippocampal astrocytes, an experimental increase of calcium leads to ATP release. This ATP release increases synaptic transmission of dentate gyrus granule cells, which leads to anxiolytic behavior in this mouse model [78]. Overexpression of astrocytic TGF-β-activated kinase 1 (Tak1) activity in the mediobasal hypothalamus (MBH) leads to anxiolytic behavior in mice. Thus, Tak1 activity in MBH astrocytes regulates anxiety-like behavior in mice, whereas changes in the calcium concentration of these MBH astrocytes have no effect on the behavior [79, 80]. These examples demonstrate a clear involvement of astrocyte function in certain brain areas and of astrocyte subpopulations in the overall expression of anxiety in mice.

There is no real evidence for a significant activation of human microglia in anxiety disorders; however, this is due to the lack of research studies and not to negative results [81, 82]. Animal models that used stress-related paradigms to mimic anxiety provided some further insight. However, in most of these models, employing chronic stress also leads to depressive behavior, as the HPA axis is affected. Still, repeated social defeat in mice increases anxiolytic behavior. It causes microglia activation within the fear circuit with upregulated inflammatory cytokines. Increased invasion of macrophages into the perivascular space was also observed. Propranolol, a β-adrenergic antagonist, blocked both microglia activation and macrophage invasion [83]. A similar repeated social defeat model in rats caused microglia activation and increased neuronal firing in the basolateral amygdala, accompanied by anxiolytic behavior [84]. Blockade of microglial activation prevented the increased neuronal firing rate and the anxiolytic behavior. CYLD (a specific lysine 63-linked deubiquitinase for various proteins including TAK1) knockout in mice leads to anxiolytic behavior, but not depression or reduced motor activity. This knockout activated microglia in the dorsal striatum and increased TNFα and IL-1β levels. The microglial activation inhibitor minocycline prevented all these changes including the anxiolytic behavior [85]. These studies are so far the best evidence for an active microglial involvement in anxiety disorders from an animal model.

Changes in oligodendrocytes and myelination have so far not been implicated in anxiolytic behavior. However, an interesting study [86] used optogenetic stimulation of OPCs (oligodendrocyte precursor cells) in the hippocampus of mice. In the hippocampal CA1 (comu ammonis 1) layer, the output of pyramidal cells is controlled by inhibitory interneurons. These interneurons are the target of presynaptic-like structures from OPCs. Stimulation of these OPCs releases GABA, which in turn reduces the activity of these interneurons. This leads to increased spiking by the pyramidal cells. The photostimulation of the OPCs causes anxiolytic behavior of these mice in vivo. Furthermore, in a chronic social defeat stress paradigm the mice developed OPC calcium increases, which were directly linked to anxiety-like behavior.

17.5 Substance Use Disorders

17.5.1 The Brain Reward System

The addiction to substances is closely related to activation of the brain reward system. The purpose of this system is to engage the animal in goal-directed behavior that is not necessarily bringing in positive results at the time of the engagement but is important for the long-term survival and fitness of the individual animal or the species. Fulfilling such a demand is often experienced as a pleasurable activity. Examples are reproduction, parental care, and hoarding of food. In humans, and other animals, learned association with extrinsic rewards can connect addictive behaviors to this system. This is the case with gambling addiction, thrill-seeking behavior, or sex addiction. Substance addiction can be the result of such a learned association (e.g., glue sniffing) or stimulation of receptors of the reward system by external chemicals (e.g., alcohol or tobacco addiction) or both.

How does this system work and what are the feedback mechanisms that prevent the individual from being overtaken by these demands and neglect necessary maintenance? Key in this system are the dopaminergic neurons in the ventral tegmental area (VTA) of the midbrain. These dopaminergic neurons have a basal tonic firing rate. When a new, unexpected, or greater-than-expected reward is presented, the neurons respond with a phasic outburst. In contrast, a predictable cue does not cause any change in the tonic pattern. If a reward is expected, but turns out to be denied or smaller than expected, the firing rate declines. Thus, there is a continuous output of these dopaminergic neurons, the firing frequency of which determines the desirability or avoidance of a cue and its related reward. There are two important output pathways: the mesocortical (to the PFC for cognitive control, emotions, and motivation) and mesolimbic (to the nucleus accumbens or NAc and other structures) pathways. The low firing frequency between 1 and 5 Hz (Hertz) activates targets within the mesolimbic pathway, especially the NAc high-affinity D2 dopamine receptors. When highly desirable reward clues are presented, the firing frequency is above 20 Hz, which in the NAc activates both high- and low-affinity D1/D2 receptors due to the higher dopamine concentration at the presynaptic release sites. The phasic increases reflect desirable situations and facilitate learning new cues connected to a pleasurable sensation. However, as soon as such a behavior is established and no longer new, the NAc is no longer involved. Instead, it is now supported by the dorsal striatum. Other pathways play a role like the mesostriatal one (dopaminergic neurons from the substantia nigra to the dorsal striatum) and the mesocortical one to the frontal cortex. The various regions (limbic system including amygdala and frontal cortex) communicate with each other to integrate the response and this integration includes emotional, motivational, and decision-making components. There are various other transmitter systems involved, including inputs and outputs into the NAc. Endocannabinoid receptors are involved in the terminal region of the mesolimbic system as are opiate and nicotinic acetylcholine receptors. For this complex reward system, the backbone is the mesolimbic-cortical dopamine system. All drugs of abuse increase the dopamine availability of the mesolimbic system in the NAc and dorsal striatum [87]. Withdrawal symptoms involve mainly forebrain areas including amygdala and habenula. During these phases, transmitters other than dopamine are involved.

In a normally functional reward system, macroglia cells fulfill their usual role as in other brain systems: astrocytes within the tripartite synapse and blood flow regulation and oligodendrocytes due to the role of the myelin sheath around axons. There is some indication, however, of microglia having a specific role in mice fed with high-caloric chocolate cafeteria diet. When the clue is new and leads to weight gain, microglia become activated and release inflammatory cytokines, namely, IL-1β and IFN-γ. This release leads to synaptic pruning and remodeling of medium spiny neurons in the NAc, leading to a spine density increase during the compulsive-seeking behavior [88]. Forced inactivation of microglia in this mouse model normalized these mechanisms and the compulsive behavior, even if the animals were exposed to the cafeteria diet. Continuous exposure to the cafeteria diet decreases the sense of the reward and microglia activation and spine density decrease.

17.5.2 Alcohol Use Disorder

It is estimated that alcohol use disorder (AUD) causes about 6% of all worldwide deaths [89]. It affects multiple organ systems and can lead to serious psychosocial dysfunctions. Alcohol (also called ethanol) is lipophilic and can easily cross biological membranes. It can enter the brain compartments without restrictions and therefore interact with receptors and other structures inside the membrane, cytoplasm, and organelles. Thus, many transmitter systems are affected by the presence of alcohol. A genetic variation of the serotonin transporter is seen as a major risk factor for alcoholism [90]. Alcohol administration leads to increased activity of VTA dopaminergic neurons and therefore increased dopamine release in the NAc [89]. In AUD, reduced dopamine receptor sensitivity causes increased motivation for alcohol intake to keep the dopaminergic response constant. Several of the dopamine receptor polymorphisms are therefore an additional risk factor for AUD. Another significant interaction is the modulation of the $GABA_A$ (gamma aminobutyric acid A) receptor by alcohol. This leads to reduced inhibition of dopaminergic neurons [91]. Due to its lipophilic nature, there are many more interactions with alcohol and transmitter systems as well as ion channel functions. Treatment options involve behavioral and group therapies for mild AUD with the addition of pharmacological approaches for severe AUD. Most common medications are disulfiram (inhibitor of ALDH or aldehyde dehydrogenase) to provoke negative systemic effects of alcohol consumption, naltrexone (opioid receptor antagonist), acamprosate (glutamate neurotransmission modulator), nalmefene (antagonist/agonist for various opioid receptors), and baclofen ($GABA_B$ receptor agonist).

The postmortem brain analysis of alcoholics exhibits increased microgliosis and astrocyte loss [92, 93]. Originally, animal models of AUD reported proinflammatory effects of alcohol in the brain. This has changed over time to a more balanced view. The effect of alcohol on the brain is now considered to be more complex and immunomodulatory rather than exclusively proinflammatory [94]. The selection and age of the animal models, the exposure route, length, dose, and pattern have a strong influence on the outcome. The various findings oscillate between proinflammatory effects and reductions of the immune response [95]. The C57BL/6J mouse was heavily used for AUD studies. However, it was later found that a mutation makes this model very susceptible to oxidative stress, obscuring the results [96]. Despite this qualification, alcohol exposure leads to astrocyte loss and microglia degeneration in these models [97, 98]. One finding, sustained by various research groups, is that microglia are primed by alcohol exposure. Prior alcohol exposure increases the microglial reaction to a second alcoholic stimulus [99–101] or another challenge like LPS exposure [102]. The microglial reaction is expressed as increased TSPO binding, upregulated complement receptor 3 and Iba1 (ionized calcium-binding adapter molecule 1) immunoreactivity. In animal models of AUD, astrocytes react with an increase of GFAP, although this might not be sustainable after long-term intake. Glutamate homeostasis by astrocytes is a critical factor. Sustained alcohol exposure reduces the glutamate regulation and contributes to damage. Experimental upregulation of astrocytic glutamate uptake decreases alcohol drinking in rats [103]. The transcriptome of astrocytes from AUD models exhibits changes in calcium signaling, gap junctions and TLRs.

MRI studies in human patients indicate a correlation between alcohol ratings and myelin integrity [104]. Even moderate alcohol consumption affects white matter microstructure. A postmortem brain analysis of severe alcoholics shows prominent demyelination. Animal models confirm these findings and point to disrupted OPC differentiation and survival as a major contributing factor.

It seems obvious that alcohol acts on the glial cells as a toxic factor with a strong dose-response dependency.

17.5.3 Opioid Use Disorder

The endogenous opioid system is a modulatory system throughout the body. It is neuromodulatory in the brain and gastrointestinal system circuits. It uses various endogenous opioids, all derived from a neuropeptide precursor and three main receptor families: delta (enkephalins), kappa (dynorphins), and mu (endorphins). The neuromodulatory roles are highly varied, but focus on pain, stress, and reward. An opioid is any substance that interacts with the opioid receptors of the body. A subgroup is opiates, which are natural products from the flowering opium poppy plant (*Papaver somniferum*), which include morphine and heroin. Some synthetic opioids are used in palliative care or to combat chronic or acute pain. Some of these, like fentanyl, are far more potent than natural ligands. Fentanyl especially is far more effective than heroin (50×) or morphine (100×). VTA and NAc dopaminergic neurons express mu receptors. Their stimulation reinforces reward effects via dopaminergic output. However, there are other mu receptors in other areas connected to the reward system, for example, amygdala and hippocampus. These G-protein (guanin nucleotide-binding protein) coupled receptors desensitize and downregulate very quickly causing the need for increased doses [105]. Abrupt cessation of opioid use precipitates very serious physical and psychological withdrawal symptoms. This is based on a loss of function in the VTA/NAc reward system and a gain of function in the amygdala-based stress system. Consequently, opioids are historically well known as addictive drugs. The best known and oldest is heroin. Due to lower cost heroin addictions increased in the last three decades. At the same time, prescriptions for opioid analgesics mushroomed and led to overuse and misuse. Due to highly potent synthetic opioids (fentanyl), this has developed into a major epidemic with high death rates in many western countries. It is estimated that, in 2016, 27 million people were living worldwide with opioid use disorder (OUD) [106]. Treatment options face the difficulty of reaching much of the addicted population. Counseling and behavioral therapies are assisted by medications: methadone (synthetic agonist), buprenorphine (antagonist/agonist depending on the receptor), and naltrexone (receptor antagonist) are the most used [107].

Astrocytes possess mu-opioid receptors. Opioid administration causes release of calcium from their intracellular stores via IP3R2 (inositol triphosphate receptor 2). This in turn provokes astrocytic glutamate release, which activates NMDA (N-methyl-D-aspartate) receptors of adjacent neurons. This activation leads to excitation of these neurons. The exposure of drug-naïve dopaminergic neurons to conditioned medium from astrocytes previously exposed to morphine enhances the rewarding effect of morphine, compared to controls [108]. VTA and NAc astrocytes also possess D1 and D2 dopamine receptors. Both receptors are expressed by distinct subgroups of astrocytes with no overlap. Experimental activation of adjacent neurons or perfusion with opioids increases dopamine levels and leads to receptor-specific calcium increase patterns in astrocytes due to the D1 and D2 activation. The astrocytes also release ATP/adenosine due to this stimulation. The subsequent activation of presynaptic A1 receptors depresses the dopaminergic transmission in the NAc [109]. Chronic stimulation with opioids causes reduction of the astrocytic glutamate transporter GLT-1 and decreased encasement of NAc dopaminergic synapses by astrocytic processes. Both mechanisms increase synaptic glutamate. The glutamate spillover enhances glutamatergic synaptic transmission and nonspecific neuronal transmission above the levels of drug-naïve animals [110]. Another, somewhat paradox finding supports this astrocyte–glutamate–drug-seeking sequence. The astrocytic xCT (cystine/glutamate) exchanger functions by shuttling one glutamate molecule out of the astrocyte into the ECS (extracellular space) in exchange for taking up one extracellular cystine. After cessation of an opioid administration period, xCT activity is impaired. This causes an opioid withdrawal-associated reduction in extracellular glutamate. This reduction of tonic glutamate has an inhibitory effect on group II mGluRs (metabotropic glutamate receptors). These receptors act to limit glutamate release from presynaptic terminals in a negative feedback loop. However, the withdrawal-associated decrease of tonic glutamate disturbs this mGluR-mediated feedback. Now the synaptic cleft is flooded with presynaptic glutamate. This in turn enhances excitation. Both mechanisms, GLT-1 and xCT inhibition, therefore shift the circuits toward drug seeking [111]. During neurodevelopment, astrocytes release thrombospondins (TSPs) to interact with neuronal receptors and promote synapse formation. In the adult, this mechanism is heavily downregulated. During cocaine consumption, NAc astrocytes increase TSP release, and this leads to silent synapse formation and spinogenesis in neighboring dopaminergic neurons. Disrupting this TSP–neuronal receptor interaction by various means prevents these formations and reduces cue-induced cocaine seeking after withdrawal [112].

Microglia also possess mu receptors, which when activated create microglia reactivity. However, there is another way in which opioid consumption affects microglia activity. In microglia, the pattern recognition receptor complex TLR4/MD2 (toll-like receptor 4/myeloid differentiation factor 2) is the main receptor for LPS and therefore for the activation of microglia toward a proinflammatory phenotype. It turns out that a breakdown product of many opioids is morphine-3-gluconide (M3G). M3G has a negligible affinity for opioid receptors, but it is an agonist for TLR4/MD2 [113]. This activation releases proinflammatory cytokines (TNFα, IL-1β) and BDNF. There is evidence that this proinflammatory reactivity of the microglia has a detrimental effect on cognitive processes, as such declines are observed in patients. Other studies suggest that the increased reactivity contributes to tolerance of the opioid reward. The mechanisms are not clear and at this point the conclusions are speculative. However, the fact remains that opioids have an off-target effect that increases microglia reactivity. With long-term use, this could lead to chronic microglia reactivity with consequences for cognitive complications. To underline this, it has been shown that LPS stimulation of the TLR4/MD2 complex promotes morphine tolerance [114]. Pharmacological inhibition of TLR4 receptors reduces dopamine release and reward behavior [115].

In summary, astrocytes seem to be heavily involved in reward seeking in the NAc mainly due to its control of glutamate homeostasis, but other mechanisms play a role. So far, the role of microglia seems relatively minor except for the interaction with a catabolic metabolite of the opioids.

17.5.4 Nicotine Dependence

Tobacco smoking leads globally to about five million deaths a year. The major psychoactive component of tobacco smoke is nicotine. This substance adheres to smoke particles and reaches the lungs via inhalation. Within the lungs, nicotine is quickly absorbed into the pulmonary venous circulation. It reaches the brain circulation within 7 s. Nicotine enters the brain within another few seconds and a yet unknown carrier-mediated transport seems to be responsible for this [116]. Nicotine interacts mainly with nicotinic acetylcholine receptors (nAChRs), which are ligand-gated cation channels. These nAChRs are primarily presynaptic receptors. Therefore, the stimulation of these receptors is mainly neuromodulatory and can activate a whole range of other transmitter systems [117]. This huge spectrum of transmitter responses complicates treatment. Since nicotine acts as nAChR agonist, its presence leads to nAChR desensitization. There is a large family of these nAChRs; therefore, the effects of nicotine are not uniform [118]. The fast kinetics of nicotine in traversing lung–blood–brain and back from the brain within seconds leads to the interesting aspect that there are brief spikes of nicotine concentration in the brain with each puff from a cigarette. This would partly disable desensitization [119].

Despite nicotine's action on many transmitter systems, the main action is thought to be on the dopaminergic neurons arising from the VTA with nerve endings in the NAc. These presynaptic endings have nAChRs, which stimulate dopamine release. Another circuit is composed of the glutamatergic neurons arising from PFC and terminating in the VTA on dopaminergic neurons. Stimulation of both nAChR targets by nicotine leads to heavily increased dopamine release in the NAc, laying the groundwork for the addictive action of nicotine. Cessation of smoking causes a powerful withdrawal response due to the previous upregulation and desensitization of nAChRs. However, this withdrawal response involves other brain regions in addition to the NAc [117].

Astrocytes possess nAChRs and their activation by nicotine increases intracellular calcium concentration, releases gliotransmitters, and causes morphological changes, but no reactive gliosis. One of these released gliotransmitter substances is D-serine, which seems to be involved in nicotine's actions. It increases synaptic transmission and long-term potentiation due to D-serine's actions as an agonist of NMDARs. If the D-serine binding sites on NMDARs are blocked, D-serine catabolized or astrocytes incapacitated, some of nicotine's actions are eliminated. They can be restored by adding excess D-serine [120]. The nicotine action decreases the function of the glutamate transporter EAAT2 in NAc astrocytes. If this decrease is restored or if its expression is upregulated, nicotine's effects on the circuits are dampened [121].

Microglia possess the α7-nAChR subunit. Two studies linked nicotine withdrawal to microglia activation and release of proinflammatory cytokines in the NAc of mice [122, 123]. However, a TSPO binding study in humans undergoing nicotine withdrawal could not replicate these results and showed a reduction in microglia activation [124].

17.6 Conclusion

The close interaction of neuronal activity with astrocytes, microglia, and OPCs means that these cells are involved in these aberrations of neuronal function. The question is, therefore, not if the glial function is affected, but how much of the neuronal aberrational functions are caused by mechanisms originally started in the glial cells. There are neurodevelopmental events where glial cell reactions are causing structural misplacements that are a risk factor for developing mental problems later. The effect of fetal stress or infection on gliogenesis as a risk factor in schizophrenia is the most prominent.

In addition, there are situations where the glial cells have a reduced function, which does not necessarily make them a cause of the mental disturbance. However, compensating this deficiency in an animal experiment seems to improve or even reverse the pathology. Examples are reduction of astrocyte density in MDD, D-serine release by astrocytes in nicotine addiction, astrocyte glutamate homeostasis, and microglial priming in AUD. Other mental problems seem to include an irreversible damaging effect on glial cells, which contributes to the pathology. An example is the effect of alcohol on OPCs.

References

1. Faden J, Citrome L. Schizophrenia: one name, many different manifestations. Med Clin North Am. 2023;107(1):61–72.
2. Hjorthøj C, Stürup AE, McGrath JJ, Nordentoft M. Years of potential life lost and life expectancy in schizophrenia: a systematic review and meta-analysis. Lancet Psychiatry. 2017;4(4):295–301.
3. Robinson N, Bergen SE. Environmental risk factors for schizophrenia and bipolar disorder and their relationship to genetic risk: current knowledge and future directions. Front Genet. 2021;12:686666.
4. Stahl SM. Beyond the dopamine hypothesis of schizophrenia to three neural networks of psychosis: dopamine, serotonin, and glutamate. CNS Spectr. 2018;23(3):187–91.
5. van Erp TGM, Walton E, Hibar DP, Schmaal L, Jiang W, Glahn DC, et al. Cortical brain abnormalities in 4474 individuals with schizophrenia and 5098 control subjects via the enhancing neuro imaging genetics through meta analysis (ENIGMA) consortium. Biol Psychiatry. 2018;84(9):644–54.
6. Gomes FV, Zhu X, Grace AA. Stress during critical periods of development and risk for schizophrenia. Schizophr Res. 2019;213:107–13.
7. de Oliveira Figueiredo EC, Calì C, Petrelli F, Bezzi P. Emerging evidence for astrocyte dysfunction in schizophrenia. Glia. 2022;70(9):1585–604.
8. Wei Y, Xiao L, Fan W, Zou J, Yang H, Liu B, et al. Astrocyte activation, but not microglia, is associated with the experimental mouse model of schizophrenia induced by chronic ketamine. J Mol Neurosci. 2022;72(9):1902–15.
9. Green IW, Glausier JR. Different paths to core pathology: the equifinal model of the schizophrenia syndrome. Schizophr Bull. 2016;42(3):542–9.
10. Levine SZ, Levav I, Yoffe R, Pugachova I. The effects of pre-natal-, early-life- and indirectly-initiated exposures to maximum adversities on the course of schizophrenia. Schizophr Res. 2014;158(1–3):236–40.
11. Sun H, Lui S, Yao L, Deng W, Xiao Y, Zhang W, et al. Two patterns of white matter abnormalities in medication-naive patients with first-episode schizophrenia revealed by diffusion tensor imaging and cluster analysis. JAMA Psychiatry. 2015;72(7):678–86.
12. Ozcelik-Eroglu E, Ertugrul A, Oguz KK, Has AC, Karahan S, Yazici MK. Effect of clozapine on white matter integrity in patients with schizophrenia: a diffusion tensor imaging study. Psychiatry Res. 2014;223(3):226–35.
13. Mauney SA, Pietersen CY, Sonntag KC, Woo TW. Differentiation of oligodendrocyte precursors is impaired in the prefrontal cortex in schizophrenia. Schizophr Res. 2015;169(1–3):374–80.
14. Schmitt A, Tatsch L, Vollhardt A, Schneider-Axmann T, Raabe FJ, Roell L, et al. Decreased oligodendrocyte number in hippocampal subfield CA4 in schizophrenia: a replication study. Cell. 2022;11(20):3242.

15. Dietz AG, Goldman SA, Nedergaard M. Glial cells in schizophrenia: a unified hypothesis. Lancet Psychiatry. 2020;7(3):272–81.
16. Liu SH, Du Y, Chen L, Cheng Y. Glial cell abnormalities in major psychiatric diseases: a systematic review of postmortem brain studies. Mol Neurobiol. 2022;59(3):1665–92.
17. Steiner J, Bielau H, Brisch R, Danos P, Ullrich O, Mawrin C, et al. Immunological aspects in the neurobiology of suicide: elevated microglial density in schizophrenia and depression is associated with suicide. J Psychiatr Res. 2008;42(2):151–7.
18. Brisch R, Wojtylak S, Saniotis A, Steiner J, Gos T, Kumaratilake J, et al. The role of microglia in neuropsychiatric disorders and suicide. Eur Arch Psychiatry Clin Neurosci. 2022;272(6):929–45.
19. Goudriaan A, de Leeuw C, Ripke S, Hultman CM, Sklar P, Sullivan PF, et al. Specific glial functions contribute to schizophrenia susceptibility. Schizophr Bull. 2014;40(4):925–35.
20. Jenkins AK, Lewis DA, Volk DW. Altered expression of microglial markers of phagocytosis in schizophrenia. Schizophr Res. 2022;251:22–9.
21. Windrem MS, Osipovitch M, Liu Z, Bates J, Chandler-Militello D, Zou L, et al. Human iPSC glial mouse chimeras reveal glial contributions to schizophrenia. Cell Stem Cell. 2017;21(2):195–208.e6.
22. Liu Z, Osipovitch M, Benraiss A, Huynh NPT, Foti R, Bates J, et al. Dysregulated glial differentiation in schizophrenia may be relieved by suppression of SMAD4- and REST-dependent signaling. Cell Rep. 2019;27(13):3832–43.e6.
23. Hill RA, Nishiyama A. NG2 cells (polydendrocytes): listeners to the neural network with diverse properties. Glia. 2014;62(8):1195–210.
24. Gibson EM, Purger D, Mount CW, Goldstein AK, Lin GL, Wood LS, et al. Neuronal activity promotes oligodendrogenesis and adaptive myelination in the mammalian brain. Science (New York, NY). 2014;344(6183):1252304.
25. Rakofsky J, Rapaport M. Mood disorders. Continuum (Minneapolis, Minn). 2018;24(3, behavioral neurology and psychiatry):804–27.
26. Czéh B, Simon M, Schmelting B, Hiemke C, Fuchs E. Astroglial plasticity in the hippocampus is affected by chronic psychosocial stress and concomitant fluoxetine treatment. Neuropsychopharmacology. 2006;31(8):1616–26.
27. Zhou X, Xiao Q, Xie L, Yang F, Wang L, Tu J. Astrocyte, a promising target for mood disorder interventions. Front Mol Neurosci. 2019;12:136.
28. Kodama M, Fujioka T, Duman RS. Chronic olanzapine or fluoxetine administration increases cell proliferation in hippocampus and prefrontal cortex of adult rat. Biol Psychiatry. 2004;56(8):570–80.
29. Cao X, Li LP, Wang Q, Wu Q, Hu HH, Zhang M, et al. Astrocyte-derived ATP modulates depressive-like behaviors. Nat Med. 2013;19(6):773–7.
30. Koizumi S. Glial purinergic signals and psychiatric disorders. Front Cell Neurosci. 2021;15:822614.
31. Böttcher C, Fernández-Zapata C, Snijders GJL, Schlickeiser S, Sneeboer MAM, Kunkel D, et al. Single-cell mass cytometry of microglia in major depressive disorder reveals a non-inflammatory phenotype with increased homeostatic marker expression. Transl Psychiatry. 2020;10(1):310.
32. Owen DR, Narayan N, Wells L, Healy L, Smyth E, Rabiner EA, et al. Pro-inflammatory activation of primary microglia and macrophages increases 18 kDa translocator protein expression in rodents but not humans. J Cereb Blood Flow Metab. 2017;37(8):2679–90.
33. Perry VH. Microglia and major depression: not yet a clear picture. Lancet Psychiatry. 2018;5(4):292–4.
34. Hellwig S, Brioschi S, Dieni S, Frings L, Masuch A, Blank T, et al. Altered microglia morphology and higher resilience to stress-induced depression-like behavior in CX3CR1-deficient mice. Brain Behav Immun. 2016;55:126–37.
35. He E, Liu M, Gong S, Fu X, Han Y, Deng F. White matter alterations in depressive disorder. Front Immunol. 2022;13:826812.

36. Abraham M, Mundorf A, Brodmann K, Freund N. Unraveling the mystery of white matter in depression: a translational perspective on recent advances. Brain Behav. 2022;12(7):e2629.
37. Tanti A, Kim JJ, Wakid M, Davoli MA, Turecki G, Mechawar N. Child abuse associates with an imbalance of oligodendrocyte-lineage cells in ventromedial prefrontal white matter. Mol Psychiatry. 2018;23(10):2018–28.
38. McGuffin P, Rijsdijk F, Andrew M, Sham P, Katz R, Cardno A. The heritability of bipolar affective disorder and the genetic relationship to unipolar depression. Arch Gen Psychiatry. 2003;60(5):497–502.
39. Pinto JV, Moulin TC, Amaral OB. On the transdiagnostic nature of peripheral biomarkers in major psychiatric disorders: a systematic review. Neurosci Biobehav Rev. 2017;83:97–108.
40. Zhang X, Alnafisah RS, Hamoud AA, Shukla R, Wen Z, McCullumsmith RE, et al. Role of astrocytes in major neuropsychiatric disorders. Neurochem Res. 2021;46(10):2715–30.
41. Sneeboer MAM, Snijders G, Berdowski WM, Fernández-Andreu A, van Mierlo HC, Berdenis van Berlekom A, et al. Microglia in post-mortem brain tissue of patients with bipolar disorder are not immune activated. Transl Psychiatry. 2019;9(1):153.
42. Dai N, Jones BDM, Husain MI. Astrocytes in the neuropathology of bipolar disorder: review of current evidence. Brain Sci. 2022;12(11):1513.
43. Tondo L, Lepri B, Baldessarini RJ. Suicidal risks among 2826 Sardinian major affective disorder patients. Acta Psychiatr Scand. 2007;116(6):419–28.
44. Brundin L, Bryleva EY, Thirtamara RK. Role of inflammation in suicide: from mechanisms to treatment. Neuropsychopharmacology. 2017;42(1):271–83.
45. Tondo L, Vazquez GH, Baldessarini RJ. Suicidal behavior associated with mixed features in major mood disorders. Psychiatr Clin North Am. 2020;43(1):83–93.
46. Arsenault-Lapierre G, Kim C, Turecki G. Psychiatric diagnoses in 3275 suicides: a meta-analysis. BMC Psychiatry. 2004;4:37.
47. Locci A, Pinna G. Social isolation as a promising animal model of PTSD comorbid suicide: neurosteroids and cannabinoids as possible treatment options. Prog Neuro-Psychopharmacol Biol Psychiatry. 2019;92:243–59.
48. Capuzzi E, Caldiroli A, Capellazzi M, Tagliabue I, Buoli M, Clerici M. Biomarkers of suicidal behaviors: a comprehensive critical review. Adv Clin Chem. 2020;96:179–216.
49. Pandey GN, Rizavi HS, Ren X, Fareed J, Hoppensteadt DA, Roberts RC, et al. Proinflammatory cytokines in the prefrontal cortex of teenage suicide victims. J Psychiatr Res. 2012;46(1):57–63.
50. Torres-Platas SG, Cruceanu C, Chen GG, Turecki G, Mechawar N. Evidence for increased microglial priming and macrophage recruitment in the dorsal anterior cingulate white matter of depressed suicides. Brain Behav Immun. 2014;42:50–9.
51. Schnieder TP, Trencevska I, Rosoklija G, Stankov A, Mann JJ, Smiley J, et al. Microglia of prefrontal white matter in suicide. J Neuropathol Exp Neurol. 2014;73(9):880–90.
52. Schnieder TP, Zhou Qin ID, Trencevska-Ivanovska I, Rosoklija G, Stankov A, Pavlovski G, et al. Blood vessels and perivascular phagocytes of prefrontal white and gray matter in suicide. J Neuropathol Exp Neurol. 2019;78(1):15–30.
53. Pandey GN, Rizavi HS, Ren X, Bhaumik R, Dwivedi Y. Toll-like receptors in the depressed and suicide brain. J Psychiatr Res. 2014;53:62–8.
54. Pandey GN, Zhang H, Sharma A, Ren X. Innate immunity receptors in depression and suicide: upregulated NOD-like receptors containing pyrin (NLRPs) and hyperactive inflammasomes in the postmortem brains of people who were depressed and died by suicide. J Psychiatry Neurosci. 2021;46(5):E538–e47.
55. Erhardt S, Lim CK, Linderholm KR, Janelidze S, Lindqvist D, Samuelsson M, et al. Connecting inflammation with glutamate agonism in suicidality. Neuropsychopharmacology. 2013;38(5):743–52.
56. Bay-Richter C, Linderholm KR, Lim CK, Samuelsson M, Träskman-Bendz L, Guillemin GJ, et al. A role for inflammatory metabolites as modulators of the glutamate N-methyl-D-aspartate receptor in depression and suicidality. Brain Behav Immun. 2015;43:110–7.

57. Oquendo MA, Sullivan GM, Sudol K, Baca-Garcia E, Stanley BH, Sublette ME, et al. Toward a biosignature for suicide. Am J Psychiatry. 2014;171(12):1259–77.
58. Chwastiak L, Ehde DM, Gibbons LE, Sullivan M, Bowen JD, Kraft GH. Depressive symptoms and severity of illness in multiple sclerosis: epidemiologic study of a large community sample. Am J Psychiatry. 2002;159(11):1862–8.
59. O'Leary LA, Belliveau C, Davoli MA, Ma JC, Tanti A, Turecki G, et al. Widespread decrease of cerebral vimentin-immunoreactive astrocytes in depressed suicides. Front Psychol. 2021;12:640963.
60. Zhao J, Lucassen PJ, Swaab DF. Suicide is a confounder in postmortem studies on depression. Biol Psychiatry. 2019;86(10):e37–40.
61. Zhao J, Verwer RW, van Wamelen DJ, Qi XR, Gao SF, Lucassen PJ, et al. Prefrontal changes in the glutamate-glutamine cycle and neuronal/glial glutamate transporters in depression with and without suicide. J Psychiatr Res. 2016;82:8–15.
62. Chandley MJ, Szebeni A, Szebeni K, Wang-Heaton H, Garst J, Stockmeier CA, et al. Markers of elevated oxidative stress in oligodendrocytes captured from the brainstem and occipital cortex in major depressive disorder and suicide. Prog Neuro-Psychopharmacol Biol Psychiatry. 2022;117:110559.
63. Tomasi J, Lisoway AJ, Zai CC, Harripaul R, Müller DJ, Zai GCM, et al. Towards precision medicine in generalized anxiety disorder: review of genetics and pharmaco(epi)genetics. J Psychiatr Res. 2019;119:33–47.
64. Maron E, Nutt D. Biological markers of generalized anxiety disorder. Dialogues Clin Neurosci. 2017;19(2):147–58.
65. Gottschalk MG, Domschke K. Genetics of generalized anxiety disorder and related traits. Dialogues Clin Neurosci. 2017;19(2):159–68.
66. Aktar E, Nikolić M, Bögels SM. Environmental transmission of generalized anxiety disorder from parents to children: worries, experiential avoidance, and intolerance of uncertainty. Dialogues Clin Neurosci. 2017;19(2):137–47.
67. Crocq MA. The history of generalized anxiety disorder as a diagnostic category. Dialogues Clin Neurosci. 2017;19(2):107–16.
68. Paulesu E, Sambugaro E, Torti T, Danelli L, Ferri F, Scialfa G, et al. Neural correlates of worry in generalized anxiety disorder and in normal controls: a functional MRI study. Psychol Med. 2010;40(1):117–24.
69. Steimer T. Animal models of anxiety disorders in rats and mice: some conceptual issues. Dialogues Clin Neurosci. 2011;13(4):495–506.
70. Picard K, St-Pierre MK, Vecchiarelli HA, Bordeleau M, Tremblay M. Neuroendocrine, neuroinflammatory and pathological outcomes of chronic stress: a story of microglial remodeling. Neurochem Int. 2021;145:104987.
71. Wegener AJ, Neigh GN. Animal models of anxiety and depression: incorporating the underlying mechanisms of sex differences in macroglia biology. Front Behav Neurosci. 2021;15:780190.
72. Garakani A, Murrough JW, Freire RC, Thom RP, Larkin K, Buono FD, et al. Pharmacotherapy of anxiety disorders: current and emerging treatment options. Front Psychol. 2020;11:595584.
73. Malivoire BL, Mutschler C, Monson CM. Interpersonal dysfunction and treatment outcome in GAD: a systematic review. J Anxiety Disord. 2020;76:102310.
74. Shigetomi E, Koizumi S. The role of astrocytes in behaviors related to emotion and motivation. Neurosci Res. 2022;187:21.
75. Martin-Fernandez M, Jamison S, Robin LM, Zhao Z, Martin ED, Aguilar J, et al. Synapse-specific astrocyte gating of amygdala-related behavior. Nat Neurosci. 2017;20(11):1540–8.
76. Wahis J, Baudon A, Althammer F, Kerspern D, Goyon S, Hagiwara D, et al. Astrocytes mediate the effect of oxytocin in the central amygdala on neuronal activity and affective states in rodents. Nat Neurosci. 2021;24(4):529–41.
77. Xiao Q, Xu X, Tu J. Chronic optogenetic manipulation of basolateral amygdala astrocytes rescues stress-induced anxiety. Biochem Biophys Res Commun. 2020;533(4):657–64.

78. Cho WH, Noh K, Lee BH, Barcelon E, Jun SB, Park HY, et al. Hippocampal astrocytes modulate anxiety-like behavior. Nat Commun. 2022;13(1):6536.
79. Zhao F, Zhang T, Shen Q, Yin K, Wang Y, Zhang G. Tak1 in the astrocytes of mediobasal hypothalamus regulates anxiety-like behavior in mice. Glia. 2021;69(3):609–18.
80. Sweeney P, Qi Y, Xu Z, Yang Y. Activation of hypothalamic astrocytes suppresses feeding without altering emotional states. Glia. 2016;64(12):2263–73.
81. Frick LR, Williams K, Pittenger C. Microglial dysregulation in psychiatric disease. Clin Dev Immunol. 2013;2013:608654.
82. Sild M, Ruthazer ES, Booij L. Major depressive disorder and anxiety disorders from the glial perspective: etiological mechanisms, intervention and monitoring. Neurosci Biobehav Rev. 2017;83:474–88.
83. Wohleb ES, Hanke ML, Corona AW, Powell ND, Stiner LM, Bailey MT, et al. β-Adrenergic receptor antagonism prevents anxiety-like behavior and microglial reactivity induced by repeated social defeat. J Neurosci. 2011;31(17):6277–88.
84. Munshi S, Loh MK, Ferrara N, DeJoseph MR, Ritger A, Padival M, et al. Repeated stress induces a pro-inflammatory state, increases amygdala neuronal and microglial activation, and causes anxiety in adult male rats. Brain Behav Immun. 2020;84:180–99.
85. Han YY, Jin K, Pan QS, Li B, Wu ZQ, Gan L, et al. Microglial activation in the dorsal striatum participates in anxiety-like behavior in Cyld knockout mice. Brain Behav Immun. 2020;89:326–38.
86. Zhang X, Liu Y, Hong X, Li X, Meshul CK, Moore C, et al. NG2 glia-derived GABA release tunes inhibitory synapses and contributes to stress-induced anxiety. Nat Commun. 2021;12(1):5740.
87. Volkow ND, Boyle M. Neuroscience of addiction: relevance to prevention and treatment. Am J Psychiatry. 2018;175(8):729–40.
88. Gutiérrez-Martos M, Girard B, Mendonça-Netto S, Perroy J, Valjent E, Maldonado R, et al. Cafeteria diet induces neuroplastic modifications in the nucleus accumbens mediated by microglia activation. Addict Biol. 2018;23(2):735–49.
89. Yang W, Singla R, Maheshwari O, Fontaine CJ, Gil-Mohapel J. Alcohol use disorder: neurobiology and therapeutics. Biomedicine. 2022;10(5):1192.
90. Seneviratne C, Franklin J, Beckett K, Ma JZ, Ait-Daoud N, Payne TJ, et al. Association, interaction, and replication analysis of genes encoding serotonin transporter and 5-HT3 receptor subunits A and B in alcohol dependence. Hum Genet. 2013;132(10):1165–76.
91. Burkhardt JM, Adermark L. Locus of onset and subpopulation specificity of in vivo ethanol effect in the reciprocal ventral tegmental area-nucleus accumbens circuit. Neurochem Int. 2014;76:122–30.
92. He J, Crews FT. Increased MCP-1 and microglia in various regions of the human alcoholic brain. Exp Neurol. 2008;210(2):349–58.
93. Korbo L. Glial cell loss in the hippocampus of alcoholics. Alcohol Clin Exp Res. 1999;23(1):164–8.
94. Carlson ER, Guerin SP, Nixon K, Fonken LK. The neuroimmune system – where aging and excess alcohol intersect. Alcohol (Fayetteville, NY). 2022;107:153.
95. Perkins AE, Varlinskaya EI, Deak T. From adolescence to late aging: a comprehensive review of social behavior, alcohol, and neuroinflammation across the lifespan. Int Rev Neurobiol. 2019;148:231–303.
96. Ronchi JA, Figueira TR, Ravagnani FG, Oliveira HC, Vercesi AE, Castilho RF. A spontaneous mutation in the nicotinamide nucleotide transhydrogenase gene of C57BL/6J mice results in mitochondrial redox abnormalities. Free Radic Biol Med. 2013;63:446–56.
97. Wilhelm CJ, Hashimoto JG, Roberts ML, Bloom SH, Andrew MR, Wiren KM. Astrocyte dysfunction induced by alcohol in females but not males. Brain Pathol (Zurich, Switzerland). 2016;26(4):433–51.
98. Marshall SA, McClain JA, Wooden JI, Nixon K. Microglia dystrophy following binge-like alcohol exposure in adolescent and adult male rats. Front Neuroanat. 2020;14:52.

99. Saba W, Goutal S, Auvity S, Kuhnast B, Coulon C, Kouyoumdjian V, et al. Imaging the neuroimmune response to alcohol exposure in adolescent baboons: a TSPO PET study using (18) F-DPA-714. Addict Biol. 2018;23(5):1000–9.
100. Barton EA, Baker C, Leasure JL. Investigation of sex differences in the microglial response to binge ethanol and exercise. Brain Sci. 2017;7(10):139.
101. Marshall SA, Geil CR, Nixon K. Prior binge ethanol exposure potentiates the microglial response in a model of alcohol-induced neurodegeneration. Brain Sci. 2016;6(2):16.
102. Qin L, Crews FT. Chronic ethanol increases systemic TLR3 agonist-induced neuroinflammation and neurodegeneration. J Neuroinflammation. 2012;9:130.
103. Erickson EK, Grantham EK, Warden AS, Harris RA. Neuroimmune signaling in alcohol use disorder. Pharmacol Biochem Behav. 2019;177:34–60.
104. Rice J, Gu C. Function and mechanism of myelin regulation in alcohol abuse and alcoholism. BioEssays. 2019;41(7):e1800255.
105. Strang J, Volkow ND, Degenhardt L, Hickman M, Johnson K, Koob GF, et al. Opioid use disorder. Nat Rev Dis Primers. 2020;6(1):3.
106. GBD 2016 Disease and Injury Incidence and Prevalence Collaborators. Global, regional, and national incidence, prevalence, and years lived with disability for 328 diseases and injuries for 195 countries, 1990–2016: a systematic analysis for the Global Burden of Disease Study 2016. Lancet (London, England). 2017;390(10100):1211–59.
107. Herlinger K, Lingford-Hughes A. Opioid use disorder and the brain: a clinical perspective. Addiction. 2022;117(2):495–505.
108. Narita M, Miyatake M, Narita M, Shibasaki M, Shindo K, Nakamura A, et al. Direct evidence of astrocytic modulation in the development of rewarding effects induced by drugs of abuse. Neuropsychopharmacology. 2006;31(11):2476–88.
109. Corkrum M, Covelo A, Lines J, Bellocchio L, Pisansky M, Loke K, et al. Dopamine-evoked synaptic regulation in the nucleus accumbens requires astrocyte activity. Neuron. 2020;105(6):1036–47.e5.
110. LaLumiere RT, Kalivas PW. Glutamate release in the nucleus accumbens core is necessary for heroin seeking. J Neurosci. 2008;28(12):3170–7.
111. Wang J, Holt LM, Huang HH, Sesack SR, Nestler EJ, Dong Y. Astrocytes in cocaine addiction and beyond. Mol Psychiatry. 2022;27(1):652–68.
112. Wang J, Li KL, Shukla A, Beroun A, Ishikawa M, Huang X, et al. Cocaine triggers astrocyte-mediated synaptogenesis. Biol Psychiatry. 2021;89(4):386–97.
113. Green JM, Sundman MH, Chou YH. Opioid-induced microglia reactivity modulates opioid reward, analgesia, and behavior. Neurosci Biobehav Rev. 2022;135:104544.
114. Eidson LN, Inoue K, Young LJ, Tansey MG, Murphy AZ. Toll-like receptor 4 mediates morphine-induced neuroinflammation and tolerance via soluble tumor necrosis factor signaling. Neuropsychopharmacology. 2017;42(3):661–70.
115. Chen JX, Huang KM, Liu M, Jiang JX, Liu JP, Zhang YX, et al. Activation of TLR4/STAT3 signaling in VTA contributes to the acquisition and maintenance of morphine-induced conditioned place preference. Behav Brain Res. 2017;335:151–7.
116. Tega Y, Yamazaki Y, Akanuma SI, Kubo Y, Hosoya KI. Impact of nicotine transport across the blood-brain barrier: carrier-mediated transport of nicotine and interaction with central nervous system drugs. Biol Pharm Bull. 2018;41(9):1330–6.
117. Kumar M, Adeluyi A, Anderson EL, Turner JR. Glial cells as therapeutic targets for smoking cessation. Neuropharmacology. 2020;175:108157.
118. Rose JE, Mukhin AG, Lokitz SJ, Turkington TG, Herskovic J, Behm FM, et al. Kinetics of brain nicotine accumulation in dependent and nondependent smokers assessed with PET and cigarettes containing 11C-nicotine. Proc Natl Acad Sci U S A. 2010;107(11):5190–5.
119. Russell MA, Feyerabend C. Cigarette smoking: a dependence on high-nicotine boli. Drug Metab Rev. 1978;8(1):29–57.

120. López-Hidalgo M, Salgado-Puga K, Alvarado-Martínez R, Medina AC, Prado-Alcalá RA, García-Colunga J. Nicotine uses neuron-glia communication to enhance hippocampal synaptic transmission and long-term memory. PLoS One. 2012;7(11):e49998.
121. Alajaji M, Bowers MS, Knackstedt L, Damaj MI. Effects of the beta-lactam antibiotic ceftriaxone on nicotine withdrawal and nicotine-induced reinstatement of preference in mice. Psychopharmacology. 2013;228(3):419–26.
122. Saravia R, Ten-Blanco M, Grande MT, Maldonado R, Berrendero F. Anti-inflammatory agents for smoking cessation? Focus on cognitive deficits associated with nicotine withdrawal in male mice. Brain Behav Immun. 2019;75:228–39.
123. Adeluyi A, Guerin L, Fisher ML, Galloway A, Cole RD, Chan SSL, et al. Microglia morphology and proinflammatory signaling in the nucleus accumbens during nicotine withdrawal. Sci Adv. 2019;5(10):eaax7031.
124. Brody AL, Hubert R, Enoki R, Garcia LY, Mamoun MS, Okita K, et al. Effect of cigarette smoking on a marker for neuroinflammation: a [(11)C]DAA1106 positron emission tomography study. Neuropsychopharmacology. 2017;42(8):1630–9.

Chapter 18
Glia in Recovery Processes and Repair

Abstract Axonal regrowth is not possible in the adult mammalian brain due to unsupportive gene expression and microenvironment. Remyelination is different. Recovery of neuronal functions is aided by the secretion of neurotrophic factors from reactive astrocytes. Uninjured and intact axons can sprout and establish new synapses. Most prominent are reparative axonal sprouting and long-distance sprouting. It occurs over long distances and connects functionally related areas. It is therefore most important for functional recovery and compensation. Reactive astrocytes are involved as they repulse and promote the sprouting. In other words, they serve as guides that direct the sprouting axons to their targets. Blocking this astrocytic activity does not lead to functional recovery. This important role of astrocytes pertains only to moderate reactive subtypes but not to scar-forming astrocytes. In addition, reactive astrocytes are known to regulate spine stabilization, plasticity, and maturation of the newly sprouted axons with target dendrites by releasing synaptogenic factors in the appropriate areas. Macrophages can induce a proinflammatory shift in microglia after an injury. This can lead to chronic inflammation which is completely detrimental to the healing process. If this macrophage influence is absent microglia shift after a short pro-inflammatory phase into an anti-inflammatory phase and assist the astrocytes in restoring function by degrading fibrotic tissue and releasing of factors that promote sprouting and synaptogenesis.

Keywords Axonal sprouting · Fibrotic tissue · Long-distance sprouting · Microglia · Macrophages · Moderate reactive astrocytes · Neuroinflammation · Neurotrophic factors · Remyelination · Reparative axonal sprouting · Scar-forming astrocytes · Spine stabilization · Spine plasticity · Synaptogenesis

18.1 Introduction

Neurons are – due to their incorporation into mostly rigid circuits – less flexible in their reactions to changing circumstances than glial cells. Glial cells have more potential to express a repertoire of functions to reach out to other cell types and exhibit therefore large phenotypical changes. The question therefore is what are the

W. Walz, *The Gliocentric Brain*, https://doi.org/10.1007/978-3-031-48105-5_18

mechanisms that these glial (and immune) cells possess to compensate, recover, or repair lost neuronal properties after insults? Microglia, astrocytes/radial glia, and OPCs (oligodendrocyte progenitor cells) play a big role in neurodevelopment to establish functional neuronal circuits and in the organization of the brain areas. Could some of these roles revive in the adult brain after a crisis to compensate for deficient neuronal function? Obviously, the answers will be dependent on the context of the injury or lesion. After such lesions the priority will be to limit damage, re-establish some kind of homeostasis, and seal blood–brain barrier leaks, if they are present. This chapter explores glial roles after lesions are stabilized and a kind of functional status is established with no further neuronal damage. This is the time point where regeneration and repair should start. Adult neurogenesis was covered in a previous chapter and is not included here.

18.2 Remyelination

Following injury, axons in the mammalian central nervous system (CNS) cannot grow back to their target. They are not able to repeat the situation which existed during neurodevelopment in the adult CNS. In adult injury, the tip of the axon will reseal but will not form a growth cone as the adult neuron has a restricted expression of the required gene activities. It instead expresses genes that restrict axonal growth, for example, phosphatase and tensin homolog gene [1, 2]. In addition, the neuronal environment is not supportive of axon regeneration. During development, the growth cone integrates various guidance cues (such as netrins, slits, semaphorins, ephrins), which attract or repulse and thus guide the cone to a target. In the adult CNS after injury, a large role falls to reactive astrocytes, which together with other nonneuronal cells (fibroblasts, macrophages, OPCs, pericytes), form a scar that is repulsive for axonal regrowth. The priorities of this scar tissue are to seal leakage from the blood and suppression of inflammatory activity and neuroprotection, but not axonal regeneration. One of the most important repulsive components of the extracellular matrix is CSPGs (chondroitin sulfate proteoglycans). Type 1 collagen is also prominent in the scar as it helps the reactive astrocytes to fuse together. If type 1 collagen is suppressed, the astrocytes are present, but do not form a scar, and axon regeneration is considerably improved [3]. However, there is at least one exception: GFAP+ bridge-forming astrocytes close to corticospinal axons in the chronic (but not acute) phase of injury. This subgroup seems to support regeneration attempts. Such GFAP+ bridge-forming astrocytes are probably derived from residential mature astrocytes [4]. Otherwise, adult mammalian axonal regeneration is – without supportive manipulation – not a major occurrence. However, as this example shows, one should not rule out a context-dependent appearance of a reactive astrocyte subtype, that is supportive of regeneration of axons, which escaped major injury [5].

In contrast, oligodendrocytes/OPCs preserve the ability to regenerate in the adult mammalian CNS. In situations where the myelin sheath is damaged or removed by disease and where the axon is in good health or only slightly damaged, remyelination can occur. Most of the new myelin comes from OPCs, which proliferate and differentiate into mature oligodendrocytes. However, some oligodendrocytes that survived the demyelination can also contribute, albeit at a much lower rate. Even partially injured surviving oligodendrocytes can contribute [6]. Microglia play an important role in facilitating remyelination by phagocytosing the myelin debris, and by interacting with OPCs to promote their proliferation and differentiation [7]. Major microglial secretions to aid in this process are IGF-1 (insulin-like growth factor-1), FGF-2 (fibroblast growth factor-2), VEGF (vascular endothelial growth factor), PDGF (platelet-derived growth factor), and several cytokines. Sustained axonal action potential traffic is an important stimulant for remyelination by both newly differentiated oligodendrocytes from OPCs and surviving mature oligodendrocytes. Especially motor learning, if it occurs in partially demyelinated axons, can improve remyelination [8]. Remyelinated axons have shorter internodal segments. This indicates that new nodes are formed during the remyelination process. It appears that during remyelination there is first a contact between an oligodendroglial process and a node-like cluster of ion channels on the axon membrane. Once this contact is established, the wrapping of the myelin around the axon starts [9]. Another key for a successful remyelination seems to be the switch of the surrounding microglia from a proinflammatory to an anti-inflammatory subtype [10]. This switch initiates remyelination. The proinflammatory microglia are reduced by necroptosis (inflammatory cell death) and replaced by microglia, which support remyelination. The main signal for this switch is type-1 interferon interaction with the microglial interferon-alpha/beta receptor subunit 2. Accordingly, depletion of microglia interferes heavily with remyelination [11]. In addition, microglia secrete several factors (IGF-1, FGF-2, HGF or hepatocyte growth factor, VEGF), which regulate OPC proliferation and differentiation. Astrocytes recruit phagocytic microglia to the sites of damaged myelin [12] mainly by secreting the chemokine CXCL10 (C-X-C motif chemokine ligand 10). In addition, astrocytes promote OPC proliferation (PDGF-AA or two A chains of PDGF, FGF-2, IL-1β, or interleukin-1 beta) and differentiation (CNTF or ciliary neurotrophic factor, LIF or leukemia inhibitory factor, IGF-1). Fibrinogen, entering the brain through a compromised blood–brain barrier, inhibits oligodendrocyte maturation and therefore interferes with remyelination [13]. However, fibrinogen also induces reactive astrocyte scar formation and therefore contributes to the resealing of the leaking blood–brain barrier. In multiple sclerosis, there is demyelination with remyelination failure and resulting subsequent axonal damage. The reason for the remyelination failure in multiple sclerosis is not clear. Some evidence points toward reduced microglial involvement, for example, microglia-specific TREM2 (triggering receptor expressed on myeloid cells 2) knockout in mouse models causes remyelination failure [14].

In the mammalian CNS, terminally differentiated neurons lost their ability to divide and therefore cannot replace themselves after an injury that involves neuronal loss. In the Zebrafish, certain areas allow glia-to-neuron conversion, but in the adult mammalian CNS, this process is not occurring naturally. There are, however, efforts under way to create such a conversion as a therapeutic avenue in certain neurological diseases [15]. The role of nonneuronal cells in recovery and repair is therefore limited to supporting and remodeling surviving neurons and circuits. Examples are sprouting and synaptogenesis.

18.3 Neurotrophic Factor Production

After injury reactive astrocytes secrete soluble factors which support not only neuronal survival but also reorganization and repair processes of neurons [2]. The most important examples are BDNF (brain-derived neurotrophic factor for neuronal survival), CNTF (axon sprouting), NGF (nerve growth factor for axon regrowth), FGF-2 (axon branching), HGF (axon regrowth), IGF-1 (stimulation of corticospinal axons), the GDNF (glial cell line-derived neurotrophic factor) family (neuronal survival, sealing of blood–brain barrier), MANF/CNDF (mesencephalic astrocyte-derived neurotrophic factor/cerebral dopamine neurotrophic factor to reduce stress on the endoplasmic reticulum and promotion of anti-inflammatory action), TGF-β (transforming growth factor beta for anti-inflammatory action), and LIF (support of remyelination). Other nonsoluble factors secreted by astrocytes are extracellular matrix proteins and MMPs (matrix metalloproteinases). These substances enhance recovery of neuronal functions [16].

18.4 Axonal Sprouting and Synaptogenesis

As pointed out, neuron replacement and axonal regeneration after injury is limited in the adult mammalian CNS. This is compensated by sprouting of surviving and uninjured axons, often far away from the injury site. The difference between regeneration and sprouting is that regeneration is undertaken by injured and cut axons, whereas sprouting involves axons, which are uninjured and are adjacent or even distant from the injury site. It may involve collateral sprouting of intact axons [17]. It may lead to the reinnervation of previously denervated but surviving neuronal targets. Obviously, this involves the establishment of new synapses. This process is a crucial component of neuroplasticity after injury. It is stimulated by synchronized low-frequency neuronal activity (<0.1 Hz). In addition, astrocytes are heavily involved in this sprouting process. As will be seen this astrocytic involvement is not only promoting sprouting, but there are inhibitory components as well. Sprouting and associated synaptogenesis occur after all brain lesions. However, most knowledge so far focuses on ischemic infarcts. Axons and synapses are lost after such an insult. With a delay of about a week or so, robust sprouting and synaptogenesis appear for several months [18]. There are several forms of sprouting. Reactive axonal sprouting occurs in the vicinity of lesion sites and is accompanied by reactive astrogliosis and activated microglia. It is a local phenomenon and a somewhat automatic cellular response connected to limited recovery. Reparative axonal sprouting is different. It works over longer distances and connects functionally related areas.

It is associated with functional recovery. This sprouting response can involve the cortical hemisphere or part of the spinal cord that are contralateral to the lesion. This long-distance sprouting is not random but involves distinct connections that facilitate the recovery of a behavior pattern, often to compensate for a pattern that has been lost. Interestingly, blocking astrocytic growth inhibitors NgR1 (nogo-66 receptor) or EphrinA5 stimulates the reparative axonal sprouting response. This indicates that astrocytes use inhibitors and activators of long-distance sprouting to direct the sprouting toward distinct targets. Indeed, if these astrocytic growth inhibitors are blocked *and at the same time* the activity of the lesioned area is increased, the sprouting is directed randomly into every direction and is not bound by functional domains [19]. Consequently, this experimental condition is called unbounded axonal sprouting. In this form of sprouting, behavioral outcome is not improved as with the other two forms. Experimental animals undergoing unbounded axonal sprouting have a behavioral performance that is worse than untreated controls [20]. It underlines the importance of astrocytic inhibitory factors in long-distance sprouting to connect functionally related areas which will compensate lost functions with compensatory actions. These sprouting responses are observed in mouse, rat, and non-human primates [18]. Despite the action of these inhibitory growth factors, there are strong indications that astrocytic factors support sprouting and play a crucial role in reactive and reparative sprouting [2]. After a cerebral cortical photothrombotic stroke in GFAP/vimentin knockout mice, the sprouting response is impaired. As GFAP and vimentin deletion interfere heavily with reactive astrogliosis, the conclusion is that reactive astrocytes promote sprouting. More important, reactive astrogliosis seems to be a prerequisite for axonal sprouting after injury [21]. However, sprouting occurs quite often away from the injury site with no reactive astrocytes in the vicinity. Interestingly, it has been shown that reactive astrocytes from scar tissue impede axonal sprouting. However, reactive astrocytes transplanted before a scar was formed can promote sprouting. Thus, scar-forming astrocytes impede sprouting, but moderately reactive astrocytes, without a surrounding scar, will support and initiate sprouting [2, 3]. Stimulation of STAT3 (signal transducer and activator of transcription 3) signaling in healthy mice creates moderately reactive astrogliosis in the gray matter. If the mice are thereafter injured by an experimental stroke, enhanced axonal sprouting compared to controls with normal astrocytes is observed. It is speculated that secretion of CNTF is the most likely astrocytic signal to promote sprouting [22].

The synaptic contacts of cortical neurons are in the form of dendritic spines. In the adult cortex, the number and location of these spines appear stable. However, due to learning the spines can remodel and change their efficacy. After a stroke, spine loss occurs around the peri-infarct areas. Surprisingly, it is also observed at a far distance away from the lesion. After the injury, the peri-infarct areas regain their lost density of spines, whereas the areas that are distant gain additional numbers of spines. These new spines are located between surviving and sprouting axons and dendrites. Astrocytes play a role in synaptogenesis in neurodevelopment (see a previous chapter). Interestingly, an astrocyte subpopulation was found in mature mouse cortex with strong gene expression for synapse-associated proteins. If co-cultured with neurons, this subpopulation promoted the generation of excess functional synapses [23]. However, after spinal cord injury such a synaptogenic phenotype could not be identified in vivo. Instead, all astrocytes express a similar density of synapse-associated proteins [24]. Despite this finding, astrocytes are known to regulate spine stabilization, plasticity, and maturation [25]. It is known that moderately reactive astrocytes at some distance from lesions promote synaptogenesis and spine stability. In a mouse injury model involving facial axotomy, when astrocytes remote from the injury site are experimentally transformed via the STAT3 pathway, they turn into moderately reactive astrocyte phenotype without proliferation and inflammation. This treatment induces the release of the synaptogenic substance TSP-1 (thrombospondin 1) from these moderately reactive astrocytes, which is a main factor in the rearrangement of synapses. A microglial contribution to this TSP-1 release was ruled out [26]. TSP-1 is released into the astrocytic matrix. Its receptor in the neuronal membrane is a subunit for a calcium channel. The interaction between both leads to a remodeling of the actin cytoskeleton and in the end to an increase in the number excitatory synapses [27]. Other astrocytic synaptogenic factors were identified, and most are upregulated after injury. APOE (apolipoprotein E) is being used to transport astrocytic cholesterol to presynaptic endings for synthesizing synaptic vesicles [28]. Other proteins released by reactive astrocytes are Hevin (also known as SPARC-like 1) together with SPARC (secreted protein acidic and rich in cysteine). Both are antagonistic, with Hevin inducing excitatory synaptogenesis and SPARC suppressing this response. Together they are used to regulate the synapse formation [29]. The proteoglycan agrin is another astrocytic extracellular matrix protein, which after injury promotes synaptogenesis [30]. Sp1 (specificity protein 1) released by astrocytes promotes neurite outgrowth and synaptogenesis [31]. In addition, astrocyte-derived BDNF is another signal to induce synaptogenesis [32]. Most interesting are the findings that astrocytes not only release substances that promote sprouting and synaptogenesis but that they also release antagonistic pairs of substances. Like Hevin/SPARC or EphrinA5/CNTF, these antagonists are a clear indication that moderately, non-scar–associated astrocytes are not only promoting axon extension or synapse number increases but also directing and regulating the location, position, and targets of these processes. Unfortunately, not too many details are yet known about these directive commands.

18.5 The Complex Role of Macrophages and Microglia

After damage to tissues outside the CNS, inflammatory and phagocytic macrophages accumulate at the injury site. After several days, however, these macrophages are replaced by other anti-inflammatory populations, which assist in the transition to the healing phase [33]. In contrast, in the CNS, this transition phase is far more complex and its extent and outcome depend on the inflammation status of the injury site. The reason for this CNS-specific response pattern is probably the isolation due to the blood–brain barrier and the interactions with components of the glial scar [34, 35].

In the CNS, after an injury existed for several days, both macrophages and microglia can have a detrimental or a healing influence on the lesion, depending on the context. A saturation of the tissue with high concentrations of TNF-α (tumor necrosis factor alpha) and persisting myelin phagocytosis mediated by macrophage scavenger receptor 1 seem to be major factors in keeping the macrophage population in an inflammatory and toxic state and preventing the transition to a healing macrophage population [36, 37]. This prolonged proinflammatory phenotype supports a cytokine storm, which sustains this detrimental situation. The microglia are affected by this continued inflammatory situation. They continue releasing proinflammatory cytokines; they trigger neuronal apoptosis, inhibit axonal growth, shift astrocytes into a more reactive state, and can cause oligodendrocyte death. This chronic inflammation will lead to secondary neuronal death instead of healing and regeneration. What factors cause this switching during the recovery of an injury? The complement component C1q can cause such a shift toward inflammatory phenotypes by tagging synapses that have to be eliminated [38]. LPS (lipopolysaccharide), indicating an infection, causes a shift toward inflammation. In contrast, CSPG, peroxisome proliferator-activated receptor-γ stimulation, some interferon response factors, and miRNA-124 (micro ribonucleic acid 124) cause shifts to an anti-inflammatory phenotype. However, other interferon response factors and miRNA-155 result in the exact opposite shift [39, 40]. In addition, experimental treatments, which interfere with inflammatory macrophage recruitment in this stage of a CNS injury, have shown beneficial effects on the recovery of neuronal functions [37, 41, 42]. If the macrophage-induced shift of the immune cells toward detrimental phenotypes is reduced or absent, the microglia shift into a phase with anti-inflammatory and restorative functions [37]. In the restoration phase, most macrophages disappear and those left are secreting metalloproteinases to degrade fibrotic tissue and IL-10 to resolve inflammation [43]. Anti-inflammatory microglia secrete BDNF and GDNF, which stimulate sprouting. Microglial-released BDNF also increases synaptophysin, GAP-43 (growth-associated protein 43), and AMPA (α-amino-3-hydroxy-5-methyl-4-isoxazolepropionic acid) receptors, which promote synaptic plasticity. VEGF and berberine, which are released from microglia, enhance angiogenesis [33].

18.6 Conclusion

OPCs and moderately reactive, nonproliferative, and noninflammatory astrocytes are an asset for remyelination, sprouting, and synaptogenesis following brain injury. There are first indications that astrocytes regulate the direction, extent, and targets of sprouting via antagonistic factors. This is a crucial process in repair and compensation after injury. It could mean that the astrocytic syncytium sets the blueprint for remodeling the CNS pathways in brain repair. If these indications can be verified in future research, a completely new field of adult neuroplasticity would open up. Strongly reactive astrocytes are not part of this process. In contrast, they hinder it. However, their roles as barriers are different and physically segregated from the reparative processes. Microglia have a response pattern that is far more flexible and can reach from driving secondary damage to assisting astrocytes and OPCs/oligodendrocytes in restoration. Macrophages, if present, have a crucial role in determining if the microglia functions are supportive of these reparative processes or detrimental.

References

1. Hilton BJ, Bradke F. Can injured adult CNS axons regenerate by recapitulating development? Development. 2017;144(19):3417–29.
2. Hemati-Gourabi M, Cao T, Romprey MK, Chen M. Capacity of astrocytes to promote axon growth in the injured mammalian central nervous system. Front Neurosci. 2022;16:955598.
3. Hara M, Kobayakawa K, Ohkawa Y, Kumamaru H, Yokota K, Saito T, et al. Interaction of reactive astrocytes with type I collagen induces astrocytic scar formation through the integrin-N-cadherin pathway after spinal cord injury. Nat Med. 2017;23(7):818–28.
4. Zukor K, Belin S, Wang C, Keelan N, Wang X, He Z. Short hairpin RNA against PTEN enhances regenerative growth of corticospinal tract axons after spinal cord injury. J Neurosci. 2013;33(39):15350–61.
5. Anderson MA, Ao Y, Sofroniew MV. Heterogeneity of reactive astrocytes. Neurosci Lett. 2014;565:23–9.
6. Duncan ID, Radcliff AB, Heidari M, Kidd G, August BK, Wierenga LA. The adult oligodendrocyte can participate in remyelination. Proc Natl Acad Sci U S A. 2018;115(50):E11807–e16.
7. Quan L, Uyeda A, Muramatsu R. Central nervous system regeneration: the roles of glial cells in the potential molecular mechanism underlying remyelination. Inflamm Regen. 2022;42(1):7.
8. Bacmeister CM, Barr HJ, McClain CR, Thornton MA, Nettles D, Welle CG, et al. Motor learning promotes remyelination via new and surviving oligodendrocytes. Nat Neurosci. 2020;23(7):819–31.
9. Auer F, Vagionitis S, Czopka T. Evidence for myelin sheath remodeling in the CNS revealed by in vivo imaging. Curr Biol. 2018;28(4):549–59.e3.
10. Lloyd AF, Davies CL, Holloway RK, Labrak Y, Ireland G, Carradori D, et al. Central nervous system regeneration is driven by microglia necroptosis and repopulation. Nat Neurosci. 2019;22(7):1046–52.
11. Sariol A, Mackin S, Allred MG, Ma C, Zhou Y, Zhang Q, et al. Microglia depletion exacerbates demyelination and impairs remyelination in a neurotropic coronavirus infection. Proc Natl Acad Sci U S A. 2020;117(39):24464–74.
12. Skripuletz T, Hackstette D, Bauer K, Gudi V, Pul R, Voss E, et al. Astrocytes regulate myelin clearance through recruitment of microglia during cuprizone-induced demyelination. Brain. 2013;136(Pt 1):147–67.
13. Petersen MA, Ryu JK, Chang KJ, Etxeberria A, Bardehle S, Mendiola AS, et al. Fibrinogen activates BMP signaling in oligodendrocyte progenitor cells and inhibits remyelination after vascular damage. Neuron. 2017;96(5):1003–12.e7.
14. Wang Y, Kyauk RV, Shen YA, Xie L, Reichelt M, Lin H, et al. TREM2-dependent microglial function is essential for remyelination and subsequent neuroprotection. Glia. 2023;71:1247.
15. Qian C, Dong B, Wang XY, Zhou FQ. In vivo glial trans-differentiation for neuronal replacement and functional recovery in central nervous system. FEBS J. 2021;288(16):4773–85.

16. Allen M, Ghosh S, Ahern GP, Villapol S, Maguire-Zeiss KA, Conant K. Protease induced plasticity: matrix metalloproteinase-1 promotes neurostructural changes through activation of protease activated receptor 1. Sci Rep. 2016;6:35497.
17. Geoffroy CG, Zheng B. Myelin-associated inhibitors in axonal growth after CNS injury. Curr Opin Neurobiol. 2014;27:31–8.
18. Carmichael ST, Kathirvelu B, Schweppe CA, Nie EH. Molecular, cellular and functional events in axonal sprouting after stroke. Exp Neurol. 2017;287(Pt 3):384–94.
19. Overman JJ, Clarkson AN, Wanner IB, Overman WT, Eckstein I, Maguire JL, et al. A role for ephrin-A5 in axonal sprouting, recovery, and activity-dependent plasticity after stroke. Proc Natl Acad Sci U S A. 2012;109(33):E2230–9.
20. Wahl AS, Omlor W, Rubio JC, Chen JL, Zheng H, Schröter A, et al. Neuronal repair. Asynchronous therapy restores motor control by rewiring of the rat corticospinal tract after stroke. Science (New York, NY). 2014;344(6189):1250–5.
21. Liu Z, Li Y, Cui Y, Roberts C, Lu M, Wilhelmsson U, et al. Beneficial effects of gfap/vimentin reactive astrocytes for axonal remodeling and motor behavioral recovery in mice after stroke. Glia. 2014;62(12):2022–33.
22. Chen M, Ingle L, Plautz EJ, Kong X, Tang R, Ghosh N, et al. LZK-dependent stimulation of astrocyte reactivity promotes corticospinal axon sprouting. Front Cell Neurosci. 2022;16:969261.
23. John Lin CC, Yu K, Hatcher A, Huang TW, Lee HK, Carlson J, et al. Identification of diverse astrocyte populations and their malignant analogs. Nat Neurosci. 2017;20(3):396–405.
24. Allahyari RV, Heinsinger NM, Hwang D, Jaffe DA, Rasouli J, Shiers S, et al. Response of astrocyte subpopulations following spinal cord injury. Cell. 2022;11(4):721.
25. Liu CY, Yang Y, Ju WN, Wang X, Zhang HL. Emerging roles of astrocytes in neuro-vascular unit and the tripartite synapse with emphasis on reactive gliosis in the context of Alzheimer's disease. Front Cell Neurosci. 2018;12:193.
26. Tyzack GE, Sitnikov S, Barson D, Adams-Carr KL, Lau NK, Kwok JC, et al. Astrocyte response to motor neuron injury promotes structural synaptic plasticity via STAT3-regulated TSP-1 expression. Nat Commun. 2014;5:4294.
27. Chiareli RA, Carvalho GA, Marques BL, Mota LS, Oliveira-Lima OC, Gomes RM, et al. The role of astrocytes in the neurorepair process. Front Cell Dev Biol. 2021;9:665795.
28. Mauch DH, Nägler K, Schumacher S, Göritz C, Müller EC, Otto A, et al. CNS synaptogenesis promoted by glia-derived cholesterol. Science (New York, NY). 2001;294(5545):1354–7.
29. Kucukdereli H, Allen NJ, Lee AT, Feng A, Ozlu MI, Conatser LM, et al. Control of excitatory CNS synaptogenesis by astrocyte-secreted proteins Hevin and SPARC. Proc Natl Acad Sci U S A. 2011;108(32):E440–9.
30. Zhang P, Yang L, Li G, Jin Y, Wu D, Wang QM, et al. Agrin involvement in synaptogenesis induced by exercise in a rat model of experimental stroke. Neurorehabil Neural Repair. 2020;34(12):1124–37.
31. Hung CY, Hsu TI, Chuang JY, Su TP, Chang WC, Hung JJ. Sp1 in astrocyte is important for neurite outgrowth and synaptogenesis. Mol Neurobiol. 2020;57(1):261–77.
32. Gómez-Casati ME, Murtie JC, Rio C, Stankovic K, Liberman MC, Corfas G. Nonneuronal cells regulate synapse formation in the vestibular sensory epithelium via erbB-dependent BDNF expression. Proc Natl Acad Sci U S A. 2010;107(39):17005–10.
33. Mesquida-Veny F, Del Río JA, Hervera A. Macrophagic and microglial complexity after neuronal injury. Prog Neurobiol. 2021;200:101970.
34. Perego C, Fumagalli S, De Simoni MG. Three-dimensional confocal analysis of microglia/macrophage markers of polarization in experimental brain injury. J Vis Exp. 2013;79:50605.
35. Koh TJ, DiPietro LA. Inflammation and wound healing: the role of the macrophage. Expert Rev Mol Med. 2011;13:e23.
36. Kroner A, Greenhalgh AD, Zarruk JG, Passos Dos Santos R, Gaestel M, David S. TNF and increased intracellular iron alter macrophage polarization to a detrimental M1 phenotype in the injured spinal cord. Neuron. 2014;83(5):1098–116.

37. Kong FQ, Zhao SJ, Sun P, Liu H, Jie J, Xu T, et al. Macrophage MSR1 promotes the formation of foamy macrophage and neuronal apoptosis after spinal cord injury. J Neuroinflammation. 2020;17(1):62.
38. Schafer DP, Lehrman EK, Kautzman AG, Koyama R, Mardinly AR, Yamasaki R, et al. Microglia sculpt postnatal neural circuits in an activity and complement-dependent manner. Neuron. 2012;74(4):691–705.
39. Bouhlel MA, Derudas B, Rigamonti E, Dièvart R, Brozek J, Haulon S, et al. PPARgamma activation primes human monocytes into alternative M2 macrophages with anti-inflammatory properties. Cell Metab. 2007;6(2):137–43.
40. Hu Y, Fryatt GL, Ghorbani M, Obst J, Menassa DA, Martin-Estebane M, et al. Replicative senescence dictates the emergence of disease-associated microglia and contributes to Aβ pathology. Cell Rep. 2021;35(10):109228.
41. Freria CM, Hall JC, Wei P, Guan Z, McTigue DM, Popovich PG. Deletion of the Fractalkine receptor, CX3CR1, improves endogenous repair, axon sprouting, and synaptogenesis after spinal cord injury in mice. J Neurosci. 2017;37(13):3568–87.
42. Morganti JM, Jopson TD, Liu S, Riparip LK, Guandique CK, Gupta N, et al. CCR2 antagonism alters brain macrophage polarization and ameliorates cognitive dysfunction induced by traumatic brain injury. J Neurosci. 2015;35(2):748–60.
43. Sindrilaru A, Scharffetter-Kochanek K. Disclosure of the culprits: macrophages-versatile regulators of wound healing. Adv Wound Care. 2013;2(7):357–68.

Chapter 19
Glial Phenotype Plasticity

Abstract This chapter summarizes the information from the previous chapters. Due to their specific position and role in a functional circuit, neurons have a restricted plasticity, when challenged. The loss or impairment of neurons can impact the circuit function negatively and must be limited to the absolute minimum. All glial cells are involved in neuronal signaling and homeostasis. Even in normal brain, they show functional properties determined by their place and space in relation to circuits. After injury, they exhibit new phenotypes appropriate for the challenges. Microglia have a major coordination role in these events, and they have properties different from other tissue-specific macrophages. In some cases, glial cells can switch cell type after injury.

Keywords Cell type switching · Concomitant activation of astrocytic and oligodendrocytic genes · Glia phenotype · Glial cell plasticity · Neuronal plasticity · Phenotype switching

19.1 Introduction

In the era of single-cell biology, the characterization of the identity of a cell, especially the distinction between different cell types and different phenotypes of an identified cell type is no longer straightforward. Gene and protein expression are major methods for this purpose. A major improvement is the emergence of new cell lineage tracking technologies. But one must accept that no accurate method to characterize the identity of a cell exists [1, 2].

The cellular dynamic of the brain has many similarities with other organs and also a few differences. The most striking difference is the organization of neurons and their processes into circuits. These circuits exhibit plasticity in a way that they react to previous activity and therefore can potentiate or depress responses. They also interact with all glial cells to modify responses. Despite this flexibility which enables memory and learning, individual neurons are fixated within such a circuit. This might be the reason most neurons are terminally differentiated. This framework suggests that almost any single adult neuron is nonexpendable and must be

W. Walz, *The Gliocentric Brain*, https://doi.org/10.1007/978-3-031-48105-5_19

protected at all costs. Neurons are under threat not just by outright pathological damage, but by any factors in their environment that change their excitability and make them unresponsive or hyperexcitable. During such pathological events, neuronal plasticity is limited. Under certain circumstances they can participate in immune responses, but these responses normally do not compromise their signaling role and therefore have to be weak. Instead, they rely on glial cells and invading cells of the immune system to exhibit enough plasticity and flexibility to counter any threats to the function of the neuronal circuits. The participation of all glial cells in neuronal signaling makes them especially sensitive to neuronal aberrations and threats. The glial plasticity is adapted to protect neurons. For example, during virus infections, microglia instruct invading T cells to avoid attacking infected neurons as much as possible. Microglia have key and dual roles in these events. In the normal adult central nervous system (CNS), they participate together with astrocytes and oligodendrocytes in neuronal signaling, by modifying and synchronizing neuronal responses. They are therefore in a prime position to sense aberrations in the CNS functioning and they can subsequently abandon their signaling and homeostatic role and turn into their original destiny as immune cells. This switch is not unknown in other organs as will be discussed. Astrocytes are not quite as flexible as they normally take their cues from microglia and these astrocytes do not completely abandon their homeostatic role but reduce and upregulate a reactive role in concert with microglia and often other immune cells. When investigating the plasticity of non-neuronal cells in the mature CNS, one must distinguish between two different events.

First, there is the now-confirmed transformation of one glial cell type into another due to pathological events. In addition, one glial cell type can – due to external, most often pathological cues – change its phenotype dramatically. The resulting new phenotype is often completely different from the original phenotype – so much so that this appears to be a transformation into a new cell type. With these cell transformations and plasticity, the question arises of what one can learn from these mechanisms about the meaning of a cell type and their conversions. Is it still justified to point to a defined cell type without the context of it is existing? Are these conversions restricted to the CNS or are there other examples known from other mature tissues?

19.2 Context Determines Glial Phenotype

Evidence points to the microenvironmental context determining the phenotype of all glial cells. Microenvironmental context is influenced by the status of functionally connected neurons including their past activity, damage-related signals in the extracellular space, antigen presentation, macrophage status, presence of T cells, neuroinflammatory factors, and the status of the neurovascular unit among others. Theoretically, it is possible that no glial cell has the same status as its neighbor despite gap junctional connections for some of them. Astrocytes are now seen as *astromeres* with specific properties in relation to their connected neuronal circuits

[3]. These circuits determine their molecular properties [4, 5]. A similar situation exists for microglia, which are far more flexible than astrocytes. This is partly due to their lineage as an immune cell and their role as a participant in neuronal homeostasis and signaling. Microglia phenotypes change with exercise, diet, and enriched environment [6]. However, the participation of microglia in healthy CNS tissue functions and maintenance is not unique to the brain. Most macrophages are involved in these functions in their resident organs and tissues [7].

Oligodendrocyte progenitor cells (OPCs) start out as a homogeneous population. With age, they diversify into functionally different phenotypes depending on brain region and age [8]. As well, mature oligodendrocyte phenotypes depend on their environmental context as was shown with transplantation studies [9].

19.3 Phenotype Switching

As was pointed out consistently in the previous chapters dealing with different challenges to the adult mammalian CNS, the potential of all glial cells and invading macrophages for phenotype plasticity to protect and repair neuronal circuits is almost limitless.

After injury astrocytes partly, but not completely abandon their homeostatic role and react to the local injury environment. The most important part of this environment is the status of the microglial cells. The astrocytes integrate these signals and the combined impact determines the reactivity status [10]. Therefore, there are many reactive astrocytic phenotypes, some of which can be grouped into categories. However, one still must evaluate the reactive status of each single astrocyte in its context.

If required, microglia will abandon their homeostatic and signal role completely, multiply and change location. They transfer into their alternate identity as a *bona vide* immune cell. In the brain, this transfer comes at a cost and there are many safety features to prevent a complete transition and additional involvement of invading T cells. This transition is not an all-or-nothing event, it is graded. Grouping of reactive microglia after injury into categories is only justified in exceptional circumstances [11]. Microglia instruct invading cells to spare neurons to protect the specificity of circuits [12]. If neurons are infected, they transfer the antigen to microglia which in turn communicate with invading T cells by presenting the antigen complex to them. These are two examples of evolving glial cell plasticity in order to isolate neurons as much as possible from external threats. The neuroinflammatory role of microglia has a powerful impact in the CNS especially if it is unrestricted or low level but chronic. The problem is to adjust the microglial inflammatory status to the threat and tune it down or terminate it whenever appropriate. For this function, microglia depend on many environmental cues from neurons to astrocytes to immune cells. This is a difficult proposition for any part of the body and it is even more so in the CNS where a substantial part of the cell types cannot be easily replaced.

However, microglia have a few properties which distinguish them from other tissue macrophages [13]. To protect precious neuronal circuits, microglia have a higher threshold for inflammatory activities. Once activated, however, microglia appear less sensitive to down-regulation than their macrophage counterparts in other tissues. In addition, the long life span of microglia in the metabolically active brain environment makes them more susceptible for "inflammaging" than other macrophages. This is a contributing factor to neurodegenerative processes.

19.4 Cell Type Switching

For the human brain, the transformation of one glial cell type into another after injury has not been clearly demonstrated. But for other mammalian brains, it is now well established that not only can embryonic OPCs create astrocytes, but also adult OPCs after stab wound injury, cryoinjury, and spreading depression [14]. The proportion and functionality of these newly created astrocytes depends on the age and injury model [15]. This adult OPC – astrocyte lineage seemed till recently the only pathway of glial cell type switching – at least without further manipulation technologies.

However, for the rodent brain, there are now increasing indications that other injury-initiated glial cell type transformations are occurring. A stab wound injury to the hippocampus of adult mice leads to the appearance of cells with specific markers for both microglia and astrocytes [16]. Furthermore, brain samples from patients with stroke, Alzheimer, and Lewy body dementia exhibited this same yet unidentified "cell type" with mixed upregulated specific markers. This might be an indication that microglia phenotypes are far more flexible in their transitions than originally thought. If this example can be verified, it would constitute a major new pathway. After all, microglia are part of the immune cell lineage whereas astrocytes are derived from the neuroepithelium.

Moreover, it has been shown for adult mice in for various hemorrhagic injury models, that a subset of oligodendrocytes is reprogrammed into a transitional AO (astro- and oligodendroglial) precursor cell type. This reprogramming is initiated and supported by microglial cytokine release. This AO cell further splits into astrocytes and OPCs (Fig. 19.1). The proportion of astrocytes depends on the conditions used but could be as high as 30% [17]. The functional properties of the newly created astrocytes were not yet evaluated.

Thus, evidence is slowly accumulating that there is not only a high degree of phenotype switching among glial cells after injuries but also the possibility of cell type switching. This cell type switch must be provoked by serious injuries or impact.

The creation of new neurons after injury was reviewed in Chap. 11. Adult glial cell–like progenitors create new neurons after injury in three brain regions. However, these are progenitor cells from restricted areas, they are not fully developed glial cells, and it is not clear if these new neurons are incorporated into fully functional circuits.

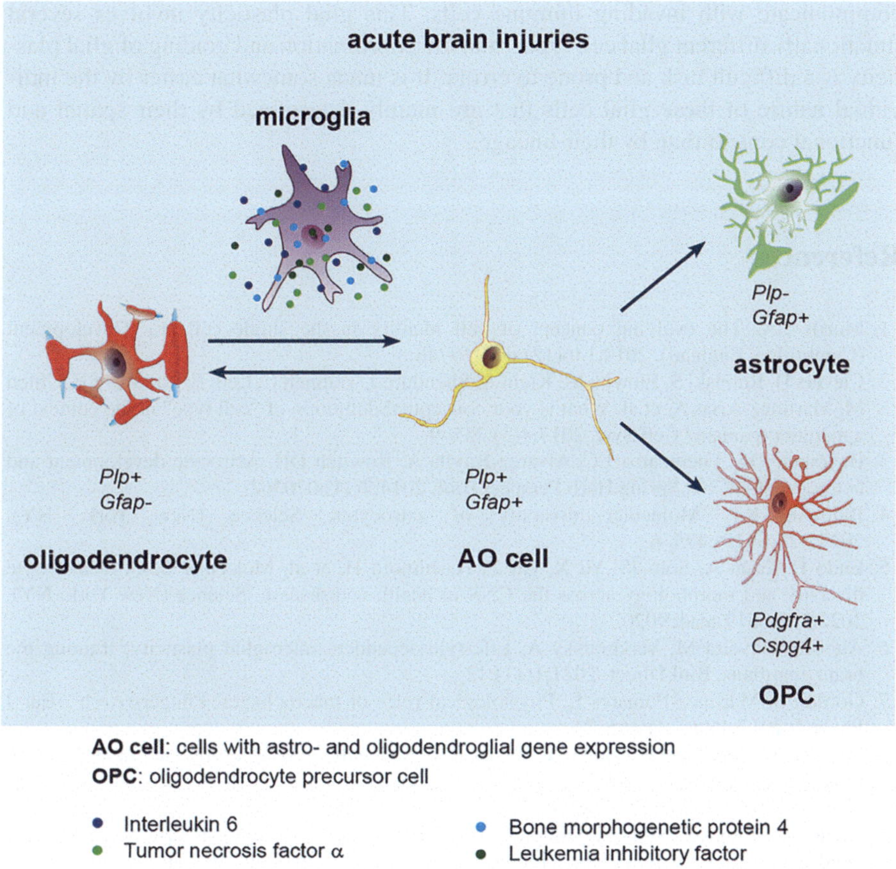

Fig. 19.1 Identification of a subset of cells in mice with concomitant activation of astrocytic and oligodendroglial genes (termed AO cells), derived from oligodendrocytes upon acute brain injury. AO cells differentiate into astrocytes, oligodendrocytes, and oligodendrocyte precursors, and microglial-derived interleukin-6 promotes the transition of mature oligodendrocytes to AO cells. (Reprinted with permission from [17]. Copyright Elsevier Inc.)

19.5 Conclusion

The brain has single neuronal cells that are completely unique and cannot easily be replaced. The role of many single neurons in a circuit is crucial for the signal integration, memory, learning, and the eventual output of the brain that affects behavior. Only the memory T cells in lymph nodes come to mind as cells with a similar importance. This role restricts the plasticity of neurons during pathological events. They cannot change shape, abandon their space, and migrate if challenged by pathological aberrations. This transition into new phenotypes to battle intrinsic and extrinsic pathological events must be taken on by the glial cells that also communicate with invading immune cells. This glial plasticity involves several functionally different glial cell types, and the coordination and grading of glial plasticity is a difficult task and prone to errors. It is made somewhat easier by the individual nature of these glial cells that are mainly determined by their spatial and functional context than by their lineage.

References

1. Morris SA. The evolving concept of cell identity in the single cell era. Development (Cambridge, England). 2019;146(12):dev169748.
2. Clevers H, Rafelski S, Elowitz M, Klein A, Shendure J, Trapnell C, Lein E, Lundberg E, Uhlen M, Martinez-Arias A, et al. What is your conceptual definition of "cell type" in the context of a mature organism? Cell Syst. 2017;4(3):255–9.
3. Bayraktar OA, Fuentealba LC, Alvarez-Buylla A, Rowitch DH. Astrocyte development and heterogeneity. Cold Spring Harb Perspect Biol. 2014;7(1):a020362.
4. Baldwin KT. Molecular diversity of astrocytes. Science (New York, NY). 2022;378(6619):475–6.
5. Endo F, Kasai A, Soto JS, Yu X, Qu Z, Hashimoto H, et al. Molecular basis of astrocyte diversity and morphology across the CNS in health and disease. Science (New York, NY). 2022;378(6619):eadc9020.
6. Augusto-Oliveira M, Verkhratsky A. Lifestyle-dependent microglial plasticity: training the brain guardians. Biol Direct. 2021;16(1):12.
7. Gordon S, Martinez-Pomares L. Physiological roles of macrophages. Pflugers Arch - Eur J Physiol. 2017;469(3–4):365–74.
8. Spitzer SO, Sitnikov S, Kamen Y, Evans KA, Kronenberg-Versteeg D, Dietmann S, et al. Oligodendrocyte progenitor cells become regionally diverse and heterogeneous with age. Neuron. 2019;101(3):459–71.e5.
9. Viganò F, Möbius W, Götz M, Dimou L. Transplantation reveals regional differences in oligodendrocyte differentiation in the adult brain. Nat Neurosci. 2013;16(10):1370–2.
10. Hemati-Gourabi M, Cao T, Romprey MK, Chen M. Capacity of astrocytes to promote axon growth in the injured mammalian central nervous system. Front Neurosci. 2022;16:955598.
11. Paolicelli RC, Sierra A, Stevens B, Tremblay ME, Aguzzi A, Ajami B, et al. Microglia states and nomenclature: a field at its crossroads. Neuron. 2022;110(21):3458–83.
12. Jeon S, St Leger AJ, Cherpes TL, Sheridan BS, Hendricks RL. PD-L1/B7-H1 regulates the survival but not the function of CD8+ T cells in herpes simplex virus type 1 latently infected trigeminal ganglia. J Immunol (Baltimore, MD: 1950). 2013;190(12):6277–86.
13. Timmerman R, Burm SM, Bajramovic JJ. Tissue-specific features of microglial innate immune responses. Neurochem Int. 2021;142:104924.
14. Tamura Y, Eguchi A, Jin G, Sami MM, Kataoka Y. Cortical spreading depression shifts cell fate determination of progenitor cells in the adult cortex. J Cereb Blood Flow Metab. 2012;32(10):1879–87.
15. Komitova M, Serwanski DR, Lu QR, Nishiyama A. NG2 cells are not a major source of reactive astrocytes after neocortical stab wound injury. Glia. 2011;59(5):800–9.
16. Wilhelmsson U, Andersson D, de Pablo Y, Pekny R, Ståhlberg A, Mulder J, et al. Injury leads to the appearance of cells with characteristics of both microglia and astrocytes in mouse and human brain. Cereb Cortex (New York, NY: 1991). 2017;27(6):3360–77.
17. Bai X, Zhao N, Koupourtidou C, Fang LP, Schwarz V, Caudal LC, et al. In the mouse cortex, oligodendrocytes regain a plastic capacity, transforming into astrocytes after acute injury. Dev Cell. 2023;58:1153.

Index

A
Acquired epileptogenesis, 187–188
Activated microglia, 49, 50, 70, 79, 152, 154, 158, 159, 161, 162, 173, 177, 189, 197, 198, 217, 234
Adaptive immune response, 92, 95–99, 104
Addiction, 218, 221, 224
Adult neurogenesis, 30, 62, 127–132, 232
Alcohol use disorder (AUD), 219–220, 224
Alzheimer's disease, 62, 88, 131, 152–158
Amyotrophic lateral sclerosis (ALS), 152, 160–161, 163, 178
Antigen presenting cells (APCs), 92–98, 110, 114, 115, 117–119, 122
Anti-inflammatory factors, 71
Anxiety disorders, 215–217
Astrocyte endfeet, 121, 145, 170, 174
Astrocytes, 2, 9, 30, 60, 68, 77, 84, 95, 102, 117, 126, 140, 156, 170, 184, 198, 208, 232, 242
Atypical reactive astrocyte, 199
Autoimmune encephalitis (AE), 113–116
Autoimmune encephalomyelitis, 116–121
Axonal sprouting, 175, 176, 185, 234–236

B
Bacterial infections, 101–110
Bipolar disorder, 211, 213–214
Blood brain barrier (BBB), 21, 92, 98, 120
Border-forming reactive astrocyte, 78, 81, 109, 171
Brain blood circulation, 34, 96, 172
Brain capillaries, 21
Brain cell proliferation, 135
Brain energy metabolism, 10, 34–36
Brain extracellular space, 11
Brain fluid dynamics, 19–23
Brain immune privilege, 92, 94
Brain injury, 13, 17, 30, 47, 70, 127, 131, 216, 237, 245
Brain innate immunity, 80
Brain lymphatics, 145
Brain reward system, 218–219
Brain swelling, 174, 177

C
Calcium waves, 37, 41, 188, 216
Cell type switching, 244
Cerebrospinal fluid (CSF), 14, 17, 18, 20–24, 92, 96–98, 115–119, 129, 131, 145, 159, 170, 171, 198, 200, 214, 215
Choroid plexus, 14, 17, 18, 20, 21, 95, 96, 102, 116, 118–119, 129
Chronic inflammation, 49, 60, 61, 70, 73, 85, 114, 118–119, 200, 237
Chronic neuroinflammation, 88
Concomitant activation of astrocytic and oligodendrocytic genes, 245
Concussion, 196–200
Cuprizone (CPZ), 122
Cytokine networks, 86
Cytokines, 50, 63, 64, 69–74, 78, 79, 83–89, 93, 94, 104–109, 115, 116, 120, 121, 131, 142, 143, 159, 161, 169, 171–173, 177, 189–191, 199, 201, 202, 210, 212–215, 222, 223, 233, 237, 244
Cytokine signaling, 84–85

W. Walz, *The Gliocentric Brain*, https://doi.org/10.1007/978-3-031-48105-5

Cytokine storm, 103, 237
Cytotoxic T cells, 93, 99, 105, 141, 156

D
Damage-associated molecular patterns (DAMPs), 72, 73, 78, 87, 88, 92, 106, 107, 158, 169, 174, 176, 179
Demyelination, 87, 116–120, 122, 156, 173, 197, 220, 233
Disease-associated microglia (DAM), 61, 62, 155, 156
Domain organization, 33, 185, 187, 188, 191, 210
D-serine, 35, 36, 216, 223, 224

E
Epigenetic changes, 145
Explosive blast injury, 196, 202

F
Fibrotic scar, 81, 171–173, 175
Fibrotic tissue, 237

G
Gastrointestinal (GI) tract, 2
Glial cell plasticity, 1, 243
Glial phenotype, 241–246
Gliogenesis, 16, 30–33, 210, 212, 223
Glioma stem cells (GSCs), 139–140, 143, 144
Glutamate homeostasis, 189, 220, 222, 224
Glutamate release, 36, 37, 45, 99, 143, 174, 178, 186, 188, 221
Glymphatic system, 22–24, 97

H
Hemorrhagic stroke, 168, 177–178
Huntington's disease, 131, 152, 159–160
Hypothalamic ventricular zone, 126, 130–131

I
Ictogenesis, 184, 188–189
Idiopathic epileptogenesis, 186
Immune-competent astrocytes, 163
Immune system, 2–4, 17, 20, 32, 60, 61, 69, 77, 79, 81, 85–87, 91–99, 102, 103, 110, 113, 114, 119, 140, 141, 174, 189–190, 242
Immunological tolerance, 114–115
Infection pathways, 101–102
Inflammatory cytokines, 49, 50, 119, 174, 178, 202, 214, 215, 217, 219
Injury-induced proliferation, 132
Innate immune response, 83, 93–95, 114, 115
Ion homeostasis, 39–40
Ischemic preconditioning, 169, 176, 179
Ischemic stroke, 168–177

J
Juxtavascular astrocytes, 201, 203

L
Latent infections, 102–105, 107
Long-distance sprouting, 235
Lymph nodes, 20, 23, 92–94, 96, 97, 104, 110, 116, 119, 141, 171, 172, 245
Lymphocytes, 80, 92–96, 99, 109, 110, 116, 117, 119–122, 127, 141, 143, 146, 159, 172, 173, 190, 201

M
Macrophages, 17, 47, 48, 61, 67–70, 77, 84–86, 88, 92–95, 102, 106–110, 117–120, 122, 127, 128, 140–142, 144, 146, 156, 161, 170, 171, 173, 177–179, 190, 199, 201, 203, 214, 215, 217, 232, 236–237, 242–244
Major depressive disorder (MDD), 211–213, 216, 224
Major histocompatibility complex (MHC), 70, 92–96, 99, 104–106, 114, 115, 117, 141, 142, 154, 158, 159, 161, 163, 171
Mechanical shearing, 197, 200
Metastatic colonization, 144–146
Microglia, 2, 14, 31, 60, 67, 77, 84, 94, 102, 116, 126, 140, 154, 169, 189, 197, 208, 232, 242
Microglia activation, 47, 49, 50, 62, 68, 69, 71–72, 86, 87, 99, 119–122, 159, 161, 162, 169–170, 190, 191, 198, 199, 201, 212, 217, 219, 223
Microglia classification, 67–69
Mild traumatic brain injury, 196–200
Moderate reactive astrocytes, 235, 236
Molecular subtype migration, 137, 138
Mood disorders, 109, 211–215
Multiple sclerosis (MS), 62, 113, 114, 116–122, 172, 215, 233
Myelination, 17, 32, 46, 109, 127, 162, 210, 217

N

Neurogenesis, 16, 31, 68, 125–132, 145, 190, 210, 211
Neuroinflammation, 4, 64, 83–85, 87–89, 95, 96, 104, 142, 172, 198, 208, 210, 212
Neuronal energy demands, 168, 197
Neuronal OFF/ON factors, 169
Neuronal plasticity, 85, 176, 242
Neuron doctrine, 1
Neurotrophic factors, 42, 87, 196, 234
Neurovascular unit, 14, 33–34, 86, 168, 242
Nicotine dependence, 222–223
NMDA receptors, 44
Nucleotide-binding oligomerization domain-like receptors, 72–74, 174, 215

O

Oligodendrocyte loss, 209
Oligodendrocyte precursor cells, 30, 32, 46, 156, 160, 209, 217
Oligodendrocyte progenitor cells, 4, 16, 61, 122, 126, 140, 197, 232, 243
Oligodendrocytes, 4, 14, 16, 17, 32, 33, 43–46, 49, 50, 61, 68, 73, 87, 88, 99, 102, 109, 116–119, 122, 126–129, 132, 140, 156, 161, 171, 175, 197, 200, 201, 208–210, 213, 215, 217, 219, 232, 233, 237, 238, 242–245
Opioid use disorder (OUD), 220–222

P

Parkinson's disease, 152, 156–159, 163
Pathogen-associated molecular patterns (PAMPs), 72, 73, 78, 87, 88, 92, 95, 104, 105
Pattern recognition receptors (PPRs), 72–75, 87, 105, 109, 120, 222
Penetrating brain injury (PBI), 200–201
Penumbra, 168–175, 179, 200
Phagocytosis, 49, 50, 69–75, 87, 88, 92–94, 102, 108, 109, 117, 120, 121, 142, 156, 158, 159, 171, 174, 190, 210, 237
Phenotype switching, 243–244
Potassium channel Kir4.1, 186
Potassium siphoning, 45
Primary brain tumors, 136
Prion disease, 162, 163
Proconvulsive astrocyte phenotype, 190, 191
Progenitor cell differentiation, 126, 197, 210
Proliferative, border-forming reactive astrocytes, 81

R

Reactive astrocytes, 49, 50, 61, 62, 78–81, 117, 121, 127, 128, 140–143, 152, 154, 156, 158, 161, 170, 171, 173, 174, 177, 178, 185–189, 191, 199, 201, 232–236, 238
Reactive astrogliosis, 63, 78, 160, 162, 173, 175, 199, 209, 211, 234, 235
Reactive gliosis, 4, 59, 61, 62, 108, 132, 143, 152, 158, 162, 185, 187–189, 191, 223
Reactive microglia, 59–64, 68–71, 80, 87, 99, 117, 122, 156, 202, 243
Reactive microgliosis, 69, 122, 128
Remyelination, 88, 117–119, 122, 127, 175, 232–234, 237
Reparative axonal sprouting, 234, 235
Retinoic acid-inducible gene-1-like receptors, 72, 74, 104

S

Saltatory conduction, 43–45
Scar-forming astrocytes, 235
Schizophrenia, 208–211, 213, 214, 223
Sensome, 60, 63, 64
Skin, 2, 3, 11, 12, 96
Spine plasticity, 236
Spine stabilization, 236
Subependymal zone (SEZ), 126, 129
Subgranular zone of the hippocampus, 30, 130
Substance use disorders, 216, 218–223
Suicide, 209–215
Synaptic plasticity, 13, 40–41, 43, 47, 84, 237
Synaptogenesis, 13, 31, 143, 175, 176, 234–237

T

T lymphocytes, 68, 88, 92–94, 96, 104, 110, 114, 117, 120–122, 141, 142, 146, 159, 161, 172–174, 179
Toll-like receptors (TLRs), 70, 72–74, 104, 106, 109, 120, 128, 158, 176, 177, 190, 215, 220, 222
Transmitter homeostasis, 37–39
Tripartite synapse, 35–37, 47, 219
Trojan horse route, 102
Tumor microenvironment, 137, 140–145

V

Vascular cognitive impairment (VCI), 178
Viral encephalitis, 107–108, 115
Viral infections, 2, 101–110, 115, 185

The manufacturer's authorised representative in the EU is Springer Nature Customer Service Centre GmbH, Europaplatz 3, 69115 Heidelberg, Germany. If you have any concerns regarding our products, please contact ProductSafety@springernature.com

Printed and bound by CPI Group (UK) Ltd, Croydon, CR0 4YY
01/10/2026
02225899-0004